AF819160

BIOFUELS

A Promising Alternate for Next Generation Fuels

MJP
PUBLISHERS

BIOFUELS

A Promising Alternate for Next Generation Fuels

Dr. B. Bharathiraja

Associate Professor,
Department of Chemical Engineering,
Vel Tech High Tech Dr Rangarajan Dr Sakunthala Engineering College,
Avadi, Chennai-600062.

J. Jayamuthunagai

Assistant Professor,
Centre for Biotechnology,
Anna University,
Chennai.

R. Praveen Kumar

Associate Professor,
Department of Biotechnology,
Arunai Engineering College,
Tiruvannamalai - 606603,
Tamil Nadu.

MJP PUBLISHERS

Chennai New Delhi Tirunelveli

MJP PUBLISHERS

ISBN 978-81-8094-434-5

MJP Publishers
No. 44, Nallathambi Street,
Triplicane, Chennai 600 005

MJP 288

Publisher : **C. Janarthanan**

Project Editor : **C. Ambica**

In Affectionate Reverence this Book is Dedicated to My Parents
J. Balasubramaniyan
&
B. Gandhimathi

Preface

Today the world is facing three critical problems: (i) High fuel prices, (ii) Climatic changes (iii) Air pollution. Currently there are several important problems to be resolved worldwide: (1) high need for energy, (2) high depletion of non-renewable energy resources and (3) high local and global environmental pollution. This book "Biofuels - A Promising alternate for next generation fuels" deals with the production of biofuels. The biggest difference between biofuels and petroleum feedstock is oxygen content. Biofuels have oxygen levels from 10% to 45% while petroleum has essentially none making the chemical properties of biofuels very different from petroleum. Oxygenates are just pre-used hydrocarbons having a structure that provides a reasonable antiknock value. Most traditional biofuels, such as ethanol from corn, wheat, or sugar beets, and biodiesel from oil seeds, are produced from classic agricultural food crops that require high-quality agricultural land for growth. The term biofuel is referred to as liquid or gaseous fuels for the transport sector that are predominantly produced from biomass. There are several reasons for biofuels to be considered as relevant technologies by both developing and industrialized countries. They include energy security reasons, environmental concerns, foreign exchange savings, and socioeconomic issues related to the rural sector. Biomass can be converted to biofuels through various methods like chemical, such as biomethanol and biodiesel production, thermochemical, such as bio-oil, bio-syngas, biohydrogen production, and biochemical, such as bioethanol, biogas, biodiesel and biohydrogen production. Bioethanol is a fuel derived from renewable sources of feedstock; typically plants such as wheat, sugar beet, corn, straw, and wood. Bioethanol is a petrol additive/substitute. Biodiesel is better than diesel fuel in terms of sulphur content, flash point, aromatic content and biodegradability. Bioethanol can be produced from plentiful, domestic, cellulosic biomass resources such as herbaceous and woody plants, agricultural and forestry residues, and a large portion of municipal solid waste and industrial waste streams. Production of bioethanol from biomass is one way to reduce both the consumption of crude oil and environmental pollution. As a renewable resource, biomass represents a potentially inexhaustible supply of feedstock for methanol production. Biomethanol can be produced from biomass using bio-sythetic gas. Biodiesel is an environmental friendly alternative liquid fuel that can be used in any diesel engine without modification. There has been renewed interest in the use of vegetable oils for making biodiesel due to its less polluting and renewable nature as against the conventional petroleum diesel fuel. Biofuel production costs can vary widely by feedstock, conversion process, scale of production and region. The major economic factor to be considered for input costs of biodiesel production is the feedstock, which is about 75–80% of the total operating cost. Other important costs are labour, methanol and catalyst, which must be added to the feedstock. In particular, the cost of producing oilseed-derived biodiesel is dominated by the cost of the oil and by competition from high-value uses like cooking.

We are thankful to MJP Publishers, Chennai, for coming forward to publish and bring the valuable materials on production of biofuels and their promising alternative as next generation fuels to the limelight for the benefits of the undergraduate, postgraduate and research students in the field of environmental engineering, chemical engineering and biotechnology.

Dr. B. Bharathiraja

List of Abbreviations

AB	Acetogenic Bacteria
AD	Anaerobic Digestion
ADP	Adenosine Diphosphate
AFBR	Anaerobic Fluidized-Bed Reactor
ATP	Adenosine Triphosphate
BDS	Bio-digested Slurry
BOD	Biological Oxygen Demand
Chl a	Chlorophyll a
Chl 700	Chlorophyll a P700, located in photosystem I, which absorbs light best at 700 nanometers (nm)
Chl 680	Chlorophyll a P680, located in photosystem I, which absorbs light best at 680 nanometers (nm)
CHP	Combined Heat and Power
C/N	Carbon to Nitrogen ratio
CNG	Compressed Natural Gas
COD	Chemical Oxygen Demand
CSTR	Continuously Stirred Tank Reactor
FD	Ferredoxin
FD (ox)	Ferredoxin oxidized
FD (red)	Ferredoxin reduced
FRP	Fibre Glass Reinforced Plastic
GAC	Granular activated carbon
HCNG	Hydrogen and Compressed Natural Gas
HPR	Hydrogen Production Rates
HRT	Hydraulic Retention Time
HSOW	High-Solid Organic Wastes

IREDA	Indian Renewable Energy Development Agency Ltd
MEC	Microbial Eelectrolysis Cell
MFC	Microbial Fuel Cell
MS	Molecular Sieves
MSW	Municipal Solid Waste
NAD	Nicotinamide Adenine Dinucleotide
NADH	Nicotinamide Adenine Dinucleotide (NAD) + hydrogen (H)
NPQR	NADP/plastoquinone oxidoreductase
OFMSW	Organic Fraction of Municipal Solid Wastes
OLR	Organic Loading Rate
PAC	Powder activated carbon
PBR	Packed-Bed Reactor
PDMS	Polydimethylsiloxane
PFOR	Pyruvate Ferredoxin Flavodoxin Oxidoreductase
PS I	Photo System I
PS II	Photo System II
PSA	Pressure Swing Adsorption
PFR	Plug Flow Reactor
RDR	Rotary Drum Reactor
TS	Total Solids
USA	United States of America
UASB	Up flow Anaerobic Sludge Blanket
USD	United States Dollars
VFA	Volatile Fatty Acids
VOC	Volatile Organic Compounds
VS	Volatile Solids

Contents

1

Trends in Biodiesel Production and Processing

1.1 INTRODUCTION

Alternate fuels for diesel engines are becoming increasingly important due to diminishing fossil reserves and studies on renewable resource have gained much importance in the present scenario. As it is known depleting petroleum resources can hinder our lifestyle, automobile industry will take a major blow and many other industry which are dependent on the fuels for energy will have to be closed. This situation will have a negative impact on the growth of the developing countries as ours. A realistic approach to the above problems will be to use an alternative fuel is a green fuel with zero sulphur content and is obtained from renewable sources. Biodiesel production has gained reputed value in the catalogue of renewable resource and is known for its reduced toxic emissions when used in conventional engines after blending with Petro Diesel [PD].

Initially in 1980, scientists considered the direct usage of oils obtained from plants or animals as a potential replacement for diesel in automobiles directly (Shag 1993) [1] or by mixing or blending them with diesel in the ratio of up to 1:2. This did not prove fruitful as long-term use in direct-injection engines caused several problems such as:

(1) Coking and trumpet formation on the injectors to such an extent that fuel automization does not occur properly or is even prevented as a result of plugged orifices,

(2) Carbon deposits,

(3) Oil ring sticking, and

(4) Thickening and gelling of the lubricating oil as a result of contamination by the vegetable oils.

The reasons for the above problems were identified as higher viscosity, lower volatility and the reactivity of unsaturated hydrocarbon chains present in the vegetable oils.

Beginning 1990, scientists found ways to reduce the viscosity and molecular weight of the vegetable oils by several ways. Notable among them was by pyrolysis and transestericationn. Pyrolysis has the disadvantages of being expensive and yielding undesirable products (Sonntag 1979b) [2]. Hence research was extensively made on the transesterification and resulting fuel was called as biodiesel. Transesterification or alcoholysis of vegetable oil is a process by which fat or oil (i.e., glyceride) is made to react with an alcohol in the presence of catalyst to form an ester and glycerol. This ester is called as biodiesel. The commonly used alcohol is methanol and in some cases ethanol. Apart from the materials mentioned above an optional organic solvent is required for increasing the mutual solubility of the reactants (Iso et al., 2001) [3]. Biodiesel is chemically known as Fatty Acid Alkyl Esters (FAAE) and the alkyl group is decided by the acyl acceptor used for the reaction. FAAE has a molecular weight that is one-third of that of the vegetable oil, viscosity of about one-eighth and contains 10-11% oxygen (w/w) which enchances its combustion process. The cetane number of FAAE is better than vegetable oils and also the calorific value of FAAE (A. Demirbas, 2005) [4]. The versatility of the biodiesel production lies in the various methods by which it can be produced commercially. Biodiesel can be produced by varying any one of the following (1) Oil/Fat source, (2) Catalyst, (3) Acyl acceptor, or (4) Solvent. Previously scientists have reported carrying out transesterification processes using homogenous catalysts like acid (Schuchardt.U et al., 1998) [5] and alkali (Al-Widyan et al., 2002) [6] which gave high conversion rates but had a disadvantage of toxic waste accumulation. Then use of heterogeneous catalyst, an enzyme known as lipase was carried out. Lipase-catalyzed transesterification gave good conversion and immobilization of the enzyme or the enzyme-producing organism was analyzed which proved the reusability of the catalyst and removed the problem of toxic wastage. Usage of novel acyl acceptors like Dimethyl Carbonate, Methyl Acetate (Wei et al., 2004)

[7] or Ethyl Acetate increased the conversion rate and decreased the enzyme deactivation which was prevalent when methanol was used as the acyl acceptor. Furthermore using novel solvents like t-butanol increased the stability and yield of the process.

Earlier, only animal fat or edible and non-edible vegetable oils or waste cooking oil were used for biodiesel production. Usage of these oils has a disadvantage of limited availability and requirement of large area for the cultivation of crops. Hence an alternative use of algae as oil source is advised. Algae are plant-like prokaryotic or eukaryotic, single cellular or multicellular organisms. They are photosynthetic and follow autotrophic or heterotrophic mode of nutrition. Algae are known for their high oil content and quick lipid productivity with less doubling time (Chisti 2007) [8] by which most of the algal species are found to have the ability to be used as sustainable oil-producing biomass.

1.1.1 Triglyceride as Diesel Fuel

Triglycerides such as vegetable oil or animal oils are the renewable sources of energy for BD fuel. However, animal oil cannot meet the existing demands, hence, fresh plant oil, waste cooking oil and microalgal oil are the expected alternate triglyceride sources that have the potential to meet the existing and future fuel demands of human kind. The use of vegetable oil as an alternative renewable fuel was proposed in the beginning of 1980s [4] and now the process is in industrial practice.

1.1.2 Chemical Composition

Glycerides are composed of hydrocarbon chains. Structure of glycerides and biodiesel is given below in Table 1.

Table 1 Chemical structure of substrates and product in biodiesel production

Chemical structure of Monoglyceride	Chemical structure of Diglyceride	Chemical structure of Fat & Oil	Chemical structure of Ester (Biodiesel)
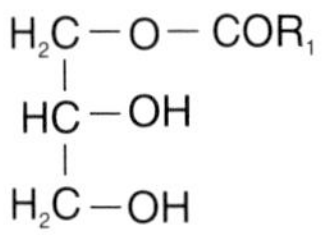	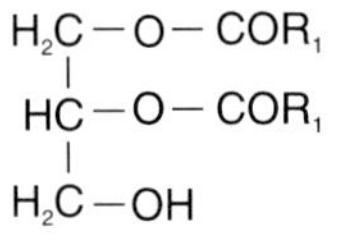	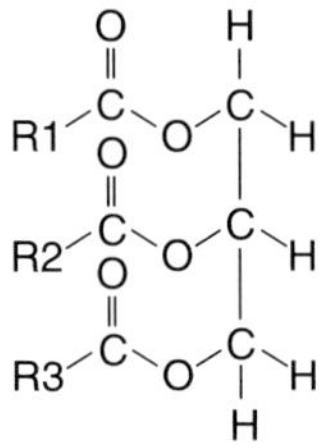	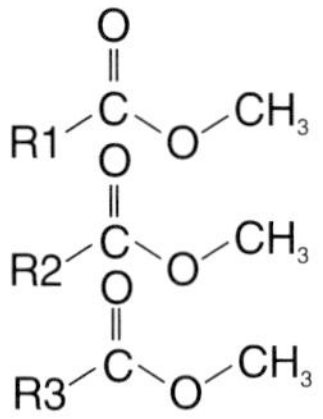

1.1.3 Properties of Vegetable Oil as Fuel

Vegetable oil has been found up to the mark satisfying the property requirements as a renewable oil resource. Reasons for opting triglyceride as alternate fuel source are [9]:

1. Liquid nature-portability
2. Ready availability
3. Renewability
4. Higher heat content
5. Lower sulphur content
6. Lower aromatic content
7. Biodegradability

Disadvantages of using triglycerides as source are:

1. High degree of viscosity
2. Low volatility
3. Reactivity of unsaturated hydrocarbon chains

1.1.4 Transesterification of Fatty Acids

Transesterification of triglycerides has been in application since the last decade. Fatty acids react with solvent or acyl acceptors to give fatty acid alkyl esters. The alkyl group of resultant FAAE and the byproduct depends on the solvent and catalyst used. Glycerol is obtained as byproduct in most cases whereas use of dimethyl carbonate, ethyl acetate, methyl acetate results in various other byproducts as discussed in latter sections. General equation for three-step formulated transesterification of fatty acid is given below:

	Catalyst	
1. Triglyceride + R'OH	$\rightleftharpoons$	Diglyceride + $R'COOR_1$
2. Diglyceride + R'OH	$\rightleftharpoons$	Monoglyceride + $R'COOR_2$
3. Monoglyceride + R'OH	$\rightleftharpoons$	Glycerol + $R'COOR_3$

1.2 PROCESS OF BIODIESEL PRODUCTION

Biodiesel production follows transesterification principle. Transesterification (also called as alcoholysis) is the reaction of fat or oil with an alcohol to form esters and glycerol. Two-step catalyzed production was also reported by Y.Wang et. al (2007) [10]. Usually a catalyst is used to mediate the reaction and bring out quicker reaction rate. After transesterification of triglycerides, the products are a mixture of esters, glycerol, alcohol, catalyst and tri-, di- and monoglycerides. Based on the process requirements, there are various methods that can be adopted for producing biodiesel from various oil sources. Some of them are described in a detailed fashion as given below.

1.2.1 Acid Catalyzed Process

Acid catalysis is the most suitable method in case of organic substrates. The transesterification process is catalyzed by acids and these catalysts give very high yields of alkyl esters, but the reactions are very slow. The homogeneous acid catalysts are H_2SO_4, HCl, BF_3, H_3PO_4 and some organic sulphonic acids. Sulphonic and sulphuric acids are mostly preferred. At the most, 99% conversion rates have been reported by using acid catalysts [11]. Freedman and Pryde [12] got the desirable product with 1 mol % sulphuric acid with a molar ratio of 30:1 at 65°C with 99% conversion rate in 51h. When excess of acid is added, better conversion of triglyceraldehyde is obtained. The main advantage of acid-catalyzed transesterification is the direct biodiesel production from low cost lipid feedstocks, such as waste cooking oil, greases, etc. [13]. These oil sources have FFAs level of >6% [14]. Though increased FFA content affects the conversion rates in acid catalyzed process, economy analysis depicts that single-step acid mediated conversion is better than two-step alkali mediated conversion where FFAs requires additional one step to be converted to methyl esters[15]. To overcome the fatty acid related problems, since the system is liquid based, liquid acid catalysts are proposed. Acid addition leads to protonation of the carbonyl group of the ester that results in carbocation which, after a nucleophilic attack of the alcohol, produces the tetrahedral intermediate, which eliminates glycerol to form the new ester and regenerates the catalyst H^+. According to the mechanism, carbocation can be allowed to undergo in the presence of water, but acid catalyzed transesterification of triglycerides are efficient in water absent systems [12, 16]. Solid heterogeneous acid catalysts have the potential to replace liquid acid catalysts. Solid catalysts are insensitive to FFA content of the oil thus eliminating the need for regular removal of biodiesel and byproduct from the reactors and enabling easy recovery and reuse of solid catalyst which in turn reduces corrosion problems. The solid catalyst is expected to have properties like interconnected system of large pores, enabled to tolerate high concentrations of acids and hydrophobic surface to promote the yield of ester conversions [13]. But unfortunately, problems rise with the glycerol recovery. This necessitates the experimental determination of optimal relations between the alcohol and raw materials [11].

1.2.2 Alkali Catalyzed Process

Most widely used homogeneous base catalysts are NaOH, CH_3ONa and KOH. Generally for homogeneous base catalysis either sodium hydroxide (NaOH) or potassium hydroxide (KOH) is used with methanol or ethanol as well as any kind of oils (refine, crude or frying). Alkaline metal oxides (CH_3ONa) are the most active catalysts, since they have very high conversion rates (>98%) in short time (30 min) even when applied at lower concentration (0.5 mol %). Effect of moisture has negative impact in case of alkali catalyzed process because of soap formation that consumes the catalyst and reduces the efficiency, as well as causing an increase in viscosity. To obtain maximum conversion rates without the use of excess alcohol, molar ratio of 6:1 is preferred [11]. The type of alcohol is usually ethanol and methanol. The first one has fewer safety problems because it is less toxic and comparatively easy to handle. Reactions with alkali catalysts are found to perform quickly than the acid catalyzed reactions [12, 17, 18]. Generally, alkali mediated reaction is a three-step formulated process. The first step is the reaction of base with the alcohol, producing an alkoxide and the protonated catalyst. The nucleophilic attack of the alkoxide at the carbonyl group of the triglyceride generates a tetrahedral intermediate, from which the alkyl ester and the corresponding anion of the diglyceride are formed. The latter deprotonates the catalyst which can react with second molecule of alcohol and start another catalytic cycle [4]. Standard value of the reaction to take place is 60°C, but depending on the oil source and catalyst, different degrees of conversion are obtained at various temperature ranging from 25 to 120°C [19-26]. Heterogeneous solid alkali catalysts are basic zeolites, alkaline earth metal oxides and hydrotalcites. They are easy to use, regenerate, supports recycling and to perform economically safer reactions. These catalysts can be directly used if the FFA content is less than or equal to 1. If, the amount of fatty acid is too big in case of waste cooking oil, then pretreatment via transesterification with alcohol but with sulphuric acid is recommended before initialising with alkali-mediated conversion [11].

1.2.3 Super Critical Methanol Process

A fluid is considered supercritical when its temperature and pressure go above its critical point. Supercritical Fluids (SCFs) posses unique transport properties. They can effuse through solids like gas and dissolve materials like liquid. The above discussed simple transesterification process is confronted with two problems, i.e., the process is relatively time consuming and it needs separation of the catalyst and saponified impurities from biodiesel. Apart from these problems, reduced efficiency of catalyst and high catalyst consumption are other problems that formed the root cause for the design of supercritical methanol process. These problems are eliminated in the non-catalytic supercritical methanol method of transesterification. This is perhaps due to the fact that the tendency of two-phase formation of oil/methanol mixture is not encountered and due to decrease in dielectric constant, single phase is encountered in the supercritical state of ethanol. Further, since the reaction is catalyst free, purification of biodiesel is easy, environment friendly and completes in 2-4 min [27]. This novel method was reported by Saka and Kusdiana [28] who demonstrated that preheating methanol at 350°C for 24 h was enough to convert rape seed oil to methyl esters. Molar ratio of methanol to rapeseed oil of 42:1 was considered as the best ratio [29]. Effect of water is an important factor in conventional catalytic conversion of vegetable oils and mostly resulted in negative results with the presence. Since, supercritical methanol have hydrophobic nature with a lower dielectric constant, non-polar triglycerides can be well solvated with supercritical methanol to form a single phase oil/methanol mixture. However, liquid methanol is a polar solvent and has hydrogen bonding between OH oxygen and OH hydrogen to form methanol clusters [30]. Thus, yield of methyl esters is high in supercritical methanol process than any other conventional techniques. Yield of methyl esters increases with increase in molar ratio of oil to methanol. Due to severe reaction conditions and high operational costs, this method suffers few disadvantages and hence it is not a viable option for industrial level commercialisation. The use of co-solvents such as carbondioxide, hexane, propane, etc., is being investigated to reduce the operational parameters and make the process economical.

1.2.4 Enzyme Catalyzed Transesterification – Lipases

In the previous sections, we discussed about the acid/alkali catalyzed transesterification reactions. Here we discuss about lipase mediated biodiesel production. Alkali catalysis mechanism is widely applied to scale up the production in industries. Though alkali catalyzed processes are economic and efficient enough, lipase dependent catalysis have certain advantages over conventional catalysis. Lipase is the most versatile industrial enzyme and known to carry out a range of bioconversions. Lipase mediated reaction offers easy removal of glycerol byproduct which is a necessary process to be carried out throughout the reaction to maintain the equilibrium in the positive domain (alkyl ester formation) in case of chemical mediated catalysis. Also, difficult glycerol recovery, removal of alkaline catalyst, energy intensiveness, treatment of highly alkaline waste water remains as a great challenge for industries. This serves as a main reason for replacing alkali catalyst with biocatalyst. Lipase mediated reactions are environmental friendly and do not promote side reactions. Extracted lipase and some lipase producing microorganisms (mostly fungi) are immobilized in biomass support particles and used as catalytic beds to obtain prolong use. It has been revealed that contribution of bacterial lipase is 45%, fungal 21%, animal 18%, plant 11%, and algae 3% [31]. Some of the lipase sources (microorganisms and plants) are listed below in Table 2.

Table 2 General sources of lipase [31, 32].

Bacteria	Fungus	Plants
Streptomyces sp.	*Mucor sp.*	*Lycopersicon esculentum*
Achromobacter sp.	*Rhizopus sp.*	*Carissa carandas* fruit
Alcaligenes sp.	*Candida sp.*	Castor bean
Arthrobacter sp.	*Geotrichum sp.*	*Cucumis melo*
Psuedomonas sp.	*Aspergillus sp.*	*Hibiscus cannabinus*
Bacillus sp.	*Pencillium sp.*	*Moringa olifera*
Staphylococcus sp.	*Humicola sp.*	Rice bran
Chromobacterium sp.		Triticum aestivum
Burkholderia sp.		

Only some species of organisms are selected based on few considerations such as activity of lipase, availability of species and amount of lipase produced to ensure the cost effectiveness and process efficiency. Microorganisms produce lipase in large quantities and hence whole organisms are chosen to act as immobilized catalytic beds. Commercial names of lipases are Novozym, Lipolase, Lipozyme, Lipomax, Lumafast, etc.[33] Some commercially available lipase producing microbial strains are given in the Table 3.

Table 3 Some commercially available lipase producing microbial strains [34]

Type	Source	Application
	P.fluorescens	Organic synthesis
	Burkholderia cepacia	Organic synthesis
Bacteria	*Chromobacterium viscosum*	Organic synthesis

Type	Source	Application
	P.aeruginosa	Organic synthesis
	Burkholderia glumae	Organic synthesis
Fungi	*Candida rugosa*	Organic synthesis
	Candida antarctica	Organic synthesis

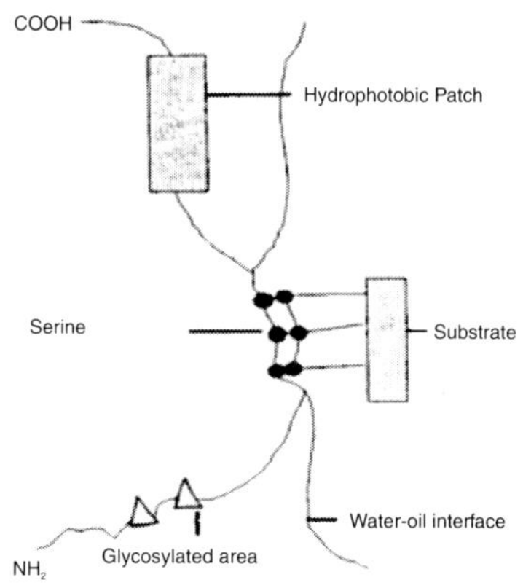

Figure 1 Diagrammatic representation of a Lipase molecule showing its main features. Substrate is Triglyceride [34].

1.2.4.1 ***Extracellular Lipase*** Extracellular lipase permits the efficient catalysis in conversion of fatty acids to alkyl esters. Extracellular enzyme catalysis requires the need of downstream processing techniques for enzyme extraction followed by immobilization to ensure repeated use. The extracted lipase from any suitable source (mostly microorganisms) is immobilized in a suitable support and used as catalytic beds. In the absence of organic solvent, lipase powders can be purchased and used for active catalysis in mediums with 4-30 wt% water without which the enzyme is nearly inactive [34]. For preparing lipase solution, lipase (0.5g) was dissolved in 5.0 ml of water, stirred for 1h,followed by centrifugation at 3500×g for 10 min. Supernatant was used as enzyme solution after dilution with water [34].

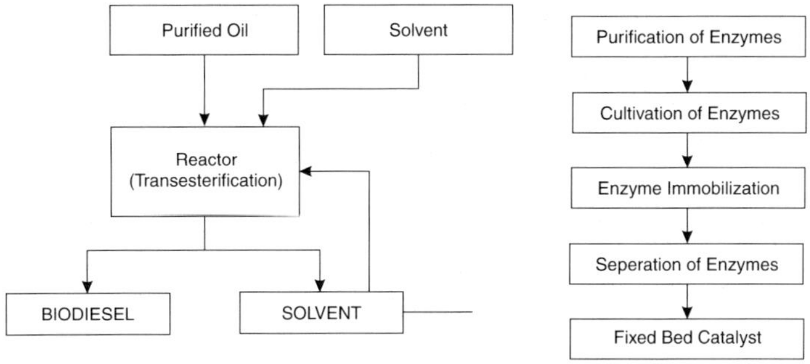

Figure 2 Process flow sheet of BDF production by extracellular lipase.

1.2.4.1.1 ***Immobilization of lipase enzyme and Optimization*** Many techniques and different carriers have been employed for immobilization of lipases to produce biodiesel. Lipases have been successfully immobilized on both hydrophilic and hydrophobic materials such as porous kaolite particle, macroporous resin and functionalized nanoscale SiO_2 spheres [3,35,36]. Immobilization of different lipases on macroporous polypropylene beds have been studied extensively which revealed that the successful adsorption of lipases on hydrophobic materials depends on structural features of "typical" lipases [37]. The catalytic activity sharply increases when the substrate forms a separate phase at which lipases are adsorbed. This phenomenon of "interfacial activation" was firstly observed by Sarda and Desnuelle [38]. The greatest part of the lipase structure seems to have a recurrent structural domain—an amphiphilic aminoaid chain called "lid" or "flap"—that buries the active site [39, 40]. It is related with the affinity of lipases on hydrophobic surfaces [39, 41]. Most lipases exist in two conformations, that is, an open and a closed conformations. The closed one dominates in water solution while open one exists in the presence of hydrophobic interface. In the closed conformation the flap exposes the hydrophilic side towards the water and hydrophobic side towards the catalytic side. In the presence of a hydrophobic substrate (Triglyceride), the hydrophobic part of lipase (catalytic site) changes in conformation, opens and adsorbs the substrate in open configuration [42]. The covalent attachment of lipase on styrene-divinyl benzene-polyglutaraldehyde support has been found to be twice stable than lipase immobilized on to styrene-divinylbenzene beds by hydrophobic interactions. The immobilized enzyme was stable and supposed to have same catalytic activity even after 30 days of storage at 4°C [43]. The optimal pH conditions for high lipase activity ranges from 5.0 to 9.0 and protein concentration from 1.14 to 11.4mg. A.Salis et al. (2008) [42] calculated the enzyme loading using the following equation.

$$L = \frac{A_i - A_r}{M_s}$$

Where L is enzyme loading; A_i is initial enzyme activity; A_r is residual enzyme activity; M_s is the mass of support (g). Loading is expressed in L (U/g support). N.Dizge et al. (2009) [43] calculated the lipase activity as follows:

$$\text{Lipase activity} = \left(\frac{U}{g}\ \text{support}\right) = \frac{\text{activity of immobilized lipase}}{\text{amount of lipase used}}$$

1.2.4.2 *Intracellular Lipase or Whole cell Biocatalyst for Biodiesel fuel Production* The use of extracellular lipase has been extensively studied and found to be expensive due to the complicated processes in the downstream techniques of enzyme extraction from microbial cells. These studies have prompted the use of microorganisms such as bacteria, yeast and fungal species as whole cell biocatalyst. Some of the species of lipase producing microrganisms having the ability to act as whole cell biocatalyst have been discussed. *Rhizopus oryzae* has been reported as an efficient whole cell biocatalyst by [44]. Catalytic activity of *Mucor meihei* has been discussed in [45]. *Candida antarctica, Candida rugosa* and *Candida cylindracea* have been noted significantly for the lipase productivity [46]. *Psuedomonas cepacia* and *Pseudomonas fluorescence* are found to have the capability to produce enough lipase and thus can be employed as catalyst in the process of methanolysis of oils [47]. Whole cell biocatalytic activity of engineered *Escherichia coli* and yeast have been discussed in detail along with the secretory mechanisms [44]. Immobilized recombinant *Aspergillus oryzae* expressing heterologous lipase is also found to be efficient for enantio-selective transesterification [48].

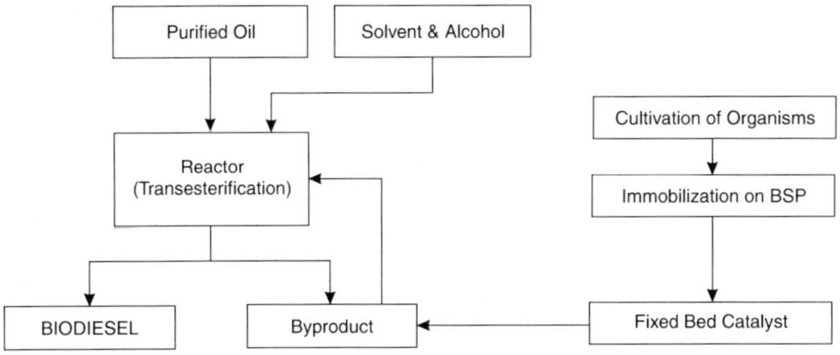

Figure 3 Process flow sheet for BDF production using intracellular lipase.

1.2.4.2.1 ***Whole cell immobilization or Bed preparation*** Immobilized *R.oryzae* cells were the first whole cell biocatalysts used in the process of transesterification that provides 1,3-positional specificity lipase. Cells were cultivated under normal conditions and then immobilized within biomass support particles (BSP). In some cases immobilized culturing of cells has been carried out [42]. For cultured cell immobilization, filamentous fungus such as *Rhizopus sp., Candida sp*. are grown in basal medium. Wei li [49] cultivated *Rhizopus oryzae* in a culture medium containing 30g soybean oil, 70g peptone, 1.2g $NaNO_3$, 1.2g $KH2PO_4$ and $MgSO_4.7H_20$ in 11 parts of tap water. Flasks (500ml) containing 100ml of basal medium were taken and spores are inoculated aseptically followed by incubation for 72h at 35°C on a reciprocator shaker (130 oscillations/min) with 80 BSPs subjected to prior sterilization. The BSPs used were 5mm cubes of reticulated polyurethane foam with a particle voidage exceeding 97% and a pore size of 50 pores per linear inch. Apart from *R.oryzae*, *R.chinensis* is also cultured and found to be active in non-aqueous system containing n-heptane [50]. For immobilized cell culturing, S.Hama [42] used Sakaguchi flasks (500ml) containing 100ml of basal media with glucose followed by sterilization and inoculated the spores of *R.oryzae* from a fresh agar slant aseptically. The inoculated flasks were incubated on a reciprocator shaker at 30°C for 24hr (150 oscillations/min; amplitude 70mm). The resultant culture broth was transferred to a 20-l air lift bioreactor, containing 10-l basal media with 30g/l of olive oil and 24,000 BSPs (6 mm × 6 mm × 3 mm cuboids). Aeration causes liquid and particle mixing and finally, temperature is maintained at 30°C. After cultivation the BSP immobilized cells were separated from the culture and stabilized. These catalytic beds are more stable and efficient in packed bed reactors than bottle flasks. Lipase can be localized using immunofluorescent detection technique using rabbit anti-ROL IgG in Phosphate Buffer Saline (PBS).

Table 4 Yield of biodiesel using lipase from various sources [50, 116]

Source of lipase	Yield (%)
Candida antarctica	30.0
Pseudomonas cepacia	27.0
Pseudomonas cepacia (immobilized on ceramic)	27.4
Candida rugosa	5.0
Porcine pancreas lipase	24.1
Rizhopus chinensis	28.4

1.2.4.2.2 ***Pretreatment and Stabilization*** Efficiency of conversion rate in whole cell catalyzed systems also depends on the cell wall fatty acid composition of the immobilized organism as discussed [44]. Pretreatment of catalytic beds triggers increased production of lipase by immobilized microorganisms. There are two types of lipases, one bound to cell wall (ROL 34) another bound to cell membrane (ROL 31). Hama et al. (2006) [51] reported that the enzyme activity increases with addition of olive oil to catalytic beds before the reaction. It has been reported that the increased catalytic activity is due to the increase in production of membrane bound lipase (ROL 31). Hence ROL 31 plays a major role in methanolysis activity. For pretreatment, substrate related compounds of lipase such as triglycerides and fatty acids acts as inducers in filamentous fungus [52]. Cells can also be preincubated in soy bean oil for some time before subjecting to methanolysis. For methyl ester pretreatment, oil is replaced with methyl ester and used. Whole cell immobilization permits prolonged use of catalytic beds thus catalyzing number of reaction cycles which depends on various other factors such as solvent used and volume considerations.

Table 5 Transesterification of oils from various sources.

Oil	Catalyst	Type	Catalyst amount (%)	Alcohol	Oil to alcohol molar ratio	Reaction conditions	Ester yield	Refs
Rice bran oil	$NaOCH_3$	Alkali, homogeneous	0.88	-	-	65°C, 1 h	83.3	[53]
	-	Acid, homogeneous	-	Methanol	1:10	60°C	<96	[54]
Cotton seed oil	$NaOCH_3$	Alkali, homogeneous	0.75	-	-	75°C, 1.5 h	97	[55]
	TiO_2/SO_4^{2-}	Acid, heterogeneous	2	-	1:12	230°C, 8 h	90	[56]
	Lipase (*Candida antarctica*)	Enzyme	1.6	t-butanol	1:4	50°C, 24 h	95	[57]
	-	Super critical condition	-	Methanol	1:41	350°C, 8 min	98	[58]
	-	Super critical condition	-	Ethanol	1:41	350°C, 8 min	75	[58]
	-	Super critical condition	-	Methanol	1:41	250°C, 8 min	85	[59]
Soybean oil	KOH	Alkali, homogeneous	0.5, 1, 1.5	Methanol	1:3, 1:6, 1:9	50-110°C, 2-50 min	99	[60]
	$Ca(OCH_2CH_3)_2$	Alkali, heterogeneous	3	Ethanol	1:12	75°C, 3 h	91.8	[61]
	Lithium-doped ZnO	Alkali, heterogeneous	-	Methanol	1:12	65°C, 3 h	96.3	[62]
	Eu(No3)3/Al2O3	Alkali, heterogeneous	-	Methanol	1:6	70°C, 6 h	63	[62]
	KF/ZnO	Alkali, heterogeneous	-	Methanol	1:10	65°C, 9 h	87	[62]
	H_2SO_4	Acid, homogeneous	3	n-Butanol	1:1	120°C, 1 h	-	[63]
	Trifluoroacetic acid	Acid, homogeneous	2	Methanol	1:20	120°C, 5 h	98.4	[64]
	S-ZrO_2 Sulfated Zirconia	Acid, heterogeneous	5	Methanol	1:20	120°C, 1 h	98.6	[65]
	$Al_2O_3/ZrO_2/WO_3$	Acid, heterogeneous	4	Methanol	1:40	250°C, 20 h	90	[66]
	SBA-15-SO_3H-P123	Acid, heterogeneous	10	Methanol	1:20	75°C, 20 h	85	[67]
	Silica-bonded N-propyl sulfamic acid	Acid, heterogeneous	7.5	Methanol	1:20	149.85°C, 60 h	90.5	[68]
	Lipase (*Thermomyces lanuginosus*)	Enzyme	15	-	1:7.5	31.5°C, 7 h	96	[69]
	Lipase (*Candida antarctica*)	Enzyme	3	Ethanol	1:2	50°C, 1.5 h	83.5	[70]

Oil	Catalyst	Type	Catalyst amount (%)	Alcohol	Oil to alcohol molar ratio	Reaction conditions	Ester yield	Refs
	-	Supercritical Condition	-	Methanol	1:40	350°C, 35 Mpa, 25 min		[71]
	-	Supercritical Condition	-	Ethanol	1:40	20 Mpa, 350°C , 15 min	80	[72]
Sun-flower oil	NaOH	Alkali, homogeneous	1	Methanol	1:6	60°C, 2 h, 600 rpm	94	[73]
	KOH	Alkali, homogeneous	-	Methanol	1:6	25°C,	90	[74]
	Magnesium-lanthanum mixed oxide	Alkali, heterogeneous	5	Methanol	1:53	65°C, 30 min	100	[75]
	Lipase (*Psuedomonas fluorescens*)	Enzyme	10	Hexane	1:4.5	40°C, 48 h	91	[76]
	Lipase (*Candida antarctica*)	Enzyme	30	Ethanol	1:5	45°C, 12 h	27	[70]
	Lipase (*Mucor meihei*)	Enzyme	20	Ethanol	1:11	45°C, 5 h	82	[70]
	-	Supercritical condition	-	Methanol	1:40	350°C, 200 bar, 40 min	96	[77]
	-	Super critical condition	-	Ethanol	1:40	400°C, 20 MPa, 40 min	100	[72]
Mahua	H_2SO_4	Acid, homogeneous	0.32	-	1:24 (v/v)	60°C, 1 h	98	[78]
Palm	NaOH	Alkali, homogeneous	1	-	-	70°C, 30 min	95	[79]
	SO_4/SnO_2	Acid, heterogeneous	3	Methanol	1:6	200°C	80.6	[80]
	ZrO_2/SO_4^{2-}	Acid, heterogeneous	1	-	1:6	200°C, 1 h	90.3, 80.6	[81]
	KF/Al_2O_3	Alkali, heterogeneous	4	Methanol	1:12	65°C, 3 h	90	[82]
	CaO/Al_2O_3	Alkali, heterogeneous	-	Methanol	-	65°C, 3 h	<95	[83]
	Lipase (*Pseudomonas fluorescens*)	Enzyme	20	Ethanol	1:18	40°C, 24 h	98	[70]
	Lipase (*Candida antarctica*)	Enzyme	30	Ethanol	1:5	45°C, 10 h	75	[70]

Oil	Catalyst	Type	Catalyst amount (%)	Alcohol	Oil to alcohol molar ratio	Reaction conditions	Ester yield	Refs
	-	Supercritical condition	-	Methanol	1:45	350°C, 60 min	60.30	[84]
	-	Supercritical condition	-	Ethanol	1:50	400°C, 20 MPa, 30 min	95	[72]
Karanja oil	KOH	Alkali, homogeneous		Methanol tetrahy-dro-furan	1:10	60°C	92	[85]
	KOH	Alkali, homogeneous	1	Methanol	1:6	65°C, 2 h, 360 rpm	98	[86]
	Li/CaO	Alkali, heterogeneous	2	Methanol	1:12	65°C, 8 h	94.8	[87]
Tobacco	H_2SO_4	Acid, homogeneous	1	Methanol	1:18	60°C,25 min	91	[88]
Refined sunflow-er oil	CH_3ONa, CH_3OK	Alkali, homogeneous	1	Ethanol	1:3, 1:6, 1:9	30-80°C, 3-60 min	46.3-99.6	[89]
Rubber seed oil	NaOH	Alkali, homogeneous	0.5	Methanol	1:9	45°C, 30 min	-	[90]
Rape-seed oil	KOH	Alkali, homogeneous	1	Methanol	1:6	65°C, 2 h, 600 rpm	96	[91]
	Styrene-divinyl benzene macropo-rous	Acid, heterogeneous	10	Methanol	1	100-140°C, 4h	16.5-55	[92]
	KF/Eu_2O_3	Alkali, heterogeneous	-	Methanol	1:12	Methanol reflux, tem-perature, 1 h	92.5	[62]
	Mg-AL HTs	Alkali, heterogeneous	-	Methanol	-	65°C, 3 h	<94	[83]
	Lipase (*Candida antarctica*)	Enzyme	3	t-butanol	1:4	35°C, 12 h	95	[93]
	-	Supercritical condition	-	Methanol	1:24	250°C, 6 MPa, 10 min	97	[94]
	-	Supercritical condition	-	Ethanol	1:42	350°C, 25 MPa, 12 min	100	[72]

Oil	Catalyst	Type	Catalyst amount (%)	Alcohol	Oil to alcohol molar ratio	Reaction conditions	Ester yield	Refs
Waste cooking oil	NaOH	Alkali homogeneous	1.1	Methanol	1:7	66°C, 33 min	88.8	[95]
	KOH	Alkali homogeneous	1	Methanol	1:6	70°C, 1 h	98.2	[96]
	H_2SO_4	Acid, homogeneous	41.8	Methanol	1:245	70°C, 4 h	99	[97]
	Zeolite Y (Y 756)	Acid, heterogeneous	-	Methanol	1:6	460°C, 37 min	26.6	[95]
	SO_4^{2-}/SnO_2-SiO_2	Acid heterogeneous	3	Methanol	1:!5	150°C, 3 h	92.3	[95]
	TiO_2/MgO	Acid, heterogeneous	-	Methanol	1:30	150°C, 6 h	92.3	[62]
	K_3PO_4	Alkali heterogeneous	4	Methanol	1:6	60°C, 2 h	97.3	[95]
	Lipase (*Candida antarctica*)	Enzyme	4	-	1:3	30°C, 50 h	90.4	[95]
Jatropha oil	KOH	Alkali, homogeneous	1	-	-	75°C, 1 h	97.6	[98]
	Alumina loaded with potassium nitrate	Alkali, homogeneous	6	Methanol	1:12	70°C, 360 min	84	[99]
	NaOH	Alkali, homogeneous	1	Methanol	1:5	65°C, 90 min	98	[99]
	KNO_3/Al_2O_3	Acid, heterogeneous	6	Methanol	1:12	70°C, 6 h	84	[100]
	Na doped SiO_2	Alkali, heterogeneous	-	Methanol	1:15	65°C, 45 h	99	[62]
	CaO	Alkali, heterogeneous	-	Methanol	-	70°C, 2.5 h	<93%	[83]
	Lipase (*Chromobactrium viscosum*)	Enzyme	10	-	1:4	40°C, 10 h	92	[101]
	Lipase (*Candida antarctica*)	Enzyme	30	Ethanol	1:5	45°C, 10 h	55	[70]
	Lipase (*Psuedomonas cepacia*)	Enzyme	10	Ethanol	1:4	50°C, 8 h	98	[70]
	-	Super critical condition	-	Methanol	1:40	350°C, 200 bar, 40 min	>90	[102]

Oil	Catalyst	Type	Catalyst amount (%)	Alcohol	Oil to alcohol molar ratio	Reaction conditions	Ester yield	Refs
	-	Super critical condition	-	Ethanol	1:50	400°C, 20 MPa, 30 min	90	[72]
Neem oil	NaOH	Alkali, homogeneous	0.7	Methanol	1:6	60°C, 6.5 h	94	[103]
Corn oil	*p*-Toluene sulphonic acid (PTSA), benzene sulphonic acid, H_2SO_4	Acid, homogeneous	1-4	Methanol	10-3	40, 60, 80°C, 20 min	90.2-97.1	[104]
Canola oil	Lewis acids (AlC_3, $ZnCl_2$)	Acid, homogeneous	1	Methanol	1:24	110°C, 18 h	98	[105]
	H_2SO_4, benzene sulphonic acid, *p*-Toluene sulphonic acid or 2,4-dimethylbenzene sulphonic acid	Acid, homogeneous	1-5	Methanol	1:15	60°C, 30min	21-99.9	[106]
	Na, K, LieBaO	Alkali, heterogeneous	-	Methanol	-	50°C, 4 h	<97.5	[83]

1.2.4.3 *Use of Novel Acyl Acceptors in Lipase Catalyzed Process* Apart from the regular acyl acceptors like methanol and ethanol, researchers have also proposed the use of other novel acyl acceptors such as methyl acetate [Wei Du et al.], ethyl acetate, [87] and dimethyl carbonate [107].

Use of alcohol for the transesterification process gives rise to practical problems such as deactivation of the immobilized lipase by lower chain alcohols like ethanol and methanol [108, 109]. Regular removal of byproduct is advised to avoid process reversion as the process is reversible [107]. Some alternative ways such as use of solvents like hexane [110] or t-butanol [111] are also reported. The use of low concentration of methanol in ratio of 0.33:1 to oil [108] was reported by Samukawa et al. (2000) for the first problem and stepwise addition of alcohols was proposed for the second problem [114]. Even though, the alternatives showed good results, they increased the operational cost and made the process complex thereby requiring more manpower.

1.2.4.3.1 *Use of methyl acetate* Du et al. (2004) [112] suggested the use of methyl acetate instead of alcohol. High yield of 92% was obtained for the molar ratio of 1:12 of oil to methyl acetate. The yield was obtained for crude and refined soybean oil. The lipase B from *Candida antarctica* was immobilized on acrylic resin and the enzymes remained active for 100 batches. The byproduct obtained is triacetyl glycerol and not glycerol and hence the loss of activity due to glycerol was eliminated.

1.2.4.3.2 *Use of ethyl acetate* Use of ethyl acetate as an acyl acceptor for lipase catalyzed transesterification was reported by [Prasad et al (2006) (87)]. Lipase B from *Candida antarctica* immobilized on acrylic resin was allowed to react with crude oil of Jatropha, Karanj and sunflower oil. A yield of 91.3%,90% and 92.7% of ethyl ester was obtained respectively when reaction was processed with molar ratio of 11:1 of ethyl acetate to oil at 50°C with reaction period of 12h. The activity of enzyme remained efficient enough till 12 repeated cycles. The byproduct obtained was triacetin, a valuable molecule which has widespread application. The advantage of ethyl ester is that the extra carbon atom increases the heat content and cetane number [87].

Ethyl esters also have lower pour and cloud points, higher flash and combustion points which improve cold starts and safety in handling. Smoke opacity and exhaust temperature are also lowered [87].

1.2.4.3.3 ***Use of Dimethyl carbonate*** Use of methyl acetate and ethyl acetate has a disadvantage that a large amount of compound is required (1:12 of oil/methyl acetate; 1:11 of oil/ethyl acetate). Hence another acyl acceptor Dimethyl Carbonate was used by E-Z Su et al. (2007) (107). DMC is a neutral, odourless, cheap, non-corrosive, non-toxic compound that exhibits good solvent properties [115]. Advantage of DMC-involved transesterification is that the reaction remains in positive product formation side (biodiesel) as the product is CO_2 which escapes as gas.

1.3 PROCESS VARIABLES

For efficient biodiesel production using enzyme catalyzed transesterification certain crucial parameters need to be addressed. The optimization of these parameters is crucial in obtaining maximum yields. The basic parameters that affect the yield obtained are selection of lipase, selection of substrates, molar ratio of substrates, temperature, pH of lipase micro environment, water content, solvent content, glycerol content, mixing intensity and purity of reactants.

1.3.1 Selection of Lipase/Organism

Selection of the lipase mainly depends on:

1) Whether the system is a solvent involved system or a solvent independent system,
2) The type of fatty acid involved, and
3) Whether the lipase to be used is intracellular or extracellular along with other reaction parameters.

An organic solvent such as n-hexane, s-butanol, petroleum ether are added to a system to increase the miscibility between the triglyceride and methanol and thus increasing the catalytic efficiency of lipase. Reactions can also be carried out in solvent free systems.

Lipase from various microbial sources gives various results in different systems and they are given in the table 5 [116].

The triacyl glycerol (TAG) and free fatty acids (FFA) present in the oil decides the activity of the lipase. Specificity of lipases for biodiesel synthesis refers to their region specificity or specificity with respect to the length of hydrocarbon chain of fatty acid [117]. The lipases hence selected for the catalysis are those that display wide substrate specificity. Examples are lipases from pseudomonas and candida species [118, 119]. Lipases from other sources are also found to show considerable yields depending on the reaction parameters. Selection of the right lipase for the particular source and condition is necessary for higher yields.

Lipase used for biodiesel production can be either used directly by immobilizing it in a suitable matrix or the whole organism can be used for the catalysis purpose. Whole cell biocatalysts are considered cheaper and hence are appropriate for industrial biodiesel manufacturing.

1.3.2 Substrates for Biodiesel production

1.3.2.1 ***Lipids*** Plant oils (refined and raw), oil from micro algae, and waste fats which remained after cooking are considered for biodiesel production. Depending on the geographical region and crops grown, the oil for the biodiesel production is considered. Watanabe et al [120] compared the effective transesterification of three forms, raw, refined and degummed soya bean oil for using lipase from *Candida antarctica*. The result showed lesser conversion in raw oil due to the presence of phospholipids. High amount of phospholipids gave low flame [112]. The phospholipid problem can be addressed by immersion pre-treatment of lipase (*C. antarctica*) in crude oil for 12h, oil degumming [122] and simultaneously dewaxing/degumming was found to be efficient [112, 122]. Even though refined oil gave higher yields, it was costly. Oil from micro algae is also potential source for biodiesel production. Hence proper selection of the source is important for industrial implementation.

1.3.2.2 ***Selection of Acyl Acceptors*** Transesterification involves two substrates, one being the oil and other being an acyl acceptor. The acyl acceptor is crucial for the production of the biodiesel. Several acyl acceptors currently under study are alcohols such as methanol, ethanol, iso propanol, iso butanol, 2-butanol and 1-butanol or other novel acyl acceptors such as methyl acetate, ethyl acetate and dimethyl carbonate.

Methanol and ethanol are industrially produced in large scale and are cheap. Use of ethanol as acyl acceptor gives biodiesel with an additional carbon atom which increases the fuel properties. One solution to denaturation by methanol is stepwise addition of the alcohol [123]. By this process 1/3 molar equivalent of methanol was added for the stoichiometric amount of the waste oil and the eluate of the first step is again treated with 1/3 molar equivalent of methanol. Immobilised *C.antarctica* was taken and the yield was reported to be more than 90%. Nelson et al [124] reported the compared yields of transesterification of methanol, ethanol and isobutanol with follow in presence of immobilized lipozyme IM 60 (M.Michel) and novozyme SP435 (*C. antarctica*); the yield was 83.8% for isobutanol in hexane solvent and in solvent free system the yield for isobutanol was highest with 97.4%. This shows that isobutanol can also be used as an acyl acceptor for lipase catalysed transesterification. The draw back with alcohols is that they inactivate the lipase at higher concentration. It has been noted that the degree of deactivation has been found to be inversely proportional to the number of carbon atoms in the linear lower alcohols [125]. The simultaneous presence of water with methanol can speed up denaturation of lipase while in systems containing ethanol, propanol, isopropanol, butanol the presence of small amounts of water is necessary [126]. Methyl acetate was used as an acyl acceptor by Du et al [121] and a yield of 92% was reported for crude soya bean oil in the presence of lipase from *C. antarctica*. Prasad et al used ethyl acetate and the yield for jatropha, karanj and sunflower oils were 91.3%, 90% and 92.7% (lipase from *C. antarctica* was used). The acyl acceptor used in both the process involve high amount of lipase (three-fold more) and molar ratio of oil to acyl acceptor is 1:12 and 1:11 respectively. Use of dimethyl carbonate by Wei et al [107] gave a yield of >90% for cotton seed oil, olive oil, sunflower oil, corn seed oil, rapeseed oil and soyabean oil in presence of organic solvent petroleum ether. Use of DNL gives an irreversible reaction.

1.3.2.3 ***Selection of Solvents*** Solvents are used in biodiesel production in order to increase the mutual solubility of hydrophobic triacyl glycerides and hydrophobic alcohols. Poor solubility of short chain alcohols such as methanol and ethanol causes inhibition on the enzyme activity. It has been observed by Laane et al [127] that the log P value was the fundamental property that described the influence of polarity-hydrophobicity of organic solvents on enzymatic catalysis. An ideal solution must ensure good solubility of oil and alcohol along with maintaining the enzyme stability. Use of a mixture of solvents can be done [128] in order to get better results.

Among the many organic solvents present, certain organic solvents were observed to give higher yields and provide more stability to the enzyme. They are hydrophobic ones such as isooctane (log P = 4.7), n-heptane (log P = 4.0), pertroleum ether 60-120 (log P=3.5-4.3), petroleum ether 20-40 (log P = 3.2), n-hexane (log P=3.5), cyclo hexane (logP=3.1) [R21 (32)] tert-butanol (logP = 0.83). The other hydrophobic solvents used are acetone (log P = -0.24), n-hexane (log P = -3.5), diesel oil (log P = -8.0). Although there are many reasons for using an organic solvent in biodiesel production by transesterification, organic solvents also have certain draw backs such as being volatile, flammable and toxic. These negative attributes make them a bane for economic biodiesel production. Also organic solvents make the biodiesel recovery difficult during downstream operations and certain organic solvents (hydrophobic solvents like n-hexane and petroleum ether) do not completely solubilize methanol and glycerol [129]. Hence their purpose is not completed. To address these issues several interesting ways have been proposed. One is using tert-butanol as solvent which is a moderate polar solvent which solubilises both methanol and glycerol [129], second being the use of diesel as organic solvent [130] which will eliminate the processes involved in solvent recovery and the third being the use of solvent free system where in the methanol is added in batches [123].

1.3.3 Glycerol Effect

In continuous and repeated batch process glycerol was formed to effect the transesterification process by inactivating the enzyme. Dossat et al [131] & Du et al [121] observed a drop in activity of lipases by glycerol when it was adsorbed in a hydrophilic matrix. This was due to the adsorbing of glycerol molecules onto the surface of the matrix which resulted in the formation of hydrophilic coating of glycerol which made the enzyme molecules inaccessible to hydrophobic substrates (reduced diffusion). Another possible mechanism proposed for inhibition by glycerol was that glycerol may cause a decrease in the water activity of the enzyme. Several novel methods were proposed to counter the inhibition caused by glycerol. Stevenson et al [132] tested various procedures for their ability to retain the initial enzymatic activity by addition of silica gel to the reaction system, which being a hydrophilic substance, partially removed glycerol from the lipase environment. Use of n-heptane amended with acetone as a reaction mixture or use of a semi-continuous process consisting of a transesterification reaction and rinsing of the catalyst with a rinsing solution amended with water, removes glycerol and also restored the thermodynamic water activity. Du et al [121] observed that washing immobilized lipase after transesterification with isopropyl alcohol restored its activity as glycerol was removed from the carrier. Li et al [129] used a hydrophilic tert-butanol as an organic solvent in the transesterification of rape seed oil using *C.antarctica* lipase and *Tidestomia lanuginosa* lipase, which not only helped in the solubilisation of methanol but also solubilised glycerol. Therefore, the coating of the immobilized lipase by glycerol was controlled. Chen & Wu [125] proposed the use of tert-butanol or 2-propanol for washing immobilized *C.antarctica* lipase. Belati-Bako et al [133] proposed the use of a membrane bioreactor using a suitable dialysis membrane. The reaction takes place in the primary side of the module , and the glycerol produced during the reaction passes through the membrane and is accumulated in the secondary (aqueous) phase. Another novel approach in countering glycerol inhibition was proposed by Xiu et al [134] where in an integrated bioprocess combining biodiesel production using lipase and the conversion of glycerol to 1,3-propane diol using bacteria *Klebsiella pneumoniae* in an hollow fiber membrane was carried out.

1.4 PRODUCTION OF BIODIESEL FROM MICROALGAE

1.4.1 Algae

Algae can be directly converted into energy [135]. Algal Biomass grows in aquatic conditions and uses light and carbondioxide for their normal metabolism and survival. There are two types of algae: Micro algae and Macro algae. Macro algae are multicellular and large usually grown in ponds and oceans characterized by varied growth patterns. Microalgae are a diverse group of species, pro karyotic and eukaryotic, photosynthetic microorganisms. Examples of prokaryotic microalgae are cyanobacteria (cyanophyceae) and eukaryotic micro algae are green algae (chlorophyta) and Diatoms (Bacillariophyta) [136, 137]. Microalgae are tiny, unicellular and multicellular organisms found growing in suspensions [138]. Microalgae are found to be rich in oil source (Triacyl Glycerols) which highlights them as a prime base for biodiesel production. The growth of Microalgae is very quick and some species have been reported to have their doubling time in 3.5 hours (Chisti 2007). Algal harvesting consists of biomass recovery from the culture medium that may contribute to 20-30% of the total biomass production cost [139]. Usually microalgae have oil content ranging from 30-70% of dry weight depending on the species of consideration. This clearly portrays that algae can replace coal and oil producing crops in the near future. Though the global fuel demand intensifies exponentially, biodiesel from oil producing crops cannot satisfy existing demand realistically and algal biofuels draws the attention of satisfying the same.

Table 6 Oil yields from various biomasses [8, 13, 138-142,144,145]

Biomass	**Oil content (% oil by wt in biomass)**	**Land use (m^2 year/ Kg biodiesel)**	**Oil Yield (lit/ ha)**	**Biodiesel productivity (kg biodiesel/ ha year**	**Uses**
Rubber Seed	50	-	80 -120	-	Nutrient rich feed for livestock, Biodiesel production etc.
Hemp	33	31	363	321	Fibre, Feed for livestock, Biodiesel production etc.
Corn/Maize	44	66	172	152	Food, Feed for livestock, Biodiesel production etc.
Soybean	18	18	446	562	Protein rich supplements, Biodiesel production etc.
Safflower	-	-	779	-	Flavouring foods, Nutritional supplements, Biodiesel production etc.
Chinese tallow	-	-	907	-	Ornamental tree and oil for Biodiesel production.
Camelina	42	12	915	809	Omega-3-supplement, Edible Oil and Biodiesel production.
Rice bran	16-32	-	-	-	Non-edible vegetable oil, Replacement of mineral diesel etc.
Sunflower	40	11	952	946	Edible oil, Biodiesel production etc.
Nahor	58-75	-	-	-	Wood for railway crossties, Boat building, Mine props ,Biodiesel fuel etc.
Peanut	-	-	1059	-	Lactose free milk like beverage, Oil for biodiesel production.
Canola// Rapeseed	41	12	1190	862	Edible oil, Biodiesel production etc.

Biomass	Oil content (% oil by wt in biomass)	Land use (m² year/ Kg biodiesel)	Oil Yield (lit/ ha)	Biodiesel productivity (kg biodiesel/ ha year	Uses
Moringa	33-41	-	-	-	Medicinal use, leaves as greens in food, Oil for biodiesel etc.
Castor	48	9	1413	1156	Adhesives, coatings, soaps, lubricants, paints and dyes etc.
Jatropha	28	15	1892	656	Biodiesel, mineral diesel substitute, live fence, to reclaim land and control erosion.
Karanj	-	-	2590	-	Tanning leather, soap, lubricant, water-paint binder, pesticide etc.
Coconut	-	-	2689	-	Edible oil, kernel, biodiesel production etc.
Mahua	35-40	-	-	-	Ointment, rheumatism, lubrication, illumination, soaps, etc.
Linseed	35-45	-	-	-	Geotextiles, oil crop, stem fibres, paper manufacturing etc.
Oil palm	36	2	5950	4747	Cosmetic, soaps, lubricants, transportation fuel and substitute for biodiesel.
Microalgae (low oil content)	30	0.2	58,700	51,927	Human nutrition, Animal feed, aquaculture, bio-fertilizers, recombinant proteins, source of poly unsaturated fatty acids, methane production etc.
Microalgae (medium oil content)	50	0.1	97,800	86,515	
Microalgae (high oil content)	70	0.1	136,900	121,104	

1.4.2 Algal Diversity and Global Collection Centres

It is estimated that 50,000 species of micro algae exists, but only 30,000 have been studied and analyzed and 15 were used commonly [143]. There are many microalgae collection centres in the world that provides mother culture for further use. Fresh water microalgae collection of Coimbra (Portugal) contains 4000 strains and 1000 species. Goettingen University has a collection of about 2213 strains and 1273 species. University of Texas Algal Culture Collection was founded in 1953 and has about 2300 different species of microalgae. National Institute for Environmental Studies Collection (NIES), in Japan holds a collection of about 2150 strains and 700 species of different algae. The CSIRO Collection of Living Microalgae (CCLM) in Australia holds about 800 strains from Australian waters [138].

1.4.3 Biodiesel from Microalgae

Microalgae require organic carbons for their normal metabolism and are categorized under autotrophs and heterotrophs. Photoautotrophic algal population finds the majority among algal species and basically requires light and carbondioxide for their growth. To satisfy the capitalization factors and justify algal biomass for biofuel source, autotrophic algal biomass helps in upholding the focus and advances the economics of algal biofuel. The table below list out the oil producing potential of few open pond algal species.

Table 7 Oil content of few open pond microalgal species. (Chisti 2007) [8]

Microalga	Oil Content (%Dry Weight)
Botryococcus braunii	25–75
Chlorella sp.	28–32
Crypthecodinium cohnii	20
Cylindrotheca sp.	16–37
Nitzschia sp.	45–47
Phaeodactylum tricornutum	20–30
Schizochytrium sp.	50–77

Although microalgae can utilize light efficiently, the phototrophic growth of algae is too slow when compared with heterotrophic algae. There are factors like photoinhibition, where algae is exposed to excessive light in sunny days and light limitations due to high cell density in large scale culture affecting the performance of phototrophic algae. Hence, heterotrophic algal cultivation in conventional fermenters is suggested [147]. Heterotrophic cultivation of algae offers several advantages such as elimination of light requirements, low cost for harvesting, good and controlled cultivation process and higher cell density [148]. Cell growth can be influenced by medium supplied, and controlled environmental factors, where heterotrophic algal cultivation superiors to phototrophic algal cultivation.

1.4.4 Microalgae Cultivation

Microalgal cultivation with suitable mediums was discussed and potential oil producing feedstocks are examined in lab scale [149,150]. Optimisation parameters such as nitrogen, phosphorous concentration and outdoor parameters such as photoperiods can make a significant change in lipid productivity [149]. Lifecycle analysis by O.Jorquera et al. (2010) [151], J.Yang et al. (2011) [152] and P.K.Campbell et al. (2011) [153] enables the screening of suitable microalgal strains and the selectivity based on various parameters. Economical production of microalgae can be done in two common methods as follows:

1.4.4.1 *Open Ponds (or) Raceways*

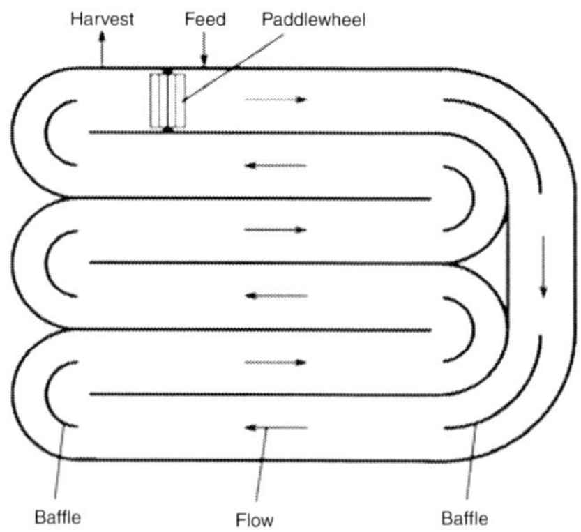

Figure 4 Open pond system or Raceways

For the production of phototrophic algae, open ponds are designed in raceway configuration usually 0.3m deep and a paddle wheel circulates and mixes the algal cells with nutrients. This system is continuous where fresh feed is added in front of the paddlewheel and the algal broth is collected behind the wheel after it has circulated the whole loop. For some marine type microalgae, seawater or high salinity water can be used. Open pond limits the biomass productivity as they do not allow microalgae to use carbondioxide efficiently [8]. In addition, optimal culture conditions are difficult to maintain in open ponds and biomass recovering methods remain expensive [154]. About 25% of biomass produced during day light may be lost during night due to respiration factors (R9). The extent of this loss depends on the light level and growth temperature at night. Besides certain disadvantages, still open ponds are the efficient ways to scale up the production of algal biomass in large quantities [154, 155].

1.4.4.2 ***Tubular Photobioreactors*** It has been reported that large quantities of microalgal biomass can be successfully produced by tubular photobioreactors [154]. Photobioreactors permit the growth of single species of microalgae for long time durations. Enclosed photobioreactors have been employed to overcome the problem of contamination and evaporation [154]. This system is made of transparent tubes and generally placed outdoors for natural light illumination. These cultivation vessels have good surface area to volume ratio to increase productivity. Tubes are generally less than 10 cm in diameter to maximize sunlight penetration. A continuous algal culture is made possible by circulating the medium broth through a pump to the tubes where it gets exposed to light and then sent back to reservoir. A portion of algae is harvested after it passes through the tubes. Spirally coiled tubes form the basic structure of helical shaped bioreactors (154). The biomass productivity of photobioreactors can average 13 times more than that of a traditional raceway pond. Biomasss harvest in this method remains easy because of the biomass concentration which is 30 times the concentration of biomass obtained via raceway method.

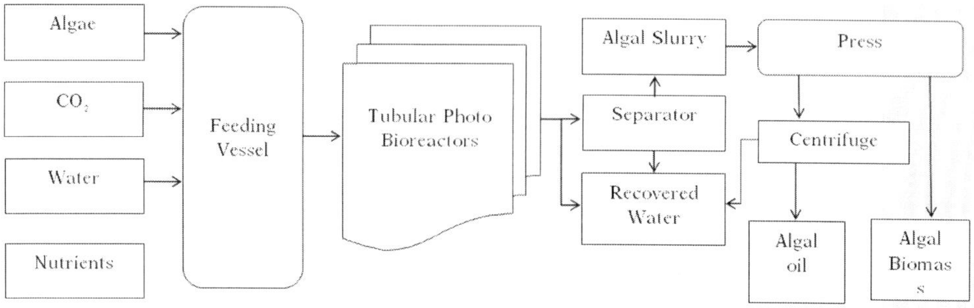

Figure 5 Cultivation of Algal biomass using Tubular Photo Bioreactors.

1.4.5 Triglyceride Synthesis

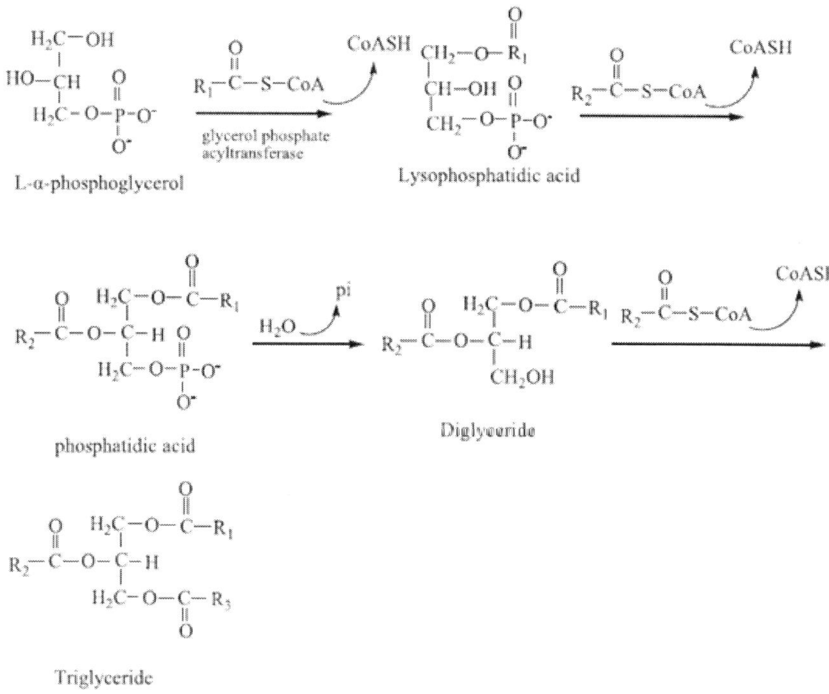

Figure 6 Synthesis of triglycerides in microalgae

The ultimate aim of cultivation is lipid production which is the main substrate for biodiesel production. Like other higher plants and animals, algal cell also produces triglycerides to store substance and energy (G.Huang et al., 2010) [156]. Generally, acetyl-coA and

L-α-phosphoglycerol are the primary compounds in triglyceride production. Acetyl-coA and phosphodihydroxyacetone, precursor of L-α-phosphoglycerol are obtained as a result of glycolysis process. Acetyl-coA reacts with the hydroxyl group of L-α-phosphoglycerol to give lysophosphatidic acid. This acid again combines with acyl-coA to form phosphatidic acid. Glycerol phosphate acyl-transferase catalyses the above two reactions. Further, lysophosphatidic acid is hydrolysed by phosphatidate phosphatase to form diglyceride. Another acetyl-coA is combined with diglyceride to give triglyceride and this is catalyzed by glyceryl diester transacylase.

1.4.6 Extraction and Analysis of Oil and Biomass

Extraction of oil from algal biomass requires the need of cell disruption techniques. Mechanical cell disruption technique involves the use of press configurations such as crushers called as oil press. Crush configurations vary with algal species as the cell physiology changes. Cavitation caused by collapsing bubbles in ultrasonicators creates shock waves that can also be employed to disrupt the cell wall. Mechanical crushing assisted by chemical extraction methods have been reported best in performance. Petroleum ether suits best for solvent criteria in case of chemical technique of extraction where extraction efficiency exceeds 90% realistically. Typical characteristics of fossil oil and bio-oil from microalgae are studied by A.L.Ahmed et al (2010) [140].

Table 8 Typical characteristics of fossil oil and bio-oil from microalgae

Properties	Unit	Standard	Bio-oil from microalgae	Fossil oil
Density	Kg/l	ASTM D-6751	1.16	0.75-1
Viscosity	Pa/s	ASTM D-445	0.1@40°C	2-1000
Heating value	MJ kg^{-1}	ASTM D-6751	29	42
Stability	-	ASTM D-6751	Not as stable as fossil fuels	-

For primary analysis of fatty acid profile of microalgal oil, approximately 50-100mg (B_1) of dried algae can be lyophilized using a vacuum freeze dryer followed by soxhlet extraction with chloroform: methanol (2:1, v/v) as the solvent (Cheung et al., 1998; Hsieh and Wu,

2009). The weight of the supernatant is calculated (B_2) and then blow dried in a fume cupboard. The supernatant was dried to a constant weight at 105°C (B_3). The lipid content (C, %) and biomass productivity (P, mg L^{-1} d^{-1}) were calculated using the following formulas [149] :

$$C(\%) = \frac{B_1 - B_2}{B_1} \times 100$$

Where,

B_1, B_2, B_3 – Weight in grams.

$$P(mg\,L^{-1}d^{-1}) = \frac{DW_2 - DW_1}{T_2 - T_1}$$

Where,

DW_1, DW_2 are the dry weight (g L^{-1}) of the algal sample at inoculation and harvesting respectively. T1 and T2 represent the inoculation and harvesting time respectively.

1.4.7 Significance of Algal Biofuel

Microalgae are a diverse group of prokaryotic and eukaryotic microorganisms that grows rapidly due to their simple structure. Micro algal biodiesel production is potentially sustainable and will be made possible to produce adequate amount of fuel to meet the fast growing fuel demands. Advancement in photobioreactor design and micro algal biomass harvesting and downstream processing techniques increases the efficiency of algal biofuel production and thereby algal fuels are likely expected to replace the conventional sources of biodiesel production. H.Xu et al. 2006 [157] have compared the properties of conventional diesel, microalgae derived biodiesel and ASTM biodiesel standards.

Table 9 Comparison of properties of microalgal oil, conventional diesel fuel and ASTM biodiesel standard [157]

Properties	Biodiesel from microalgal oil	Diesel fuel	ASTM biodiesel standard
Density (kg l^{-1})	0.864	0.838	0.84-0.90
Viscosity (mm^2 s^{-1}, cSt at 40°C	5.2	1.9-4.1	3.5-5.0
Flash Point (°C)	115	75	Min 100
Solidifying point (°C)	-12	-50 to 10	-
Cold filter plugging point (°C)	-11	-3.0 (max -6.7)	Summer max 0
Acid value (mg KOH g^{-1})	0.374	Max 0.5	Winter max <-15
Heating value (MJ kg-1)	41	40-45	-
H/C ratio	1.81	1.81	-

Combining microalgal farming and the production of biofuels using biorefinery strategy is expected to significantly enhance the overall cost effectiveness of biodiesel fuels from microalgal approach. Many researches have been undertaken and some are found to be in progress in making the algal biofuels significant in the biodiesel catalogue with main targets such as, increasing oil content of existing strains or selecting new strains with high oil content, increasing the growth rate of algae and developing robust algal-growing systems in either open-air or enclosed environments. Developing co-products other than oil-like usage of residual biomass for development of economically valuable products, using algae in bioremediation and developing an efficient oil-extraction method will make this concept of algal biodiesel a sustainable one.

1.5 CONCLUSION

The problem of oil crisis is the hotly debated topic worldwide since the mid-1970s which resulted in the search for new sustainable oil source. Though many plants and microbes are known to produce oil, most of them are not able to produce oil sustainably. Trends in microlagae biotechnology have contributed hopes for oil revolution in the field of alternate fuels, whereas some of the algal species with low lipid content and low yield of biomass are the stoppage for industrial production. Though microalgae have much significance both economically and environmentally, macroalgae are found to attract the researchers because of its easy-to-grow and eco-friendly nature. Hence macroalgae biodiesel stocks may become a very attractive oil source as the latest explorations about their oil content turns confidence on investment in the area of macroalgae cultivation. Research in macroalgae and microalgae such as growth optimization for high yield of biomass and oil and good downstream processing practices coupled with intellectual selection of transesterification methods may roll to bring advancements in the algal biodiesel production.

1.6 REFERENCES

1. Shay EG. Diesel fuel from vegetable oils: status and opportunities. Biomass Bioenergy 1993;4: 227-42.
2. Sonntag NOV. Reactions of fats and fatty acids . In: Swern D , editors. Bailey's industrial oil and fat products, volume 1. New York : John Wiley & Sons; 1979. p 99.
3. Iso M, Chen B, Eguchi M, Kudo T, Shrestha S. Production of biodiesel fuel from triglycerides and alcohol using immobilized Lipase. J Mol Catal B: Enzym 2001;16:53-58.
4. Demirbas A. Biodiesel production from vegetable oils via catalytic and non-catalytic supercritical methanol transesterification methods. Prog Energy Comb Sci 2005;31:466-487.
5. Schuchardt U, Ricardo Sercheli R,Vargas RM. Transesterification of vegetable oils: a review. J Braz Chem Soc 1998;9:199-210.
6. Al-Widyan MI,Al-Shyoukh AO. Experimental evaluation of the transesterification of waste palm oil into biodiesel. Bioresour. Technol. 2002;85:253-6.
7. Wei Du, Yuanyuan Xu, Dehua Liu, Jing Zeng. Comparitive study on lipase-catalysed transformation of soyabean oil for biodiesel production with different acyl acceptors. J Mol. Catal. B: Enzym. 2004;30:125-9.
8. Chisti Y. Biodiesel from microalgae. Biotechnology advances 2007;25:294-306.

9. Demirbas A. Biodiesel fuels from vegetable oils via catalytic and non-catalytic supercritical alcohol transesterifications and other methods: a survey. Energy Convers Manage 2003;44:2093-109.

10. Wang Y, Ou S, Liu P, Zhang Z. Preparation of biodiesel from wastecooking oil via two-step catalyzed process. Energy Convers Manage 2007;48:184-188.

11. Marchetti JM, Miguel VU, Errazu AF, Possible methods for biodiesel production. Renew Sust Energy Rev 2007;11:1300-311.

12. Freedman B, Butterfield R, Pryde E. Transesterification kinetics of soybean oil. JAOCS 1986;63(10):1375–80.

13. Borugadda.V.B, Goud.V.V. Biodiesel production from renewable and sustainable feedstocks: Status and oppurtunities. Renew Sust Energy Rev 2012;16:4763-4784.

14. Mangesh KG, Ajay DK, Charan ML, Narayan NS. Transesterification of karanja (Pongamia pinnata) oil by solid basic catalysts. European Journal of Lipid Science and Technology 2006;108:389–97.

15. Zhang Y, Dube MA, McLean DD, Kates M. Biodiesel production from waste cooking oil: 1. Process design and technological assessment. Bioresource Technology 2003;89:1–16.

16. Noureddini H, Zhu D. Kinetics of transesterification of soybean oil. JAOCS 1997;74(11):1457–63.

17. Ma F, Hanna MA. Biodiesel production: A Review. Bioresour. Technol 1999;70:1-15.

18. Grossley TD, Heyes TD, Hudson BJF. The effect of heat on pure glycerides. J Am Oil Chem Soc 1962;39:9-14.

19. Darnoko D, Cheryan M. JAOCS 2000;77(12):1263–7.

20. Harrington KJ, DA´ rcy-Evans C. Ind Eng Chem Prod Res Dev 1985;24:314.

21. Stern R, Hillion G, Eur P. Appl EP 1990;356:317 [Cl.C07C67/56]; Stern R, Hillion G, Eur P. Chem Abstr 1990;113:P58504k.

22. Aksoy HA, Kahraman I, Karaosmanoglu F, Civelekoglu H. JAOCS 1988;65:936–8.

23. Perez G. Analysis of enzymatic alcoholisis reaction with vegetables oils. Master thesis, February 2003.

24. Nelson LA, Folgia TA, Marmer WN. JAOCS 1996;73(8):1191–5.

25. Shimada Y, Watanabe Y, Sugihara A, Tominaga Y. J Mol Catal B: Enzymatic 2003;17:133–42.

26. Paiva AL, Balca˜o VM, Malcata FX. Enzyme Microb Technol 2000;27:187–204.

27. Barnwal BK, Sharma MP. Prospects of biodiesel production from vegetable oils in India. Renew Sust Energy Rev 2005;9:363-378.

28. Saka S, Kusdiana D, Biodiesel fuel from rape seed oil as prepared in supercritical methanol. Fuel 2001;80:225-231.

29. Kusdiana D, Saka S. Kinetics of transesterification in rapeseed oil to biodiesel fuel as treated in supercritical methanol. Fuel 2001;80:693-698.

30. Fukuda H, Kondo A, Noda H. Biodiesel fuel production by Transesterification of Oils. J Biosci Bioeng 2001;92:405-416.

31. Kishore JP, Manojkumar ZC, Raghunath TM. Lipase Biodiversity. Indian J.Sci.Technol. 2011;4:971-982.

32. Gupta R, Gupta N, Rathi P. Bacterial lipases: an overview of production, purification and biochemical properties. Appl Microbiol Biotechnol 2004;64: 763–781.

33. Vijay G, Debabrata D. Lipase Fermentation: Progress and prospects. Appl Microbiol Biotechnol 2004;64: 763–781.

34. Kaieda M, Samukawa T, Matsumoto T, Ban K, Kondo A, Shimada Y et al. Biodiesel fuel production from plant oil catalyzed by Rhizopus oryzae Lipase in a water-containing System without an Organic Solvent. J Biosci Bioeng 1999;88:627-631.

35. Wang F, Gao Y, Tan TW, Nie KL. Immobilization of lipase on macroporous resin ans its application in synthesis of biodiesel in low aqueous media. Chin J Biotechnol 2006;22:114-118.

36. Bai YX, Li YF, Yang Y, Li LX, Covalent immobilization of triacylglycerol lipase onto functionalized nanoscale SiO_2 spheres. Process Biochem 2006;41:770-771.

37. Salis A, Svensson I, Monduzzi M, Solinas V, Adlercreutz P. The atypical lipase B from Candida antarctica is better adapted for organic media than the typical lipase from Thermomyces lanuginosa. Biochim Biophys Acta 2003;1646:145.

38. Sarda L, Desnuelle P. Action de la lipase pancréatique sur les esters en émulsion. Biochim. Biophys. Acta 1958;30: 513-521.

39. Grochulski P, Li Y, Schrag JD, Bouthillier F, Smith P, Harrison D, Rubin B, Cygler M et al. *JBiolChem* 1993;*268*: *12843*-12847.

40. Grouchulski P, Li Y, Schrag JD, Cygler M. Two conformational states of Candida rugosa lipase. Protein Sci. 1994;3: 82–91.

41. Derewenda ZS, Derewenda U, Dodson GG. The crystal and molecular structure of the Rhizomucor miehei triacylglyceride lipase at 1.9 A resolution. J Mol Biol. 1992 Oct 5;227(3):818-39.

42. Salis A, Pinna M, Monduzzi M, Solinas V. Comparision among immobilized lipases on macroporous polypropylene toward biodiesel synthesis. J Mol Catal B: Enzym 2008;54:19-26.

43. Dizge N, Keskinler B, Tanriseven A. Biodiesel production from canola oil using lipase immobilized onto hydrophobic microporous styrene-divinylbenzene copolymer. Biochem Eng J 2009;44:220-225.

44. Fukuda H, Hama S, Tamalampudi S, Noda H. Whole cell biocatalysts for biodiesel fuel production. Trends Biotechnol 2008;28:668-673.

45. Goh SH, Yeong SK, Wang CW. Transesterification of cocoa butter by fungal lipase: effect of solvent on 1,3- specificity. J Am Oil Chem Soc 1993;70:567-70.

46. Wong DWS. Lipase In: Whitaker JR,editor. Handbook of food enzymology. Newyork, USA. Marcel Dekker Inc, 2002:667-80.

47. Ban K, Hama S, Nishizuka K, Kaieda M, Matsumoto T, Kondo A. Repeated use of whole cell biocatalysts immobilized within biomass support particles for biodiesel fuel production. J Mol Catal B: Enzym 2002;17:157-65.

48. Tamalampudi S, Hama S, Tanino T, Talukder MR, Kondo A, Fukuda H. Immobilized recombinant Aspergillus oryzae expressing heterologous lipase: An efficient whole-cell biocatalyst for enanioselective transesterification in non-aqueous medium. J Mol Catal B: Enzym 2007;48:33-37.

49. Li W, Du W, Liu D. Rhizopus oryzae IFO 4697 whole cell catalyzed methanolysis of crude and acidified rapeseed oils for biodiesel production in tert-butanol system. Process Biochem 2007;42:1481-1485.

50. Qin H, Yan X, Yun T, Dong W, Biodiesel Production catalyzed by whole-cell lipase from Rhizopus chinensis.Chin J Catal 2008;29:41-46.

51. Hama S, Tamalampudi S, Fukumizu T, Miura K, Yamaji H, Kondo A, Fukuda H et al. Lipase localization in *Rhizopus oryzae* cells immobilized within biomass support particles for use as whole-cell biocatalysts in biodiesel-fuel production. J Biosci Bioeng. 2006;101:328 333.

52. Long, K, Ghazali HM, Ariff A, Ampon K, Bucke C. Mycelium-bound lipase from a locally isolated strain of Aspergillus flavus link: Pattern and factors involved in its production. J Chem Technol Biotechnol 1996;67: 157-163.

53. Rashid U, Anwar F, Ansari TM, Arif M, Ahmad M. Optimization of alkaline transesterification of rice bran oil for biodiesel production using response surface methodology. Journal of Chemical Technology and Biotechnology 2009;84:1364–70.

54. Zullaikah S, Lai CC, Vali SR, Ju YH. A two-step acid-catalyzed process for the production of biodiesel fromricebranoil. BioresourTechnol2005 1889– 1896;96.

55. Rashid U, Anwar F, Knothe G. Evaluation of biodiesel obtained from cottonseed oil. Fuel Processing Technology 2009;90:1157–63.

56. He C, Peng B, Wang D, Wang J. Biodiesel production by the transesterifica- tion of cottonseed oil by solid acid catalysts. Frontiers of Chemical Engineering in China 2007;1:11–5.

57. Royon D, Daz M, Ellenrieder G, Locatelli S. Enzymatic production of biodiesel from cotton seed oil using t-butanol as a solvent. Bioresource Technology 2007;98:648–53.

58. Ayhan D. Studies on cottonseed oil biodiesel prepared in non-catalytic SCF conditions. Bioresource Technology 2008;9:1125–30.

59. Raman JK, Sariah A, Denis P, Seng CE, Ravindra Pogaku. Production of biodiesel using immobilized lipase—a critical review. Critical Reviews in Biotechnology 2008;28:253–64.

60. Encinar JM, Gonza´ lez JF, Martı´nez G, Sa´nchez N, Pardal A. Soybean oil transesterification by the use of a microwave flowsystem Fuel 2012;95:386–93.

61. Xuejun L, Xianglan P, Yujun W, Shenlin Z. Calcium ethoxide as a solid base catalyst for the transesterification of soybean oil to biodiesel. Energy Fuels 2008;22:1313–7.

62. Boro J, Deka D, Thakur AJ. A review on solid oxide derived from waste shells as catalyst for biodiesel production. Renewable and Sustainable Energy Reviews 2012;16:904–10.

63. Bradley WD, Brett BM, Lance SC. Synthesis of biodiesel from mixed feed- stocks and longer chain alcohols using an acid-catalyzed method. Energy Fuels 2008;22:4223–8.

64. Miao X, Li R, Yao H. Effective acid-catalyzed transesterification for biodiesel production. Energy Conversion and Management 2009;50:2680–4.

65. Martins GC, Sergio T, Ledo ML, Schuchardt U. Transesterification of soybean oil catalyzed by sulfated zirconia. Bioresource Technology 2008;99: 6608–13.

66. Furuta S, Matsuhashi H, Arata K. Biodiesel fuel production with solid amorphous-zirconia catalysis in fixed bed reactor. Biomass and Bioenergy 2006;30:870–3.

67. Lin V.S.Y., Radu D.R. Use of functionalized mesoporous silicates to esterify fatty acids and transesterify oils. United State Patent 7,122,688 B2, 2006.

68. Xie W, Yang D. Silica-bonded N-propyl sulfamic acid used as a hetero- geneous catalyst for transesterification of soybean oil with methanol. Bioresource Technology 2011;102:9818–22.

69. Costa RR, Giandra V, Zachia AMA, Keiko W. Lipase-catalyzed ethanolysis of soybean oil in a solvent-free system using central composite design and response surface methodology. Journal of Chemical Technology & Biotech- nology 2008;83:849–54.

70. Stamenkovic OS, Velickovic AV, Veljkovic VB. The production of biodiesel from vegetable oils by ethanolysis: current state and perspectives. Fuel 2011;90:3141–55.

71. Huayang H, Tao W, Shenlin Z. Continuous production of biodiesel fuel from vegetable oil using supercritical methanol process. Fuel 2007;86: 442–7.

72. Raman JK, Sariah A, Denis P, Seng CE, Ravindra Pogaku. Production of biodiesel using immobilized lipase—a critical review. Critical Reviews in Biotechnology 2008;28:253–64.

73. Umer R, Farooq A, Bryan MR, Samia A. Production of sunflower oil methyl esters by optimized alkali-catalyzed methanolysis. Biomass & Bioenergy 2008;32:1202–5.

74. Tomasevic AV, Siler-Marinkovic SS. Methanolysis of used frying oil. Fuel Processing Technology 2003;81:1–6.

75. Babu SN, Rekha S, Prasad SPS, Lingaiah N. Room-temperature transester- ification of edible and non edible oils using a heterogeneous strong basic Mg/La catalyst. Energy Fuels 2008;22:1965–71.

76. Mohamed SM, Uwe BT. Lipase-catalyzed alcoholysis of vegetable oils. European Journal of Lipid Science and Technology 2003;105:656–60.

77. Giridhar M, Chandana K, Rajnish K. Synthesis of biodiesel in supercritical fluids. Fuel 2004;83:2029–33.

78. Ghadge SV, Raheman H. Biodiesel production from mahua (Madhuca indica) oil having high free fattyacids. Biomass and Bioenergy 2005;28:601–5.

79. Lubes ZIZ, Zakaria M. Analysis of parameters for fatty acid methyl esters production from refined palm oil for use as biodiesel in the single- and two- stage processes. Malaysian J Biochem Mol Biol 2009;17:5–9.

80. Shuit S.H., Ong Y.T., Lee K.T., Subhash B., Tan S.H. Membrane technology as a promising alternative in biodiesel production: a review. Biotechnology Advances 2012, in press.

81. Jitputti J, Kitiyanan B, Rangsunvigit P, Bunyakiat K, Attanatho L, Jenvanit- panjakul P. Transesterification of crude palm kernel oil and crude coconut oil by different solid catalysts. Chemical Engineering Journal 2006;116: 61–6.

82. Xu B, Xiao G, Cui L, Wei R, Gao L. Transesterification of palm oil with methanol to biodiesel over a $KF/Al2O_3$ heterogeneous base catalyst.Energy Fuels 2007;21:3109–12.

83. Ndalew AK, Kiros Y, Zanzi R. Inorganic heterogeneous catalysts for biodiesel production from vegetable oils. Biomass and Bioenergy 2011;35:3787–809.

84. http://tisd.en.kku.ac.th/tisd2008/data/TISD 2006%20Proceedings/CHE/CHE_6I. pdf.

85. Sarma AKA. Comprehensive analysis of fuel properties of biodiesel from koroch seed oil. Energy Fuels 2005;19:656–7.

86. Meher LC, Vidya SSD, Naik SN. Optimization of alkali-catalyzed transesterification of Pongamia pinnata oil for production of biodiesel. Bioresource Technology 2006;97:1392–7.

87. Mangesh KG, Ajay DK, Charan ML, Narayan NS. Transesterification of karanja (Pongamia pinnata) oil by solid basic catalysts. European Journal of Lipid Science and Technology 2006;108:389–97.

88. Veljkovic VB, Lakicevic SH, Stamenkovic OS, Todorovic ZB, Lazic ML. Biodiesel production from tobacco (Nicotiana tabacum L.) seed oil with a high content of free fatty acids. Fuel 2006;85:2671–5.

89. Richard R, Li Y, Dubreuil B, Roux ST, Prat L. On-line monitoring of the transesterification reaction between triglycerides and ethanol using near infrared spectroscopy combined with gas chromatography. Bioresource Technology 2011;102:6702–9.

90. Ramadhas AS, Jayaraj S, Muraleedharan C. Biodiesel production from high FFA rubber seed oil. Fuel 2005;84:335–40.

91. Umer R, Farooq A. Production of biodiesel through optimized alkaline- catalyzed transesterification of rapeseed oil. Fuel 2008;87:265–73.

92. Galia A, Scialdone O, Tortorici E. Transesterification of rapeseed oil over acid resins promoted by supercritical carbon dioxide. Journal of Supercritical Fluids 2011;56:186–93.

93. Lilin L, Wei D, Dehua L, Li W, Zebo L. Lipase-catalyzed transesterification of rapeseed oils for biodiesel production with a novel organic solvent as the reaction medium. Journal of Molecular Catalysis B: Enzymatic 2006;43:58–62.

94. Sawangkeaw R, Bunyakiat K, Ngamprasertsith S. A review of laboratory- scale research on lipid conversion to biodiesel with supercritical methanol (2001–2009). Journal of Supercritical Fluids 2010;55:1–13.

95. Shahid EM, Jamal Y. A review of biodiesel as vehicular fuel. Renewable and Sustainable Energy Reviews 2008;12:2484–94.

96. Agarwal M, Chauhan G, Chaurasia SP, Singh K. Study of catalytic behavior of KOH as homogeneous and heterogeneous catalyst for biodiesel production. Journal of the Taiwan Institute of Chemical Engineers 2012;43: 89–94.

97. Lam MK, Lee KT, Mohamed AR. Homogeneous, heterogeneous and enzy- matic catalysis for transesterification of high free fatty acid oil (waste cooking oil) to biodiesel: a review. Biotechnology Advances 2010;28: 500–18.

98. Miao X, Wu Q. Biodiesel production from heterotrophic microalgal oil. Bioresource Technology 2006;97:841–6.

99. KohMY, Ghazi TIM. A review of biodiesel production from Jatropha curcas L. oil. Renewable and Sustainable Energy Reviews 2011;15:2240–51.

100. Amish VP, Subrahmanyam N, Payal PA. Production of biodiesel through transesterification of jatropha oil using KNO_3/Al_2O_3 solid catalyst. Fuel 2009;88:625–8.

101. Mohamed SM, Uwe BT. Lipase-catalyzed alcoholysis of vegetable oils. European Journal of Lipid Science and Technology 2003;105:656–60.

102. Rathore V, Madras G. Synthesis of biodiesel from edible and non-edible oils in supercritical alcohols and enzymatic synthesis in supercritical carbon dioxide. Fuel 2007;86:2650–9.

103. Duz MZ, Saydut A, Ozturk G. Alkali catalyzed transesterification of safflower seed oil assisted by microwave irradiation. Fuel Processing Technology 2011;92:308–13.

104. Guan G, Kusakabe K, Sakurai N, Moriyama K. Transesterification of vege- table oil to biodiesel fuel using acid catalysts in the presence of dimethyl ether. Fuel 2009;88:81–6.

105. Soriano Jr NU, Venditti R, Argyropoulos DS. Biodiesel synthesis via homo- geneous Lewis acid-catalyzed transesterification. Fuel 2009;88:560–5.

106. Furukawa S, Uehara Y, Yamasaki H. Variables affecting the reactivity of acid- catalyzed transesterification of vegetable oil with methanol. Bioresource Technology 2010;101:3325–32.

107. Er-Zheng Su,Min-Jie Zhang,Jian-Guo Zhang,Jian-Feng Gao ,Dong-Ziu Wei . Lipase-catalyzed irreversible transesterification of vegetable oils for fetty acid methyl esters production with

dimethyl carbonate as the acyl acceptor. Biochem Eng J 2007;36:167-73.

108. Samukawa T,Kaieda M,Matsumoto T,,Ban K,Kondo A,Shimada Y,Noda H,Fukuda H . Pretreatment of Candida antartica lipase for biodiesel fuel production frm plant oil. J Biosci Bioeng 2000;90:180-3.[C14 (13)]

109. Chen JW, Wu WT. Regeneration of immobilized *Candida antartica* lipase for transesterification. J Biosci Bioeng 2003;95:466-9.

110. Nelson LA,Foglia TA,Marmer WN. . Lipase-catalyzed production of biodiesel.J Am Oil Chem Soc 1996;73:1191-5.

111. [R19]

112. Du W,Xu Y,Zeng J,Liu D . Novozym 435-catalyzed transesterification of crude soya bean oil for biodiesel production in a solvent-free medium.Biotechnol Appl Biochem 2004;40:187-90.

113. Prasad et al.

114. Shimada Y,Watanabe Y,Sugihara A,Tominaga Y. Enzymatic alcoholysis for biodiesel fuel production and application of the reaction to oil processing. J Mol. Catal. B:Enzym2002;17:133-42.

115. Pioch D,Lozano P,Graille J . First lipase-catalyzed synthesis of fatty carbonate esters. Biotechnol. Lett.1991;13:633-5.

116. Mirolawa SA, Aneta K, Tadeusz A, Stanislaw B. Enzymatic biodiesel synthesis-Key factors affecting efficiency of the process. Renew energy 2009;34:1185-94.

117. Maragoni AG. Lipases: structure function and properties. In :Kuo TM, Gardner HW. Lipid biotechnology. New York: Marcel Dekker Inc;2002.p.357-84.

118. Wong DWS.Lipase In:Whitaker JR,editor.Handbook of food enzymology. New York,USA:Marcel Dekker Inc;2002.p.667-80.

119. Bana K, Hamab S, Nishizuka Kaieda M, Matsumoto T, Kondo A, et al. Repeated use of whole-cell biocatalysts immobilized within

biomass support particles for biodiesel fuel production. J Mol Catal B Enzym 2002;17:157-65.

120. Watanabe Y, Shimada Y, Sugihara A, Tominaga Y. Conversion of degummed soybean oil to biodiesel fuel with immobilized *Candida antartica* lipase. J Mol Catal B Enzym 2002;17:151-5.

121. Haas MJ, Piazza GJ, Foglia TA, Enzymatic approaches to the Production of biodiesel fuels.In:Kuo TM, Gardener HW, editors – Lipid biotechnology. Newyork : Marcel Dekker Inc ; 2002.587-98.

122. Lai CC, Zullaikah S, Vali RS, Ju YH. Lipase catalyzed production of biodiesel from rice barn oil. J Chem Technol Biotechnol 2005;80:331-7.

123. Shinada Y, Watnabe Y, Suggirhara A, Tominaga Y. Review – Enzymatic alcoholysis for biodiesel fuel production and application of the reaction to oil processing. J.Mol.Catal B EEnzyme 2002;19:133-42.

124. Royon D, Daz M, Ellenrieder G, Locatelli S . Enzymatic production of biodiesel from cotton seed oilusing *t*-butanol as a solvent. Bioresour. Technol. 2007;98:648-53.

125. Chen JW, Wu WT. Regeneration of immobilized Candida antartica lipase for transesterification. J Biosci Bioeng 2003;95:466-9.

126. Haas MJ, Piazza GJ, Foglia TA, Enzymatic approaches to the Production of biodiesel fuels. In: Kuo TM, Gardener HW, editors – Lipid biotechnology. Newyork : Marcel Dekker Inc ; 2002.587-98.

127. Lanne C, Boern S, Vor K, Veeger C. Rules for optimization of biocatalysis in organic solvents. Biotechnol Bioeng 1987;30:81-7.

128. Antczak T, Patura J, Szczesna-Antczak M, Hier D, Bielecki S.Sugar ester synthesis by mycelium bound mucor circinelloides lipase in a micro-reactor equipped with water activity sensor. J.Mol Catal B Enzyme 2004;29:155-61.

129. Lilin Li, Wei Du, Dehua Liu, Li Wang, Zebo Li. Lipase-Catalyzed transesterification of rapeseed oils for biodiesel production with

a novel organic solvent as the reaction medium. J Mol Catal B Enzym 2006;43:58-62.

130. Seiji K, Du D, Sato M, Park EY. Efficient production of fatty acid Methyl ester from waste activated bleaching earth using diesel oil as organic solvents. J Biosci Bioenng 2004;98 :420-4.

131. Dossat V,Combes D,Marty A. Continuous enzymatic transesterification of high oleic sunflower oil in a packed bed reactor: influence of the glycerol production. Enz Microb Technol 1999;25:194-200.

132. Stevenson DE,Stanley RA,Furneaux RH. Near-quantitative production of fatty acid alkyl esters by lipase-catalysed alcoholysis of fats and oils with adsorption of glycerol by silica gel. Enz Microb Technol 1994;16:478-84.

133. Bélafi-Bakó K,Kovács F,Gubicza L,Hancsók. Enzymatic biodiesel production from sunflower oil by *Candida antartica* lipase in a solvent-free system. Biocatal Biotransfer 2002;20:437-9.

134. Mu Y, Xiu ZL, Zhang DJ.A combined bioprocess of biodiesel production by lipase with microbial production of 1,3-propanediol by *Klebsiella pneumoniae*. Biochem Eng J 2008;40:537-41.

135. Najafi G, Ghobadian B, Yusaf.TF. Algae as a sustainable energy source for biodiesel production in Iran: A case study. Renew Sust Energy Rev 2011;15:3870-3876.

136. Canakci M, Sanli H. Biodiesel production from various feedstocks and their effects on the fuel properties. Journal of Industrial Microbiology and Biotechnology 2008;35:431–41.

137. Li Y, Wang B, Wu N, Lan CQ. Effects of nitrogen sources on cell growth and lipid production of Neochloris oleoabundans. Applied Microbiology and Biotechnology 2008;81(4):629–36.

138. T.M. Mata, A. AMartins, N.S. Cetano. Microalgae for biodiesel production and other applications: A review. Renew Sust Energy Rev 2010;14:217-232.

139. Grima ME, Belarbi EH, Fernandez FGA, Medina AR, Chisti Y. Recovery of microalgal biomass and metabolites: process options and economics. Biotechnology Advances 2003;20(7-8):491–515.

139. Demirbas A, Fatih Demirbas M. Importance of algae oil as a source of biodiesel. Energy Convers Manage 2010.

140. Ahmad AL, Mat Yasin NH, Derek CJC, Lim JK. Microalgae as a sustainable energy source for biodiesel production: a review. Renew Sustain Energy Rev 2010.

141. Singh J, Gu S. Commercialization potential of microalgae for biofuels production. Renew Sustain Energy Rev 2010;14:2596–610.

142. Janaun J, Ellis N. Perspectives on biodiesel as a sustainable fuel. Renew Sustain Energy Rev 2010;14:1312–20.

143. Foidl N, Foidl G, Sanchez M, Mittelbach M, Hackel S. Jatropha curcas as a source for the production of biofuel in Nicaragua. Bioresource Technology 1996;58:77–82.

144. Lele S. Indian Green Energy Awareness Center, http://www.svlele.com karanj.htm.

145. Gui MM, Lee KT, Bhatia S. Feasibility of edible oil vs. non edible oil vs. waste edible oil as biodiesel feedstock. Energy 2008;33(11):1646–53.

146. Foidl N, Foidl G, Sanchez M, Mittelbach M, Hackel S. Jatropha curcas as a source for the production of biofuel in Nicaragua. Bioresource Technology 1996;58:77–82.

147. Chen F. High cell density culture of microalgae in heterotrophic growth. Trends Biotechnol 1996;14:421-6.

148. Chen F. Johns MR, Effect of C/N ratio and aeration on the fatty acid composition of heterotrophic chlorella sorokiniana. Appl Phycol 1991;3:203-9.

149. Zhou XP, Ge HM, Xia L, Zhang D, Hu CX. Evaluation of oil-producing algae as potential biodiesel feedstock 2013;134:24-29.

150. Song M, Pei H, Hu w, Ma G. Evaluation of the potential of 10 microalgal strains for biodiesel production. Renew Sust Energy Rev. 2013;141:245-251.

151. Jorquera O, Kiperstock A, Sales EA, Embrirucu M, Ghirardi M. Comparative lifecycle analyses of microalgal biomass production in open ponds and photobioreactors. Renew Sust Energy Rev. 2010;101:1406-1413.

152. Yang J, Xu M, Zhang X, Hu Q, Milton S, Chen Y. Lifecycle analysis on biodiesel from microalgae: Water footprint and nutrient balance. Renew Sust Energy Rev. 2011;102:159-165.

153. Campbell PK, Beer T, Batten D. Lifecycle assessment of biodiesel from microalgae in ponds. Renew Sust Energy Rev. 2011;102:50-56.

154. Molina Grima E, F. Ancien Fernandez, F.Garcia Camacho and Y.Chisti. Photobioreactors: Light regime, mass transfer and scaleup. Journal of Biotechnology 1999;70:231-47.

155. Terry KL,Raymond LP. System design for autotrophic production of microalgae. Enzyme Microb. Technol 1985;7:474-87.

156. Huang GH, Chen F, Wei D, Zhang XW, Chen G. Biodiesel production by microalgal biotechnology. Renew Sust Energy Rev. 2010;87:38-46.

157. Xu H, Miao XL,Wu QY. High quality biodiesel production from a microalga chlorella protothecoides by heterotrophic growth in fermenters. J. Biotechnol 2006;126:499-507.

2

Bioethanol Production from Lignocellulosic Materials

2.1 INTRODUCTION

Global demand for energy continues to grow due to the expanding human population and industries. The major energy demand is still supplied from conventional fossil fuels such as oil, coal and natural gas. Petroleum-based fuels can be replaced by renewable biomass fuels such as bioethanol, bio-diesel and biohydrogen. Ethanol has been used by humans as intoxicating ingredient in alcoholic beverages. The isolation of a relatively pure alcoholic compound was first obtained by the Islamic alchemist who developed the act of distillation. Ethanol has long been considered as a suitable alternative to fossil fuels either as a sole fuel in cars with dedicated engines or as an additive in fuel blends with no engine modification requirement when mixed up to 30%. Henry Ford designed a car model called the model T to run on ethanol and quoted ethanol as "the fuel of the future". Ford and others continued to promote the use of ethanol as fuel. By 1938, an alcohol plant in Atchison, Kansas was developed with the capacity of producing 18 million gallons of ethanol a year [1-4]. Ethanol production in the United States grew from 175 million gallons to 1.4 billion gallons in a span of 18 years. Worldwide production capacity of ethanol in 2005-06 was about 45-49 billion liters per year, and is expected to reach 115 billion liters in 2015 [5]. The use of fuels blended with ethanol gave two mandatory types namely (E10) and (E85) which contain 10% and 85% of ethanol respectively. Most E85 using vehicles fall under the US government vehicles. The transportation of this ethanol has also become of great concern. When blended ethanol is run through pipelines, the ethanol gathers the moisture content from the pipelines. Hence the ethanol and gasoline need to be brought through different tankers and are to be blended at another terminal before loading into trucks [1-4].

Various other substrates such as cellulosic bio-mass [1], cassava [2-5], sago [6], sorghum [7, 8], blackstrap molasses [9], and maize [10] are tested for their potential to be efficient alternate feedstock for bioethanol production. In particular, ethanol produced from biomass by fermentation is expected to find use as fuel and in the production of various chemicals [11]. Three main reasons for the production of bioethanol from cellulosic biomass are: i) it is renewable, (ii) doesn't emit harsh gases like CO_2, SO_2, NO_2 into the environment, and (iii) it holds the key role in a country's international economy. The only low cost fermentation substrate that can meet the demands of oil of the future is lignocellulosic biomass. The cellulose biomass consists of mainly waste of inedible cellulose fibers that form the stems and branches of most plants. Grain crops, switch grass, crop residues like cornstalks, wheat straw, rice straw, grass dipping and wood residues are the many forms of cellulosic biomass. Cellulose has been deemed to be comprised of half of all the organic carbon on the planet. An analysis by NRDC (National Research Defense Council) has found that cellulosic ethanol can provide a solution for the US transportation fuel needs by 2050 without decreasing the production of food and animal feed. Cellulosic biomass has been a challenge for scientists to convert it into ethanol. Researchers use harsh acids like sulfuric acid at high temperatures to destabilize and breakdown the matrix of cellulose into sugars. An economical process has never been developed using traditional chemistry. Developing a cost effective process for the production of bioethanol is the hotly debated topic in the field of alternate energy [8, 9]. Numerous biotech firms are in the process of perfecting enzymes for cellulosic ethanol. Assorted bio refineries are being constructed to intake the cellulosic biomass, convert them to quality sugars for fermentation of ethanol. There have been great dramatic breakthroughs in the production of cellulosic ethanol till date. In 2001, the cost of cellulase enzyme was the inhibiting factor for the production of the cellulosic ethanol at a cost of \$5 per gallon. It was then highly reduced to a mere amount of 10–18 cents per gallon which was an enormous 30-fold decrease. Reputed companies like Diversa Corporation, Dyadic International have invested large in the research of bioethanol production and came out successfully. America is the largest continental producer of bioethanol constituting 22.3 billion litres/year. Asia stands second with 5.7 billion/year. Europe, Africa and Oceanic countries follow the list. The first commercial production of cellulosic ethanol was produced by the Iogen Corporation in 2004. Abengoa bioenergy was the first to construct cellulosic ethanol production plant in commercial scale. The plant could process 70 tonnes of agriculture waste to produce 1 million gallon of ethanol on an annual basis. The DOE (Department of Energy) of the USA has granted $10 million to develop a next generation dry mill corn ethanol plant. These plants will be able to produce ethanol from corn and residual fibers, which require processing of the cellulase enzymes. 80% of the world's biomass consists of forest plants. Hence, the raw material cellulose is the highly preferred choice for the production of ethanol. Lignocellulosic biomass comprises mainly of compounds cellulose, hemicellulose and the adjoining molecule lignin [12]. Cellulose is the 6-carbon sugar, glucose arranged in bundles. Hemicellulose primarily consists of 5-carbon sugars and xylose. Five distinct steps involved in converting the lignocellulosic biomass to bioethanol are:

(1) biomass delignification to release cellulose and hemicellulose from the lignin-cellulose-hemicellulose complex,

(2) depolymerization of the cellulose and hemicellulose carbohydrates to produce their respective free sugars (pentoses and hexoses),

(3) fermentation of the free sugars to produce ethanol,

(4) ethanol recovery, and

(5) effluent treatment [13].

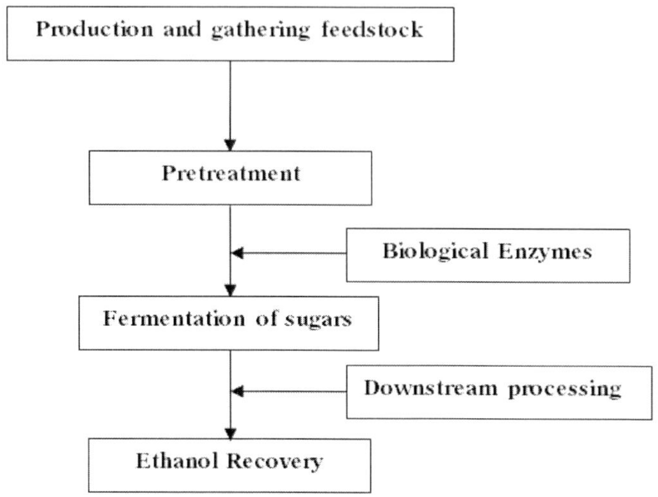

Figure 1 Recovery of ethanol from lignocellulosic materials

Raw materials of lignocelluloses obtained from chemical pulps consist of large percent of cellulose content of 60-80. These cellulosic materials being away from the human food chain add up to its positives. The maximum theoretical yield of ethanol from 42% cellulose and 21% hemicellulose in wood can be calculated to be 0.32 grams of ethanol per gram of wood. This is based on complete conversion of cellulose and hemicellulose to sugars, and conversion of sugars to bioethanol at theoretical yield of 0.51 g/g [14]. Lignocellulosic materials may be hydrolyzed with the emission of radiation by gamma ray or microwave irradiation or electron beam irradiation. However, these processes are commercially unimportant. Abundant non-edible waste from the field can be screened based on the lignocellulose content and can be used as substrates for efficient production of ethanol. [15].

Significantly used microbial strains for the production of ethanol were mostly fungal and bacterial species. Researchers showed improved bioethanol production from lignocellulosic waste materials by *Saccharomyces sp.* [16-19]. It converts the reducible sugar glucose by fermentation into ethanol. The fermentation conditions depend on the microbes used. Pure culture techniques are used to culture the selective species of microbe and grown in culture medium. Then the strains are extracted and cultured in the substrate medium for the process of fermentation [20]. A bulk fraction of the money invested for the production of bioethanol is used for the pre-treatment purpose which involves breaking of lignin and extracting cellulose, glucose and other mono sugars to increase the yield of bioethanol [21-22].

Another major processing step involved is the saccharification of cellulose. Pentose and hexoses are obtained from hemicellulose and cellulose by breaking down the H_2 bond. Conc. sulphuric acid or HCl is used for the hydrolysis of lignocellulosic materials [23, 24]. Though these acids may reduce or degrade certain amount of these sugars, the use of dilute acids will take increased time comparatively. The important two factors needed to make the process economically viable are to optimize sugar recovery and recovery of acid for recycling. Use of concentrated acid increases the % sugar recovery. Acid and sugar are separated by ion exchange and the concentration of the acid is revived using multiple effect evaporates. There are also other means of pretreatment that includes steam explosion, ammonia fiber explosion, CO_2 explosion, ozendysis, alkaline hydrolysis, oxidative delignification, organosolvent process and other biological treatment [25]. Various resources and strains used for the production of bioethanol are given in table 1. This chapter reviews the possibility of using agro wastes as sustainable substrate in ethanol production and recent drifts in microbial fermentation of various lignocellulose rich biomass for cost efficient production of ethanol.

Table 1 Fermentation of bioethanol from different sources

Strain/Media formulation	**Raw material/ Substrate**	**Pre-treatment method**	**Fermentation conditions**	**Author/ Year**
Zymomonas mobilis 25g xylulose, 5g yeast extract 10mg/ml tetracycline culture -70°C frozen culture	Wheat–acid hydrolysate supplemented with glucose in 3 seed tank	Dilute acid pre-treatment for hemicellulose recovery. Sulphuric acid -0.25–2% heated for specific temperature (180°C, 100°C, 120°C) time 0.5–5.5 hours	Fermentation media - pH controlled 3M NaOH. Temperature -30°C, agitation – 200 rpm.	[2]

Strain/Media formulation	Raw material/ Substrate	Pre-treatment method	Fermentation conditions	Author/ Year
Trichoderma reesei Media – MGYP medium at 33⁰C for 12 hours in orbital shaker. Inoculum conc. 1:5x10^8 yeast cell/ml	Raw wheat straw	Conventional alkaline pretreatment. 160ml of 1% Sodium Hydroxide boiled in 500 ml beaker for 60 minutes. Dried at 65⁰C for 2 days. (MW assisted alkali pretreatment) MW microwave (700W in microwave oven for 25 minutes)	Fermentation media in the wheat (pre-treated), Temperature - 35⁰C– 45⁰C pH - 4.8–5.8 160 rpm/min HRT – 72 hours fermentation 96 hours conventional.	[32]
Saccharomyces cerevisiae (ATCC 200062) *Pichia stipitis* (ATCC PTA 3717)	Wet distiller's grains (Wet – DG)	Acid pretreatment Time: 5.40 minutes. Temp - 140–185⁰C (depends on substrate) H_2SO_4 - 1.1%–3.27%	Fermentation media with wet distillers grain (Wet DG) temperature - 32⁰C, pH – 5, rpm – 150, HRT – 24 hours preferred	[73]
Saccharomyces cerevisiae (normal standard media for culture strength improvisation)	*Sorghum bicolor* (Sweet sorghum of China)	Crushing and extraction of juice for fermentation	Fermentation media with extract from sweet sorghum. Temperature - 33⁰C–35⁰C pH – 5 rpm – 160	[4]
Kluyveromyces spp. (Crude whey, 3.4% lactose (MTCC 1288) Culture strength 24 hours	Crude whey	-	Fermentation media with crude whey temp. 34⁰C, pH-4.5 (maintained by 6N NaOH)	[20]
Kluyveromyces marxianus (Y01070) media – glucose 50.0 g/l, yeast extract 2.5g/l, peptone 5.0 g/l,KH_2PO_4, $MgSO_4$, NH_4Cl (0.2g/l) Temp - 30⁰C.	Old conjugated card board (OCC) and paper sludge, paper waste	Enzyme pretreatment cellulase and Iogen cellulase	Fermentation media with OCC and paper waste separately. Temp – 40⁰C pH – 4.5–5.3 (maintained by 10% NaOH or H_2SO_4) HRT – 96 hours.	[6]

Strain/Media formulation	Raw material/ Substrate	Pre-treatment method	Fermentation conditions	Author/ Year
Baker yeast (*Saccharomyces cerevisiae*) standard yeast media 30⁰C / 24 hours, culture pH – 5, culture strength	Corn stover (chopped and dried in room temp)	Physico chemical pretreatment. SO_2 treatment by impregnation with steam. Enzymatic hydrolysis by cellulase enzyme.	Fermentation media with corn stover pretreatment by physicochemical and enzymatic method. Temperature 30⁰C pH – 5 HRT – 72 hours Aerobic fermentation (batch and fed batch)	[25]
Fusarium oxysporum potato dextrose agar. Temp - 30⁰C 200 rpm orbital shaker pH – 6 strength - 3–4 days	Glucose basal media		Aerobic potato dextrose agar. Temp - 30⁰C 200 rpm pH – 6 HRT – 2 days O_2 concentration– 30%min	[3]
Saccharomyces cerevisiae Zymomonas mobilis Pichia spp. (Yeast extract, urea, KH_2PO_4, $MgSO_4.7H_2O$, hydrate)	Cellulose pyrolysate	Neutralization method followed by dilution with distillation H_2O	Yeast – 30⁰C /140 hours/ 150 rpm Same even for *Pitichia spp.*	[9]
Bakers yeast pH – 5.5 30⁰C 24 hours culture	Wood chip (salix)	Steam pretreated treated with dilute H_2SO_4 for excess for 90 min.	Media with wood chip Temperature -30⁰C pH-5 (auto maintained by 10% NaOH) 350 rpm, 24 hours.	[74]
Saccharomyces cerevisiae (CFTRI101) Malt extract peptone yeast extract MPYD (liquid medium broth)	Finger millet (Ragi) germinated	Pulverization (Physical pretreatment method)	MPYD media (broth) + finger millet (germinated) Temperature -30⁰C 100 rpm pH – 5 HRT – 48 hours	[75]
Schizosaccharomyces pombe	Reducing sugar containing biomass	General pretreatment methods	Fermented at Temperature - 30⁰C pH – 4.5 at rpm - 300	[24]

Strain/Media formulation	Raw material/ Substrate	Pre-treatment method	Fermentation conditions	Author/ Year
Rhizopus spp, Mucor spp, Rhizomucor spp, (totally 12 strains) Used pH-5.6 Temp - 50^{0}C (PDA slant potato dextrose agar)	Spruce tree wood hydrolysate	Impregnated with H_2SO_4 (acid pre-treatment)	Fermentation media with spruce tree wood hydrolysate. (Pre treated) Temperature -37^{0}C pH – 5.5 HRT – 7 days (Controlled by 2M NaOH)	[76]
Liriodendron tulipifera	Saw dust	Acid pretreated with dilute sulphuric acid	General fermentation condition prevailed	[77]
Kluyveromyces marxianus, 42^{0}C 150 rpm 16 hours culture	Plant biomass	Steam explosion (appropriate time duration)	Fermentation media with plant biomass (pre treated) Temperature -42^{0}C HRT-160 hours culture	[78]
Genetically modified yeast strains	Cellulosic materials	Enzyme pretreatment	Aerobic condition with Temperature - 30^{0}C pH – 5.0 72 days	[80]
Klebsiella oxytoca, Zymomonas mobilis, (variety of species)	Sources of reducing sugars	General pretreatment methods	Maintained at a pH of 5.0 and Temperature -35^{0}C – 37°C.	[7]
Saccharomyces pastorianeus	Porous cellulose carriers	General pretreatment methods	Fermented for 10 days at 30°C with 50 rpm	[79]
Clostridium thermocellum In CMS medium	Paddy straw, sorghum, deshelled corn cobs.	General pretreatment followed	Fermentation conditions maintained for the specific substrates.	[81]
Kluyveromyces oxytoca	Assorted raw materials	Enzyme pretreatment	Maintained at a pH of 4.8 at 50°C	[82]

2.2 STRAINS USED

Microorganisms are the harvesting units of the substrate that converts biomass to valued product (biofuel). Most commonly, fungal and bacterial cultures are widely used for bioethanol fermentation. However fungi play an alarming role in the fermentation as it is known that the entire fermentation setup started with fungi *Saccharomyces cerevesieae* traditionally [26]. Most of the fermentation procedure involves the yeast as the biological agent converting the monosaccharide to ethanol in an anaerobic condition. In the recent days, thermo-tolerant organisms such as *Cryptococcus tepidarius, Candida acidithermophilium, Candida brassicae, Saccharomyces uvarum* and *Kluyveromyces* species are reported as high yielding strains [26, 27]. *Escherichia coli* and *Saccharomyces spp.* are also genetically modified for the production of biofuel. The cellulase-gene cloned transformant *Streptomyces sp.* T3-1was used for fermentation of agro wastes by Hung-Der Jang which resulted in efficient release of saccharides from corncob [28]. For producing ethanol from glucose rich substrates such as sugar cane syrup, *Zymomonas sp.* was found to be effective that gives yield greater than 90% [29, 30, 31]. However, the efficiency of conversion is mainly influenced by the nature of substrate used and the pretreatment method employed along with the fermentation process.

2.3 SUBSTRATES USED FOR BIOETHANOL FERMENTATION

Various kinds of substrates are at present being used for biofuel production, but to be economical in fermentation, lignocellulosic agro wastes will be good choice of substrate. Lignocellulosic materials generally include all the kind of crop residues, agriculturally waste products and the biomass constituents. After substrate selection, it is highly regarded to find the sort of pretreatment method that should be adopted for the efficient hydrolysis of cellulosic biomass. To set the production process in a cost effective and trend setting track, microbial fermentation is the only intelligent choice existing in current scenario. Some of the biomass that are widely being used are crop residues, corn and corn wastes like corn stover, barley and barley waste material, oats, oat grain, rice, rice waste, wheat, wheat hydrolysate, sorghum, sugar cane bagasse, etc. Recently, for the past couple of decades, woods, salix, millet, lactose waste, lactose, whey are also being widely used as the raw material for microbes-assisted bioethanol fermentation. Wood, millet and other crop residues require a different sort of pretreatment like acid hydrolysis, alkaline hydrolysis and also enzyme based hydrolysis [32].

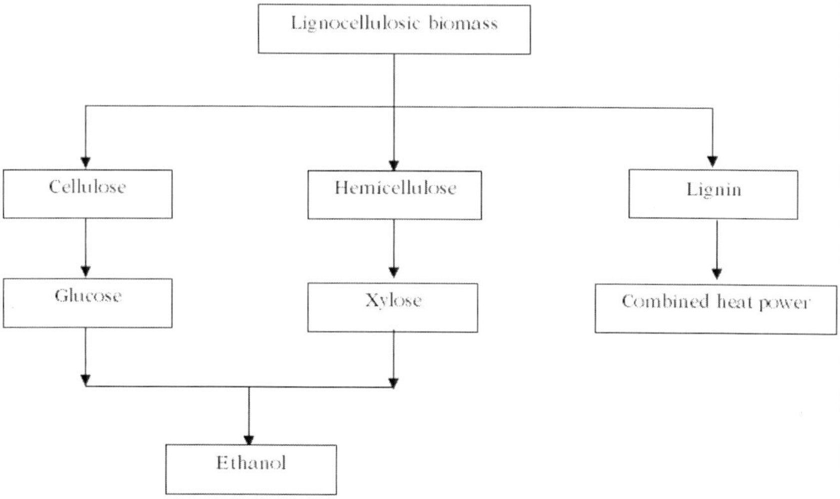

Figure 2 Different composition of lignocellulosic biomass

2.3.1 Corn

Corn is the major food crop of North America, Africa (26%), Europe (12%) and South America (9%). North America is regarded as the highest producing area of corn. Most of the corn is used as animal feed. About 64% is secondly used as human food. Nearly 5% of the corn is wasted as logistic waste. Hence, if this wasted part is being used as the substrate for bioethanol production, then it has been calculated statistically that 9.3 gallons of bioethanol can be produced. Some of the waste products in corn milling are dry distiller's grain, non-edible waste, etc. Lignin is rich in corn stover. Mild alkaline pretreatment coupled with fungal saccharification offers a promising bioprocess for ethanol production from corn stover without the addition of commercial enzymes. Saccharification can be achieved by cultivating either *Phanerochaete chrysosporium* or *Gloeophyllum trabeum* on the treated stover, and fermentation was then performed by using either *Saccharomyces cerevisiae* or *Escherichia coli* K011 [33]. Mild alkaline pretreatment coupled with fungal saccharification offers a promising bioprocess for ethanol production from corn stover without the addition of commercial enzymes. For corn grains, an integrated corn ethanol-methane fermentation system was proposed to solve the problem of stillage handling where the energy assessment indicated that application of this novel process enhanced the net energy balance ratio from 1.26 (conventional process) to 1.76 and had great technical advantage [34]. The use of distiller's dried grains with solubles has an important positive environmental impact [35]. Due to ethanol subsidy, correlation between corn and ethanol has increased. Increasing grants for ethanol production reduces the production risk and based on all these statistics available it is judged that corn can be used as the efficient substrate for the biofuel production [36].

2.3.2 Barley

Barley is an important crop with annual production of 62% in Europe and Asia (15%). Apart from food to animals and humans, 4% of barley is wasted. 4% of barley can be effectively converted into 1.5 gallons of bioethanol which can effectively replace 1.1 gallons of gasoline. Barley also produces DDGS (Distillers Dry Grain and Solubles) which can be effectively used as the substrate for bioethanol production. There is not much information available right now regarding the DDGS. In EDGE (Enhanced Dry Grind Enzymatic) process β-glucanases is added to reduce the viscosity of the mash and β-glucosidase is also added to completely hydrolyze the oligomers obtained during the hydrolysis of β-glucans to glucose which resulted in low cost ethanol production from barley [37]. Endo-xylanases improves glucan conversion and ethanol yield markedly at low substrate concentration and the positive effect of xylanase was most evident in hydrolysis that resulted in 70% ethanol yield [38]. Mutation of barley for low phytic acid content in the grain decreases lignin content modestly, preferentially decreases the guaiacyl (G) lignin counterpart, moreover, enhances the ethanol yield by up to 21.1% using conventional biomass fermentation processes after diluted acid pretreatment [39]. Cellulose and hemicellulose are critically important for the feasibility of biofuel production. 1,3-1,4 glucan 4-gluconohydrolyase-mediated saccharification of barley β-glucan (BG) was performed for ethanol production using thermotolerant

Saccharomyces sp. where maximum hydrolysis of 71% was obtained in 24 h [40]. Hence barley competes as one among the best biomass for ethanol production by fermentation.

2.3.3 Oats

Oat is also an important crop variety aimed to feed the animal stock and to less present humans. Highest producer is the Europe (64%) followed by North America (21%), Oceania (5%). The lost oat is very less of the order 2%. If this wasted oat is being converted to biofuel it can replace 161 million litres of gasoline and there by producing 225 million litres of bioethanol. Oat straw is globally available which could produce bioethanol. Oat straw is a waste crop residue with major percent of cellulose and lignin with certain other polysaccharides. On the whole, oat and oat straw can together produce about 3.16 gallons of bioethanol. Sequential ethanol fermentation and biogas digestion of steam exploded oat straw converts 85–87% of the higher heating value of holocellulose (cellulose and hemicellulose) in the oat straw into biofuel energy [41]. Excessive viscosity development during mashing of oats can be prevented by hydrolyzing -glucan with crude preparations of ' -glucanase' or 'biocellulase'. Ethanol yields of 353.2±3.7 L and 317.6±1.3 L were obtained per tonne (dry weight basis) of hulless (59.8% starch) and hulled (50.8% starch) oats respectively which gives a significant yield [42]. Hence, bioethanol and fuel production using oat is gaining importance

2.3.4 Rice

Rice is yet another important food crop, mainly concerned for human needs than animals. Asia is the largest producer of rice with more than 90%. Wastage in rice is around 4.8%. Hence, research works are being carried out in most countries to effectively convert this substrate into biofuel. Already 12.3 gallons of bio alcohol have been produced by the western countries and other parts of Asia. Tribals of eastern part of India and Tibetan regions are known to produce alcohol for drinking purpose under the name 'sake'. Hydrolysate obtained by solubilizing rice hulls with sulphuric acid fermented by *Escherichia coli* KO11 with provided additional nutrients produced over 46 g ethanol/L in

72 h [43]. Rice bran extract serves as a nitrogen source for improved hemicellulosic ethanol production from sugarcane bagasse by xylose-fermenting yeast strains [44]. Even the rice straw just like other waste agricultural byproducts can be used as the effective means to produce alcohol as a fuel.

2.3.5 Wheat

Wheat and wheat straw with wheat grains that are being wasted can be potentially counted into the effective means of biofuel through fermentation. Asia (43%), Europe (32%) are the primary producers of the wheat. Second largest producer is India while first being China. The use of substrate as wheat is limited and at the same time, the production of ethanol as biofuel from wheat is also less. Wasted wheat grain and wheat dry milling together constitutes the substrate for biofuel production. Wheat straw is potentially being used as substrate for the biofuel production. Europe is currently producing bioethanol to a large extent. Steam-pretreated wheat straw can be mixed with pre-saccharified wheat meal and converted to ethanol in simultaneous saccharification and fermentation (SSF) for good yield [45]. Pretreatment and fractionation of wheat straw with acetic acid can enhance enzymatic hydrolysis and ethanol fermentation [46]. Hydrolysates of washed alkaline peroxide-pretreated wheat biomass provides high sugar concentrations, 31.82 g/L glucose, and 13.75 g/L xylose, their fermentation yield is promised to reach17.37 g/L which figures wheat to be one of the best substrate for ethanol fermentation [47].

2.3.6 Sorghum

Sorghum is another primary crop in Africa (33%) and North America (23%). Sorghum is mainly used as animal waste and almost equally for humans. 6% of the sorghum produced is being wasted annually. Sorghum grain waste, sorghum dry milling, sorghum straw are collectively used for producing biofuel. The sorghum usage for bioethanol production is carried out only in North America and not in Africa, because of low yield of the straw and grain. Nearly 4.9 gallons of bioethanol is being produced. Modified method of ensiling to produce ethanol directly in the silo can be adopted to get good ethanol production from sorghum based biomass [48]. Combined ethanol distillation, NaOH pretreatment and simultaneous saccharification and co-fermentation (SSCF) were adopted to get theoretical ethanol conversion of 69.49% using bagasse of sweet sorghum [49].

2.3.7 Sugarcane Bagasse

Sugarcane bagasse is the straw-like substance that is left over as waste after extracting juice. This sugarcane bagasse is extensively being used as the substrate for the ethanol production. However, in recent times, bagasse is directly being used for the production of the biogas or gobar gas. The above mentioned sources are the most common sources for producing bioethanol. Apart from these substrates wood, salix, millets, are also intensively being used in many cases. The overall view of the biofuel and bioethanol producing industries is that any substrate which has polysaccharide material with lignin in it can be potentially utilized and adapted for the bioethanol production. However for the effective and enhanced production of the bioethanol, foremost thing to be considered is the pretreatment method. This is because the percent of ethanol conversion from substrate is directly related to the hydrolysis of the polysaccharide substrate. Hence based on this criterion the suitable pretreatment method can be adapted for biofuel production in a proficient way. Ethanol yield of 0.45 $g\ g^{-1}$ and volumetric productivity of 0.33 $g\ l^{-1}\ h^{-1}$ from sugarcane bagasse hemicellulose hydrolysate detoxified by neutralization and activated charcoal treatment was found to have potential application in practical process of ethanol production from lignocellulosic hydrolysate [50]. Galactose-adapted *Pichia kudriavzevii* cells for ethanol production from sugarcane juice holds potential for scale-up studies and ethanol productivity of 4g/l/h from sugarcane juice could be achieved [51].

2.4 PRETREATMENT METHODS

A potential and efficient way to produce ethanol is through the utilization of lignocellulosic material like crop residues, grasses, saw dust, wood chips and solid animal waste. Research works have almost been so concrete in the preparation methods for producing ethanol using lignocellulosic materials. However, it is high time to increase the percent of ethanol being produced by this substrate usage. Among the methods, for increasing the percent, the steps needed to be improvised in the pretreatment methods. Hence, now we can deal a bit with the various pretreatment methods being applied in this regard [10, 52, 53]. Hence, various methods being applied are broadly classified into four categories—physical pretreatment, physico-chemical pretreatment, chemical pretreatment and biological pretreatment.

2.4.1 Physical Pretreatment

2.4.1.1 ***Mechanical Disruption*** Agro-waste biomass of the order lignocellulosic materials can be pretreated by steps like chipping, grinding, milling, etc. The length of cellulose chains in biomass is reduced by mechanical disruption techniques to facilitate continuous release of sugar molecules for fermentation in the later stage. Ball milling and vibratory ball milling are the most satisfactory methods recommended for mechanical pretreatment of cellulosic biomass [54]. These methods record good concentrations of free sugar in the pretreated biomass.

2.4.1.2 ***Pyrolysis*** It employs the use of temperature of the order greater than 250^{0}C roughly. At temperature of 300 ^{0}C, the cellulose reduces to gaseous products and residual char [55, 56]. At lower temperature the decomposition is quite at a low rate. Pyrolysis at low concentration and high concentration of chemicals and heat is being used as the efficient form of pyrolysis. The pyrolysed substrate is heated with mild acid for hydrolysis in which nearly 80–85% conversion of cellulose to reduced sugar is being obtained [57]. The process can be further improvised by treating in the presence of oxygen [31]. At lower temperature also the decomposition of cellulose can be obtained by using catalysts. Zinc chloride and sodium carbonate are the catalysts used in this method [58]. The physical methods of pretreatment are the primitive cases of pretreatment methods.

2.4.2 Physico-Chemical Pretreatment

The streams adapted in this technique are steam explosion (auto-hydrolysis), ammonia fiber explosion (AFEX) and CO_2 explosion.

2.4.2.1 ***Steam Explosion*** This is the most commonly used method for pretreatment and is also referred to as auto hydrolysis [59]. Chipped biomass is treated with high pressure steam and then the pressure is swiftly reduced due to which the cellulose material undergoes explosive decompression. Usually recommended temperature is between 160^{0}C–260^{0}C. This condition is continued with the biomass for few seconds to several minutes before it is introduced into atmospheric pressure. Hemicellulose degradation and lignin transformation is obtained at high temperature and hence the degree of cellulose hydrolysis is increased by treatment with dilute sulphuric acid. Under conditions of steam explosion for optimal treatment of cellulosic biomass, two conditions prevail: High temperature and low residence time or low temperature and high residence time [60]. However, lower temperature and longer residence time can be employed beneficially and has high percentage substrate pretreatment [61]. Steam explosion technique is merited because of its low energy requirements than physical methods of pretreatment. Steam explosion stands as a cost effective method for hard woods, agricultural wastes and other high cellulose rich residues. Newly formed residues during the steam explosion process are found to have highly inhibiting effect over the fermentation process which makes the method unreliable for industrial scale establishment. Using this technique needs careful selection of the substrate, temperature and residence time to make the process suitable and feasible at conversion.

2.4.2.2 ***Ammonia Fiber Explosion (AFEX)*** When instead of steam, if liquid ammonia at high pressure is being used as the gaseous medium for a particular period of time it is called as ammonia fiber explosion (AFEX). AFEX concept is similar to steam explosion. This pretreatment method can be employed satisfactory for grass straw, herbaceous crop residues and several lignocellulosic materials like alfalfa, wheat straw, wheat chaff [62], barley straw, corn stover, rice straw [63], municipal solid waste, softwood, news paper [64] and bagasse [65]. AFEX is found not to solubilize hemicellulose like acid pretreatment and steam hydrolysis [62, 63]. Various percentage of hydrolysis has been obtained for various substrates mentioned above. However, some of the major disadvantages include reducing the cost and protecting the environment, ammonia need to be recycled after pretreatment. It is done using a process of superheated ammonia at 200^{0}C passed through the AFEX-treated substrate which vaporizes the rest of ammonia in the treated substrate. The ammonia is collected if a pressure controller is used for the recovery. The main assert of this technique is it does not produce any inhibitory products as in the previous technique. Also, small particle size is not compulsory for this sort of pre-treatment technique.

2.4.2.3 *CO_2 Explosion* This is similar to steam and ammonia pretreatment methods. CO_2 explosion is used for pretreatment of lignocellulosic materials. In this, it is found that during pretreatment, carbonic acid is formed from CO_2 which increases the hydrolysis rate. However it is found that, this method yields relatively low percentage conversion to glucose compared to steam or ammonia explosion. CO_2 explosion was found to be an efficient and cost effective way of pretreatment compared to steam and AFEX with no inhibitory product being obtained finally [66]. However, this treatment method can be efficiently modified by using other gases for treatment in such a way that it produces no toxic products at the end, and has high percentage of substrate treatment. It should also be noted that any new method employed under physico-chemical pretreatment technique is cost effective and efficient.

2.4.2.4 *Chemical Pretreatment* Chemical pretreatment involves wide range of choice of selection. Five methods are available in this technique.

2.4.3 Ozonolysis

Ozone can be used to degrade lignin and hemicellulose in many lignocellulosic materials. Ozone can act on substrates like wheat straw, bagasse, green hay, peanut, pine [67], cotton straw [68], and saw dust [69]. Lignin is not effectively treated by means of ozone but has following advantages: i) It produces no toxic residues ii) It removes lignin to a good extent iii) It is cheap for processing as it requires no much increases or decrease in temperature pressure. Disadvantage mainly includes the ozone requirement in large amounts which makes the process expensive.

2.4.3.1 ***Acid Hydrolysis*** Concentrated acids like sulphuric acid and hydrochloric acid are used in acid hydrolysis. They are toxic, corrosive and hazardous in the reactors being used. These are highly efficient of pretreatment methods. But care to be taken so that the reactor paves favorable way for the process to be carried. In case, if acid of high concentration is being used, then it should be made a note that, after treatment of substrate, the acid needs to be removed for economical and feasible processing. At present, acid with lower concentration that is dilute acid hydrolysis is carried out most successfully. Dilute sulphuric acid can be used for pretreatment of cellulosic materials. At moderate temperature it has low conversion rate of substrate to sugar molecules. However, in the case of high temperature dilute acid treatment increases the rate of cellulose hydrolysis. Two processes are involved in acid hydrolysis. They are i) High temperature continuous flow process—in this process low solid loading and high temperature of the order 1600^{0}C is used. The solid concentration is around 10% weight of substrate/weight of reaction mixture [70, 71]. ii) Batch process—low temperature—temperature maintained is less than 1600^{0}C. The solid concentration is of the order 10%–40% [72]. For ethanol, the production is most feasible at pH greater than 5 to 8. Hence in this case, the acid used need to be neutralized. Hence for this reason this method is highly cost effective than the physico chemical treatment methods.

2.4.3.2 ***Alkaline Hydrolysis*** Similar to acids, some bases can also be used for pretreatment methods of lignocellulosic materials. Its effect depends on the lignin content of the materials [57, 59]. The porosity of the lignocellulosic material increases when the cross linking xylan and hemicellulose bonds are being removed. This is primarily done by the saponification process by the alkaline materials. Generally NaOH of various strengths are being employed in the pretreatment process. Ammonia which is an alkaline chemical is also used in this regard. Hence apart from acids, alkali solutions can also be used to a greater extent in the pretreatment processes.

2.4.3.3 ***Oxidative Delignification*** In oxidative delignification process hydrogen peroxide is used. It is a strong corrosive agent. Lignin in the substrate is greatly degraded by peroxidases class of enzyme in the presence of H_2O_2 [73]. This method of pretreatment is not satisfactory and not widely applied.

2.4.3.4 ***Organo Solvent Process*** Organic or aqueous organic solvents are used in this process. The various organic acids used in this method are the methanol, ethanol, acetone, ethylene glycol, triethylene glycol and tetra hydrofurfuryl alcohol. In this process inorganic acid catalysts are also incorporated for the organosolvent process. But when the organosolvent process is carried at high temperature, inorganic acid catalyst addition is necessary. In this process after hydrolysis of the organosolvent, it is highly essential to remove the solvent as it may be an inhibitory factor in the process of fermentation and might be toxic to the microorganisms used in most cases. Thus, of the various methods employed for pretreatment under chemical methods, the most efficient are the acid hydrolysis and alkaline hydrolysis which are of great importance even today.

2.4.4 Biological Pretreatment

Biological pretreatment method uses the microorganisms that have the capability to degrade lignin and hemicellulose in the substrate. Generally fungal strains are used exhaustively as they can break the lignin and hemicellulose effectively [74].

2.4.4.1 ***Various Microbial Strains used in Pretreatment*** Brown, white and soft rot fungi are most widely used. Some of the sub spp. used are *Pleurotus ostreatus, Phanerochaete sordida, Pycnoporus cinnabarinus 115, Sporotrichum pulmententum, Ceriporiopsis subvermispora, Cyathus stercoreus, Penicillium chrysosporium*, etc. White rot fungus produces lignin peroxidase and manganese-dependent peroxidase during secondary metabolism for the degradation of lignin. Enzymes like polyphenol oxidases, laccases, H_2O_2 producing enzymes and quinone reducing enzymes can also degrade the lignin and hemicellulose. The main advantage of biological pretreatment is that it is a low energy process with mild environmental conditions. But, the rate of hydrolysis in most of biological pretreatment processes is very low. Many research works in the lignin and the hemicellulose degradation are carried for effective and fast degradation of substrate.

2.4.4.2 ***Enzymatic Hydrolysis of Cellulose*** It involves usage of enzyme cellulase for the pretreatment of cellulose. The cellulase enzyme degrades the cellulose into the reduced sugar.

$$\text{Cellulose} \xrightarrow{\text{Cellulase}} \text{Reduced Sugar}$$

The cellulose degradation by cellulase can also involve the usage of the biological pretreatment method which employs microorganism that produces enzyme cellulase which in turn degrades the cellulose. Bacteria producing cellulase enzyme are *Clostridium spp., Cellulomonas spp., Bacillus spp., Thermomonos porafusca, Bacteroides cellulosolvens*, etc. Fungal strains producing cellulase enzyme are *Scterotium rolfsii, Penicillium chrysosporium, Trichoderma spp., Aspergillus spp., Schizophyllum spp.* and other *Penicillium* strains. *Trichoderma spp.* is most extensively used in the lignocellulosic fermentation process. Thus there are various pretreatment methods being employed for enhancing hydrolysis of the substrate. The most widely applied methods are chosen depending on the nature of the substrate, the hydrolysis rate, time of reaction, cost effectiveness and above all the environmental issues related to the methods. Thus, based on the substrate and its nature, pretreatment methods are selected. Thus, on the whole, various researchers in this field have produced varying results of ethanol produced using variety of substrate and also diversified strains of microorganisms. The various conditions like temperature, aeration, agitation, HRT and the microorganisms used are given as a consolidated account from the past five year (2001–2006) research data for better reference.

Substrate

(Cellulose biomass or lignocellulosic material)

↓

Pretreatment method

(Converts the polymer to monomer form)

↓

Pretreated substrate

↓

Highly efficient microorganism

(Conversion of ethanol)

Figure 3 Conversion of substrate into bioethanol

2.5 CONCLUSION

The world's crude oil is high on its decreasing phase and the need for an alternative is in its exponential phase. Developed countries like USA, Europe and other countries are on the stake to find an alternative fuel since their consumption of fuel is high and their import of crude oil from the gulf countries is increasing. Though it has been cued out that ethanol can be one among the apt alternative, the commercialization of it in a successful phase may take a little time. If the cost of the cellulase enzyme is reduced even more, there is a great scope for the commercialization of the bioprocess industries. Engineering of microorganisms by gene cloning experiments has gained huge watch in the genetics sector for the future of ethanol based fuels. As far as substrate is concerned, lignocellulosic biomass is present in abundance and distributed throughout the global habitat. Sustainable existence of substrate is a promising factor for continuous fuel production which has to be armed with more technical findings and optimized results. The world is striving for a quick breakthrough in the field of bioethanol production to provide an economical and eco friendly fuel and possibilities of ethanol existence as alternate fuel is in the near future with lots more research to be done. Promoting more number of researches in the field of alternate fuels can have fruitful results for the fuel problems of future generation.

2.6 REFERENCES

1. Licht, F.O. 2006. World Ethanol Market: The Outlook to 2015, Tunbridge Wells, Agra Europe Special Report, UK

2. L. Davis, Y-J. Jeon, C. Svenson, P. Rogers, J. Pearce, and P. Perris, *Evaluation of wheat stillage for ethanol production by recombinant Zymomonas mobilis*, Biomass Bioenergy 29 (2005), pp. 49 – 59.

3. G. Panagioton, S.G. Villas Boar, P. Cheristatkopoulous, J. Nielsen, and N. Olsson, *Intracellular metabolite profiling of Fusarium oxysporum converting glucose to ethanol*, J of Biotechnol. 115 (2005), pp. 425 – 434.

4. E. Gnansounou, A. Dauriat, and C.E Wyman, *Refining sweet sorghum to ethanol and sugar: economic trade-offs in the context of North China*, Bioresource Technol. 96 (2005), pp. 985 – 1002.

5. M. Ballesteros, J.M. Oliva, M.J. Negro, P. Manzanares, and B. Allesteros, *Ethanol from lignocellulosic materials by a simultaneous saccharification and fermentation process (SFS) with Kluyveromyces marxianus CECT 10875*, Process Biochemistry 39 (2004), pp. 1840 – 1848.

6. Z.S. Kadar, Z.S. Szengyel, and K. Reczey, *Simultaneous saccharification and fermentation (SSF) of industrial wastes for the production of ethanol*,Industrial crops and products. 20 (2004), pp. 103 – 110.

7. H. Golias, G.J. Dumsday, G.A. Stanley, and N.B. Pammet, *Evaluation of a recombinant Klebsiella oxytoca strain for ethanol*

production from cellulose by simultaneous saccharification and fermentation: comparison with native cellobiose-utilising yeast strains and performance in co-culture with thermotolerant yeast and Zymomonas mobilis, J Biotechnol. 96 (2002), pp. 155 – 168.

8. Y. Sun, and J. Cheng, *Hydrolysis of lignocellulosic materials for ethanol production: a review*, Bioresource Technol, 83 (2002), pp. 1 – 11.

9. Z. Yu, and H. Zhang, *Pretreatments of cellulose pyrolysate for ethanol production by Saccharomyces cerevisiae, Pichia sp. YZ-1 and Zymomonas mobilis*, Biomass Bioenergy. 24 (2003), p.p. 257 – 262.

10. A. Demirbas *Biomass and Wastes: Upgrading Alternative Fuels*, Energy Sources, 25 (2003), p.p. 317-329.

11. H. Shigechi, K. Uyama, Y. Fujita, T. Matsumoto, M. Uedac, A. Tanaka, H. Fukuda, and A. Kondo, *Efficient ethanol production from starch through development of novel flocculent yeast strains displaying glucoamylase and co-displaying or secreting -amylase*, Journal of Molecular Catalysis B: Enzymatic, 17 (2002), p.p. 179–187.

12. A. Demirbas, *Ethanol from Cellulosic Biomass Resources*, Int J Green Energy, 1 (2004), p.p. 79-87.

13. C.A.C. Alzate, O.J.S. Toro, *Energy consumption analysis of integrated flow sheets for production of fuel ethanol from lignocellulosic biomass*, Energy, 31 (2006), p.p. 2447-2459.

14. Tahrzadeh,M.J. *Ethanol from lignocellulose: Physiological effects of inhibitors and fermentation strategies*, Chemical reaction Engineering. (1999).

15. J.D. Murphy, K. McCarthy, *Ethanol production from energy crops and wastes for use as a transport fuel in Ireland*, Applied Energy, 82 (2005), p.p. 148 – 166.

16. C. Lei, C.K. Tim, Y.L. Joo, and Z. Xiang, *Dilute acid pretreatment for cellulosic alcohol production*, Biomass Conversion and Biorefinery, 2 (2012), p.p. 169-177.

17. K. Saravanakumar, P. Senthilraja, and K. Kathiresan, *Bioethanol production by mangrove-derived marine yeast Sacchromyces cerevisiae,* Journal of King Saud University – Science, 25 (2013), p.p. 121–127.

18. M. Saritha, A. Arora, and Lata, *Biological pretreatment of Lignocellulosic substrates for enhanced delignification and enzymatic digestibility,* Ind J. Microbiol, 52 (2012), p.p. 122–130.

19. M. Yasuyuki, I. Toyoki, T. T. H. Yasuo, and F. Kazuhiko, *Conversion of sulfuric acid lignin generated during bioethanol production from lignocellulosic materials into polyesters with ε-caprolactone* J. Wood Sci. 57 (2011), p.p. 214–218.

20. S. Zafar, M. Owais, *Ethanol production from crude whey by Kluyveromyces marxianus,* Biochemical Engg J, 27 (2006), p.p. 295 - 298.

21. H. Jorgensen, J.B. Kristensen, and C. Felby, *Enzymatic conversion of lignocellulose into fermentable sugars: challenges and opportunities,* Biofuel Bioprod. Bior. 1 (2007), p.p. 119–134.

22. C. Munoz, J. Baeza, J. Freer, and T.R. Mendonça. *Bioethanol production from tension and opposite wood of Eucalyptus globulus using organosolv pretreatment and simultaneous saccharification and fermentation,* J. Ind. Microbiol. Biotechnol. 38 (2011), p.p. 1861–1866.

23. M. Lewandowska, W. Kujawski, *Ethanol production from lactose in a fermentation/pervaporation system,* J Food Eng, 79 (2007), p.p. 430–437.

24. X.M. Ge, F.W. Bai, *Intrinsic kinetics of continuous growth and ethanol production of a flocculating fusant yeast strain SPSC01,* J Biotechnol, 124 (2006), p.p. 363-372.

25. K. Karimis, G. Emtiazi, and M.J. Taherzadeh, *Production of ethanol and mycelial biomass from rice straw hemicellulose hydrolyzate by Mucor indicus,* Process Biochemistry. 41 (2006), p.p. 653-658.

26. D.D. Spindler, C.E. Wyman, A. Mohagheghi and K. Grohman, *Thermotolerant yeast for simultaneous saccharification and fermentation*

of cellulose to ethanol, Applied Biochemistry and Biotechnology 88 (1988), 279-293.

27. M. Takashima, T. Sugita,Y. Toriumi and T. Nakase,*Cryptococcus tepidarius sp. nov., a thermotolerantyeast species isolated from a stream from a hotspringarea in Japan,* International Journal of Systematic and Evolutionary Microbiology 59 (2009), 181–185.

28. H.D. Jang, C.L. Hsu, K.S. Chang, M.Z. Lai, T.C. Chang, Y.H. Chang. *Pretreatment and hydrolysis of cellulosic agricultural wastes with a cellulase-producing Streptomyces for bioethanol production*,biomass and bioenergy 35(2011), 1878-1884.

29. P.K. Bajpai and A. Margaritis, *Kinetics of ethanol production by immobilized cells of ZymomonasnroMi55 at varying D-glucose concentrations*, Enzyme Microb. Technol., 7 (1985), 462- 464.

30. M.B. Doelle and H.W. Doelle, *Ethanol production from sugarcane syrup using Zymomonas mobilis,* Journal of biotechnology, 11 (1989), 25-36.

31. J. Baratti and E.F. Torres, *Continuous ethanol production by Zymomonas mobilis from an equimolar mixture of glucose and fructose,* Biomass 13 (1987), 75-85.

32. S. Zhu, Y. Wu, Z. Yu, X. Zhang, C. Wang, and S. Jin, *Production of ethanol from microwave-assisted alkali pretreated wheat straw,* Process biochemistry. 41 (2006), p.p. 869-873.

33. M. Vincent, A. L. Pometto III, J. (Hans) van Leeuwen,*Ethanol production via simultaneous saccharification and fermentation of sodium hydroxide treated corn stover using Phanerochaete chrysosporium and Gloeophyllum trabeum,*Bioresource Technology, 158 (2014), p.p. 1-6

34. K. Wang, J. Zhang, L. Tang, H. Zhang, G. Zhang, X. Yang, P. Liu, Z. Mao, *Establishment and assessment of a novel cleaner production process of corn grain fuel ethanol,* Bioresource Technology148 (2013), p.p. 453–460

35. C. Pieragostini, P. Aguirre, M.C. Mussati, *Life cycle assessment of corn-based ethanol production in Argentina,* Science of The Total Environment, 472 (2014), p.p. 212-225.

36. C. Maxwell, M. Davison, *Using real option analysis to quantify ethanol policy impact on the firm's entry into and optimal operation of corn ethanol facilities*, Energy Economics 42 (2014), p.p. 140-151.

37. N.P. Nghiem, E.C. Ramírez, A.J. McAloon, W. Yee, D.B. Johnston, K.B. Hicks, *Economic analysis of fuel ethanol production from winter hulled barley by the EDGE (Enhanced Dry Grind Enzymatic) process*, Bioresource Technology102 (2011), p.p. 6696-6701

38. M.J. Negro, M.P. García-Aparicio, J.M. Oliva, P. Manzanares, M. Ballesteros, I. Ballesteros, A. González, *Second-generation ethanol production from steam exploded barley straw by Kluyveromyces marxianus CECT 10875*, Fuel 90 (2011), p.p. 1624-1630.

39. S. Chen, Jing Ke, Dhrubojyoti D. Laskar, Mackenzie T. Ellison, Robert S. Zemetra, *Modulation of lignin deposition/composition via phytic acid reduction in seed improves the quality of barley straw for sugar release and ethanol production*, Biomass and Bioenergy 46 (2012), p.p. 584-592

40. M. Rao, R. Divate, V. Menon, *Approach towards biocatalytic valorisation of barley β-glucan for bioethanol production using 1,3-1,4 β-glucanase and thermotolerant yeast*, International Biodeterioration & Biodegradation 82 (2013), p.p. 81–86.

41. J. Stahlberg,D.Y. Dererie, S. Trobro, M.H. Momeni, H.Hansson, J. Blomqvist, V. Passoth, A. Schnürer, M. Sandgren,*Improved bio-energy yields via sequential ethanol fermentation and biogas digestion of steam exploded oat straw*, Bioresource Technology 102 (2011), 4449 – 4455.

42. K C Thomas, Dr W M Ingledew, *Production of fuel alcohol from oats by fermentation*, Journal of Industrial Microbiology15 (1995), 125 – 130

43. Mohammed Moniruzzaman,Lonnie O. Ingram, *Ethanol production from dilute acid hydrolysate of rice hulls using genetically engineered Escherichia coli*, Biotechnology Letters 20 (1998), p.p.943-947.

44. S.E. Martiniano, R.R. Philippini, A.K. Chandel, C.A. Rosa, F.C. Pagnocca, S.S. da Silva, *Evaluation of Rice Bran Extract as a Nitrogen Source for Improved Hemicellulosic Ethanol Production from Sugarcane Bagasse by New Xylose-Fermenting Yeast Strains Isolated from Brazilian Forests*, Sugar Tech16 (2014), p.p. 1-8.

45. B. Erdei, Z. Barta, B. Sipos, K. Réczey, M. Galbe, G. Zacchi, *Ethanol production from mixtures of wheat straw and wheat meal,* Biotechnology for Biofuels 3 (2010).

46. L. Wang, Z. Luo,S. Xiu,A. Shahbazi, *Pretreatment and Fractionation of Wheat Straw with Acetic Acid to Enhance Enzymatic Hydrolysis and Ethanol Fermentation,*Energy sources 33 (2011),p.p. 1230-1238

47. S Bolado, C. Toquero, *Effect of four pretreatments on enzymatic hydrolysis and ethanol fermentation of wheat straw. Influence of inhibitors and washing,* Bioresource Technology 157 (2014), p.p. 68–76

48. L.L. Henk, J.C. Linden, *Solid-state production of ethanol from sorghum,* Applied Biochemistry and Biotechnology57 – 58 (1996), p.p.489 – 501.

49. S. Li, Menghui Yu, J. Li, R. Du, Y. Jiang, G. Fan, G. Zhao, S. Chang,*A cost-effective integrated process to convert solid-state fermented sweet sorghum bagasse into cellulosic ethanol,* Applied energy 115 (2014), p.p. 331 – 336.

50. A. Hande, S. Mahajan, A. Prabhune,*Evaluation of ethanol production by a new isolate of yeast during fermentation in synthetic medium and sugarcane bagasse hemicellulosic hydrolysate,* Annals of Microbiology 63 (2013), p.p.63-70.

51. H.S. Oberoi, S.S. Dhaliwal, S.K. Sandhu, D. Nanda, D. Kumar, S.K. Uppal, *Enhanced ethanol production from sugarcane juice by galactose adaptation of a newly isolated thermotolerant strain of Pichiakudriavzevii,* Bioresource Technology 102 (2011), p.p. 5968 – 5975.

52. N. Fujii, A. Sakurai, K. Onjoh, and M. Sakakibara, *Influence of surface characteristics of cellulose carriers on ethanol production by immobilized yeast cells,* Process Biochemistry, 34 (1999), p.p. 147-152.

53. T. Lebeau, T. Jouenne, and G.A Junter, *Simultaneous fermentation of glucose and xylose by pure and mixed cultures of Saccharomyces cerevisiae and Candida shehatae immobilized in a two-chambered bioreactor,*Enzyme Microbial Technol, 21 (1997), p.p. 265-272.

54. M.A. Millet, A.J. Baker, and L.D. Scatter, *Physical and chemical pretreatment for enhancing cellulose saccharification*. Biotech. Bioeng. Symp, 6 (1976), p.p. 125–153.

55. F.J. Kilzer, A. Broido, *Speculations on the nature of cellulose pyrolysis* Pyrodynamics, 2 (1965), pp. 151–163.

56. F. Shafizadeh, A.G.W. Bradbury *Thermal degradation of cellulose in air and nitrogen at low temperatures* J. Appl. Poly. Sci., 23 (1979), pp. 1431–1442

57. L.T. Fan, M.M. Gharpuray, and Y.H. Lee (1987). In: Cellulose Hydrolysis Biotechnology Monographs. Springer, Berlin.

58. F. Shafizadeh, Y-Z. Lai, *Thermal degradation of 3-deoxy-D-erythro-hexosulose* Carbohydrate Research 40 (1975), p.p. 263–274

59. J.D. McMillan, Pretreatment of lignocellulosic biomass M.E Himmel, J.O Baker, R.P Overend (Eds.), Enzymatic Conversion of Biomass for Fuels Production, American Chemical Society, Washington, DC (1994), pp. 292–324

60. S.J.B. Duff, W.D. Murray, *Bioconversion of forest products industry waste cellulosics to fuel ethanol: a review.* Bioresour. Technol, 55 (1996), p.p. 1–33.

61. Wright JD. *Ethanol from lignocellulosics: an overview.* Energy Progress, 84 (1988), p.p. 71-80.

62. M. Mes-Hartree, B.E. Dale, and W.K. Craig. *Comparison of steam and ammonia pretreatment for enzymatic hydrolysis of cellulose,* Appl. Microbiol. Biotechnol. 29 (1988), p.p. 462–468.

63. E.Y. Vlasenko, H. Ding, J.M. Labavitch, and S.P. Shoemaker. *Enzymatic hydrolysis of pretreated rice straw,* Bioresour. Technol. 59 (1997), p.p. 109–119.

64. Holtzapple MT, Lundeen JE, Sturgis R. *Pretreatment of lignocellulosic municipal solid waste by ammonia fiber explosion (AFEX),* Appl. Biochem. Biotechnol. 34 (1992), p.p. 5–21.

65. M.T. Holtzapple, J-H. Jun, G. Ashok, S.L. Patibandla, and B.E. Dale, *The ammonia freeze explosion (AFEX) process: a practical*

lignocellulose pretreatment, Appl. Biochem. Biotechnol. 28 (1991), p.p. 59–74.

66. Y.Z. Zheng, H.M. Lin, and G.T. Tsao. *Pretreatment for cellulose hydrolysis by carbon dioxide explosion,* Biotechnol. Prog. 14 (1998), p.p. 890–896.

67. W.C. Neely. *Factors affecting the pretreatment of biomass with gaseous ozone,* Biotechnol. Bioeng. 20 (1984), p.p. 59-65.

68. D. Ben-Ghedalia, G. Shefet. *Chemical treatments for increasing the digestibility of cotton straw,* J. Agric. Sci. 100 (1983), p.p. 393–400.

69. P.F. Vidal, J. Molinier, *Ozonolysis of lignin – improvement of in vitro digestibility of poplar sawdust.* Biomass, 16 (1988), p.p. 6:1-17.

70. A.H. Brennan, W. Hoagland, and D.J. Schell, *High temperature acid hydrolysis of biomass using an engineering-scale plug flow reactor: result of low solids testing,* Biotechnol. Bioeng. Symp. 17 (1986), p.p. 53-70.

71. A.O. Converse, R. Matsuno, M. Tanaka, and M. Taniguchi. *A model for enzyme adsorption and hydrolysis of microcrystalline cellulose with slow deactivation of the adsorbed enzyme.* Biotechnol. Bioeng. 32 (1988), p.p. 38–45.

72. A. Esteghlalian, A.G. Hashimoto, J.J. Fenske, and M.H. Penner, *Modeling and optimization of the dilute-sulfuric-acid pretreatment of corn stover, poplar and switchhgrass,* Bioresour. Technol. 59 (1997), p.p. 129–136.

73. M.P. Tucker, N.J. Nagle, E.W. Jennings, K.N. Iosen, H.K. Young, H. Kim, and S.L. Noll, *Conversion of distiller's grain into fuel alcohol and a higher-value animal feed by dilute-acid pretreatment,*Applied Biochemistry Biotechnol, 113-116 (2004), p.p. 1139-1159.

74. P. Sassner, M. Galbe, and G. Zacchi, *Bioethanol production based on simultaneous saccharification and fermentation of steam-pretreated Salix at high dry-matter content,* Enzyme Microbial Technol, 39 (2006), p.p. 756-762.

75. L.V.A. Reddy, O.V.S. Reddy, *Rapid and enhanced production of ethanol in very high gravity (VHG) sugar fermentation by Saccharomyces cerevisiae: Role of finger millet (Eleusine coracana L.) flour,* Process Biochemistry, 41 (2006), p.p. 726-729.

76. R. Millati, L. Edebo, and M.J. Taherzadeh, *Performance of Rhizopus, Rhizomucor, and Mucor in ethanol production from glucose, xylose, and wood hydrolyzates,* Enzyme and Microbial Technol, 36 (2005), p.p. 294-300.

77. Q. Xiang, Y.Y. Lee, and R.W. Torget, *Kinetics of glucose decomposition during dilute-acid hydrolysis of lignocellulosic biomass,* Applied Biochemistry Biotechnol,113-116 (2004), p.p. 1127-1138.

78. M. Ballesteros, J. Oliva, P. Manzanares, M. Negro, I. Ballesteros, *Ethanol production from paper material using a simultaneous saccharification and fermentation system in a fed-batch basis.* World J. Microbiol. Biotechnol., 18 (2002) p.p. 559–561.

79. N. Fujii,A. Sakurai,K. Onjoh, and M. Sakakibara, *Influence of surface characteristics of cellulose carriers on ethanol production by immobilized yeast cells,* Process Biochemistry, 34 (1999), p.p. 147-152.

80. Y. Fujita, S Takahashi, M. Ueda,A. Tanaka, Y. Morikawa, T. Kawaguchi, M. Arai, H. Fukuda, and A. Kondo, *Direct and Efficient Production of Ethanol from Cellulosic Material with a Yeast Strain Displaying Cellulolytic Enzymes,* Applied and Envronmental Microbiology, 68 (2002), p.p. 5136-5141.

81. K. Sudharani, M.V. Swamy, and G. Seenayya, *Production of ethanol from various pure and natural cellulosic biomass by Clostridium thermocellum strains SS21 and SS22,* Process Biochemistry, 33 (1998), p.p. 435-440.

82. L.O.Ingram, J.B.Doran, *Conversion of cellulosic materials to ethanol,* FEMS Microbiology Reviews, 16 (1995), p.p. 235-241.

3

Biobutanol Production Strategies

3.1 INTRODUCTION

High consumption of fuels in the form of petroleum products as energy source for working of various engines has brought uphill depletion of natural resources and petroleum products which lead to increasing demand for energy source for industrialization and motorization [1]. Currently 80% primary energy source consumed by the world is taken up by fossil fuels of which transport sector alone consumes around 58% of the fossil fuels [2]. Though fossil fuel serves as an efficient energy source around the globe, they greatly become the root cause for the emission of green house gases (GHG) due its combustion, for the loss in biodiversity, rise in sea level, receding of glaciers, etc. By foreseeing the issues caused by energy security climatic changes and climbing prices of raw materials, alternative energy sources have got increasing attention universally [3-9].

The alternative energy source has made the world to capsize as a sustainable, competent and financially viable energy source. Alternative source drawn from biomass acts as the significant energy source to mitigate the increasing demand for the fuel energy source and the CO_2 emitted during the combustion of this fuel can be recycled in synthesizing biomass. Among the many alternative energy source reinstating the present situation obtained from biomass, bio-fuels have proven to be the most eco-friendly and have increased in popularity worldwide due to the rising oil prices and need for energy security. Bio-fuels are the predominantly converted products of biomass. Due to its renewability, biodegradability, spawning acceptable quality exhaust gas, biofuels puts back fossil fuels [10]. A variety of bio-fuels are being produced from the biomass such as ethanol, butanol, methanol, biodiesel, green diesel, syngas and methane [11]. Therefore bio-fuels have emerged as one of the most decisive sustainable fuel source which has less GHG emission, improves air quality and keeps the surrounding pollution free. Bio-fuels from biomass are either in the form of *solid, liquid* or *gas*. *Solid fuels* refers to various types of solid materials that are used as fuels that produce energy through combustion which include wood, charcoal, peat, coal and pellets. These solid fuels fall under the *primary bio-fuels* [12-30] whereas the *secondary fuels* are liquid or gaseous fuels used for transportation and these are further classified as *1st, 2nd* and *3rd generation fuels* [12].

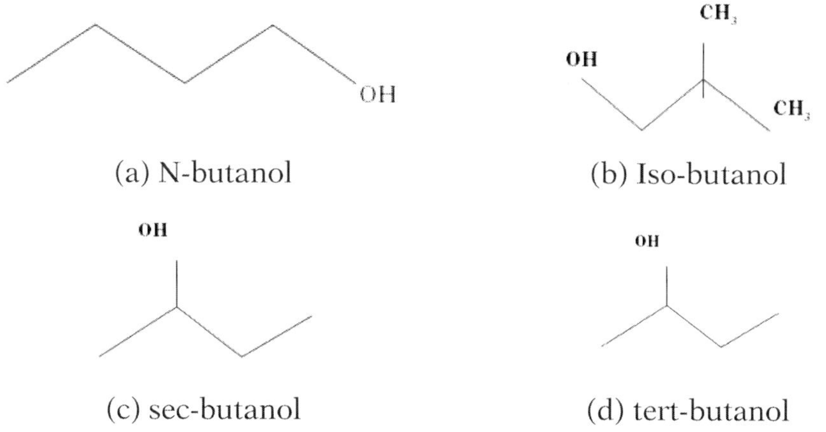

Figure 1 Chemical structure of butanol in various forms performing its unique activity

First generation secondary bio-fuels (or) *conventional fuels* are made from substrates such as sugars, vegetable oil, starch, grains and seeds, etc., but this brought high competence with the food production sector which in turn raised the cost of crops and food. *Second generation secondary bio-fuels* are sustainable bio-fuels from feedstocks (lignocellulosic biomass or agricultural wastes). Fuel from feedstocks follows a complicated production process which is not truly economical. Biobutanol, bioethanol, mixed alcohol, and Fischer–Tropsch gasoline by thermochemical and biochemical processes plunges under this group of bio-fuels. *Third generation secondary bio-fuels* are fuels that are obtained from substrates such as algae, seaweeds, etc. [12, 31]

Anticipating the challenges and benefits like energy security, economic stability, environmental benefits, technological issues and substrate availability and maintenance of the major bio-fuels mentioned above, biobutanol is receiving swelling attention as a proficient bio-fuel.

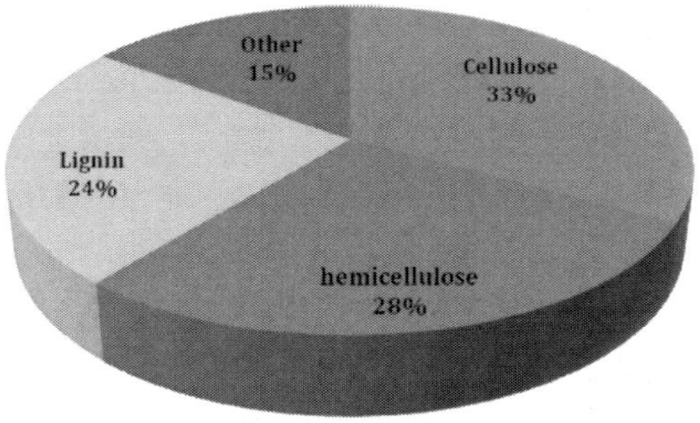

Figure 2 Universal composition of Lignocellulose

3.1.1 Biobutanol's History as a Sustainable Biofuel

Butanol, a versatile four carbon alcohol (C_4H_9OH) found to be colourless and flammable [32] is largely produced by the petrochemical industry and has been recognized in history as a chemical and solvent used in various industries. During the 20th century, it was particularly used as a raw material for butadiene used for synthetic rubber [33]. Currently butanol is considered to be the highly developed biofuel which is used as a supplement for gasoline, diesel and kerosene, superior to ethanol almost in all aspects also competing well with other major bio-fuels [33] and chiefly butanol contributes for the biological production of butyl-t-butyl ether like ethanol for ethyl-t-butyl ether (BTBE & ETBE respectively) [34]. Being a biofuel, butanol enhances the tolerance to water contamination, blends at higher concentration without retrofitting vehicles, has higher energy density, reduces viscosity of vegetable oils and has the ability to dissolve in vegetable oils at any ratio. Butanol is convertible to jet fuel, blends with gasoline and also acts as a rich fuel extender. As an additive, butanol when mixed with fuel can be used in internal combustion that yields only carbon dioxide making it a more environmental friendly biofuel [32]. Butanol is used much more than ethanol due to the following reasons: its high calorific value 29.2 $MJdm^3$, melting point -89.5^{0}C, boiling point 117.2^{0}C, flash point 36^{0}C, the self-ignition is said to be 340^{0}C, less flammable, less volatile and has a low vapour pressure,less corrosive, relative low heat of vaporizationand, and highly explosive over bioethanol. Physical properties comparison of butanol with other fuels is discussed in Table1 [33-42].

Butanol exists in other forms of butanol isomers such as n-butanol or normal-butanol ($CH_3CH_2CH_2CH_2OH$), 2-butanol or secondary butanol ($CH_3CH_2CHOHCH_3$), iso-butanol ($(CH_3)_2CH_2CHOH$) and t-butanol or ter-butanol ($(CH_3)_3COH$) [32]. Structures of these above mentioned butanol are shown in Figure 1(a, b, c, d). Iso-butanol, a branched isomer of butanol blends efficiently up to 15% on volumetric basis with petrol which replaces ethanol that blends only 10% with petrol and n-butanol is said to be the fermentation product that is more complicated both in upstream and downstream processing which is dicussed in this chapter. Butanol contains 110,000 BTUs per gallon, closer to gasoline's 115,000 BTUs, and is safer to handle with Reid value of 0.33 psi, which is a measure of a fluid's rate of evaporation when compared to gasoline at 4.5 and ethanol at 2.0 psi [43, 44]. An 85% butanol/gasoline blend can be used in unmodified petrol engines and the presence of 22% oxygen makes it a beneficiary fuel that ensures cleaner burning than ethanol [45, 46].

Table 1 Physical property comparison of various fuels

Properties	Butanol	Gasoline	Ethanol	Methanol
Melting point (°C)	- 89.3	-57.1 to -56.6	-114	-97.6
Specific gravity	0.810– 0.812	0.713	0.79	0.7866-0.7960
Ignition temperature (°C)	35–37	422.7	276-456	464
Auto-ignition temperature (°C)	343–345	246-280	422	470
Flash point (°C)	25–29	13	12.77	15.6
Relative density (water: 1.0)	0.81	0.71-0.77	0.805-0.812	0.8
Critical pressure (hPa)	48.4	-	4.60 MPa	6.14 MPa
Critical temperature (°C)	287	241	239.85	239
Explosive limits (vol. % in air)	1.4–11.3	1.4-7.6	3.3-19	5.5-44.5

Properties	Butanol	Gasoline	Ethanol	Methanol
Relative vapor density (air: 1.0)	2.6	3-4	1.59	1.1
Vapor pressure (kPa at 20°C)	0.5	7.0-7.8	5.95	12.8
Boiling point (°C)	117–118	27–221	78	64.7
Density at 20°C (g/ml)	0.8098	0.7–0.8	0.7851	0.7866
Solubility in 100 g of water	immiscible	immiscible	miscible	miscible
Energy density (MJ/l)	27–29.2	32	19.6	16
Energy content/value (BTU/gal)	110000	115000	84000	76000
Air-fuel ratio	11.2	14.6	9	6.5
Heat of vaporization (MJ/kg)	0.43	0.36	0.92	1.2
Liquid Heat capacity (Cp) at STP (kJ/k-mol.°K)	178	160–300	112.3	81.14
Research octane number	96	91–99	129	136
Motor octane number	78	81–89	102	104
Octanol/Water Partition Coefficient (as logPo/w)a	0.88	3.52±0.62	-0.31	-0.77
Dipole moment (polarity)	1.66	NA	1.7	1.6
Viscosity (10^{-3}Pa.s)	2.593	0.24–0.32	1.078	0.5445

Industrial biosynthesis of biobutanol was carried out with *Clostridium* species. Under the batch conditions of ABE fermentation, the process of solvent producing Clostridia proceeds with the biphasic process, acidogenesis and solventogenesis. Other than *Clostridium*, heterologous microbial hosts including *Escherichia coli, Pseudomonas sp., Saccharomyces sp. and Bacillus subtilis* when genetically engineered are used as biocatalysts for the enhanced butanol biosynthesis. Also production of bio-fuels requires a good substrate based on the microorganism that would be food crops, lignocelluloses and plant biomass. Starch is the fermented crop product that is not economically viable. To be more economical, algal cultures are used as it does not require lots of work and cost. Using agricultural and food crops as a major source for butanol production puts forth major challenges that compete with the food industry. Hence there is a need of perpetual hunt for cost-effective and easily available raw material for economic production of biobutanol. In this case, lignocellulosic materials were chosen to be the best, cheaper and readily available substrates [47-60].

3.2 *Clostridial Species* – Kingmaker of Butanol Production

Butanol production is carried out by the ABE fermentation process that employs bacterial strain of *Clostridium* family. *Clostridium* strains are saccharolytic butyric acid-producing bacteria under appropriate operating conditions to ferment a starch substrate (pentoses, hexoses, mono-, di- and polysaccharides), sugar beets, glucoses, cellulosic-based materials and other biomass feedstocks. These strains have the ability to secrete profuse enzymes that abet the metabolism of polysaccharides into monosaccharides. Few species that produce butane effectively include *Clostridium acetobutylicum, C. beijerinckii, C. saccharoperbutylacetonicum, C. saccharoacetobutylicum, C. aurantibutyricum, C. pasteurianum, C. sporogenes, C. cadaveris,* and *C. tetanomorphum*. Providentially, *Clostridium acetobutylicum, C. beijerinckii, C. saccharoperbutylacetonicum* and *C. saccharo acetobutylicum* [17] showed the evidence for significant butanol production with high yields. The selection of the bacterial strains for butanol production depends on the following factors, (1) type of substrates selected, (2) nutrient requirement, (3) tolerance of butanol, (4) steered yield and concentration, and (5) resistance to bacteriophage and antibiotics. Novel organisms being employed for the butanol production should have coveted characteristics and the enhancement in production of butanol by these novel organisms is taken up by the genetic modification.

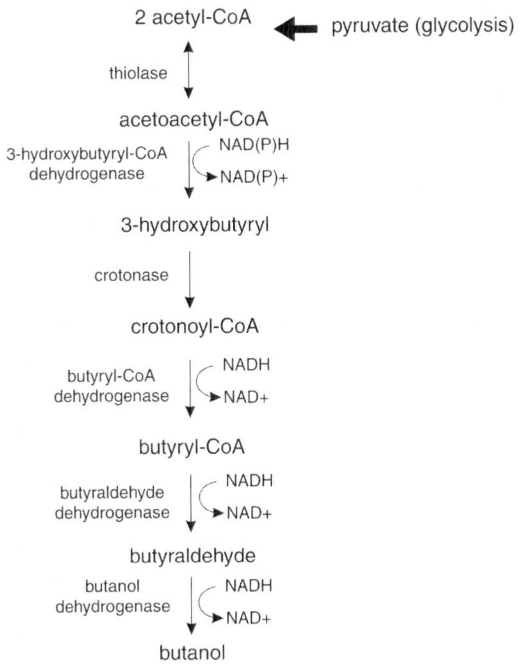

Figure 3 Metabolic pathway that produces butanol by *Clostridium sp.*

In 1914, Wiezman isolated the *Clostridium acetobutylicum* strain to produce the Acetone-Butanol-Ethanol precisely known as the ABE fermentation which gave higher yield of acetone and butanol better than the isolated organism by Fernbach. Later, butanol was produced industrially using the Wiezman process employing *Clostridium acetobutylicum* [45]. *C. acetobutylicum* is a widespread popular bacterial species used for ABE fermentation process as they are amylolytic in nature [32, 61, 62]. Conversely three other strains of *Clostridium species* were identified to grow collectively in the mixed culture—*C. beijerinckii, C. saccharoperbutylacetonicum, C. saccharoacetobutylicum* along with *Clostridium acetobutylicum*. Each strain exhibited its own unique properties in the production of biobutanol. Butanol production also used coculture for enhanced production; one such coculture is *C. beijerinckii* and *C. tyrobutyricum* which showed high yield when compared with other cultures. *C. tyrobutyricum* has the ability to produce butyric acid which is then used as a substrate to produce butanol by ABE fermentation. This process was proved to be an efficient and cost effective process [61, 62].There is yet another organism *Kluyvera* species whose strain is isolated from spent mushroom substrate which has the ability to produce elevated level of cellulase free xylanase (5.12 U/ml). This species when used as a coculture with *Clostridium species* will readily convert lignocellulosic xylan to butanol. This showed a comparable result to the separate hydrolysis of xylan and fermentation using *Clostridium species* [63]. A hydrolysate of green seaweed *Ulvalactuca* was used for the production of acetone, butanol and ethanol (ABE) by *Clostridium acetobutylicum* and *Clostridium beijerinckii*. Hydrolysate-based media were fermentable without nutrient supplementation.*C. beijerinckii* utilized all sugars in the hydrolysate and produced ABE at high yields (0.35 g ABE/g sugar consumed), while *C. acetobutylicum* produced mostly organic acids (acetic and butyric acids) [64]. In the new era, genetic engineering startled the bioprocess technology by using genetically modified species in the fermentation process to enhance the productivity and also favourable results in ABE fermentation. *Bacillus subtilis* and *E.coli* are the two organisms that are genetically modified by manipulating the expressed genes which will be discussed later in this chapter. *Clostridium species* also undergo manipulation in order to increase the productivity. *Bacillus subtilis* is an aerobic bacterium which consumes the oxygen and maintains strict anaerobic conditions for the anaerobe *Clostridium acetobutylicum* without any anaerobic pre-treatment for ABE fermentation process [65]. Further, *Bacillus subtilis* is also a high amylase producing bacteria and does not require any additional anaerobic pre-treatment by reducing agent and N_2 flushing. The amylase activity was increased to 10 folds without anaerobic pre-treatment and enhanced ABE fermentation to 5.4 and 6.5 folds. The advantages of aerobic *Bacillus* in a co-culture with anaerobic *Clostridium* is that, there was no additional requirement of any reducing agent in the medium or flushing with N_2 to ensure anaerobic condition and making the anaerobic fermentation more economical and cost effective [66]. N-butanol fermentation by *Clostridium beijerinckii* ATCC 10132 involves nitrogen source and metal ions for titre of n-butanol production. This resulted in about 6-fold increase and an unprecedented titre of 20 g L^{-1} of n-butanol in 72 h incubation period. The strain was found to be tolerant to 25 g L^{-1} n-butanol under optimized conditions and the tolerance could also be demonstrated by the strain's ability to accumulate rhodamine 6 G [67].

3.3 IDENTIFICATION OF SUBSTRATES

Feedstock for fermentation process is an important parameter which relies both on its economic feasibility and its availability [68-71]. At present, biomass accounts for 9.8% of the world's primary energy use annually, among which 30% is used in modern forms and 70% is used traditionally (combustion for domestic heating with the energy density of 15-20 MJ/kg). Butanol from biomass is environmental friendly; the digested materials from bio refinery can be readily used as an excellent and sustainable fertilizer for cultivation and crops (a true recyclable process); the energy density (MJ/kg) of biofuel will be higher than that of biomass, and the common problem of combustion, e.g., fly ash disposal and super heater corrosion, can be eliminated via biofuel production [72-76]. The increasing demands for the raw materials in the production of biobutanol have bought threat to the future production and use of agricultural crops, to overcome all this the future biofuel should have a high energy density on a mass; be produced at yields near the stoichiometric maximum from a given biomass feed, be compatible with existing fuel distribution infrastructures; have a minimum impact on environment; and should not affect the global food supplies [77-79] which led to the search of agricultural wastes, cellulosic biomass, lignocellulosic biomass which was classified as 1st generation biofuel and 2nd generation bio-fuels [25, 26, 59] among which lignocelluloses were chosen to be best foreseeing food security.

Table 2 Composition of cellulose, hemicellulose and lignin in % for various lignocellulosic materials

Lignocellulosic material	Cellulose (%)	Hemicellulose (%)	Lignin (%)
Hardwood stems	40-55	24-40	18-25
Softwood stems	45-50	25-35	25-35
Nut shells	25-30	25-30	30-40
Corn cobs	45	35	15
Grasses	25-40	35-50	10-30
Paper	85-99	0	0-15
Wheat straw	30	50	15
Sorted refuse	60	20	20
Leaves	15-20	80-85	0

Lignocellulosic material	Cellulose (%)	Hemicellulose (%)	Lignin (%)
Cotton seed hairs	80-95	5-20	0
Newspaper	40-55	25-40	18-30
Waste papers from chemical pulps	60-70	10-20	5-10
Primary wastewater solids	8-15	NA	24-29
Swine waste	6	28	NA
Solid cattle manure	1.6-4.7	1.4-33	2.7-5.7
Coastal Bermuda grass	25	35.7	6.4
Switchgrass	45	31.4	12.0

Table 3 Methods of pre-treatment implied on various substrates

Raw material	Bacterial strain	Hydrolysis treatment	References
Palm Empty Fruit Bunches	*C. acetobutylicum*	Acid and enzymatic hydrolysis	[95]
Waste water algae	*C. saccharoperbutylacetonicum* N1-4	Acid and enzymatic hydrolysis	[252]
Fibre-Enhanced DDGS	*C. beijerinckii*	Electrolyzed water and enzymatic hydrolysis	
Birchwood Xylan	*Kluyvera* species and *Clostridium sp.* BOH_3	Enzymatic hydrolysis	[88]
Sugar maple hemicellulose	*C. acetobutylicum*	Alkali hydrolysis	[256]
Ulvalactuca	*C.acetobutylicum* and *C beijerinckii*	Hot-water and enzymatic hydrolysis	[64]
Decanter cake	*C.beijerinckii*	Acid hydrolysis	[28]
2-year-old willow biomass	*C.beijerinckii*	Acid hydrolysis	[253]
Barley straw	*C.beijerinckii* P260	Acid hydrolysis and over liming	[57]
Corn stover and switchgrass	*C.beijerinckii* P260	Hot-water with wheat straw hydrolysate (1:1) and overliming	[53]

Studies on lignocelluloses showed that, it is composed of carbohydrate polymers (cellulose, hemicellulose), and an aromatic polymer (lignin). Carbohydrate polymers contain different sugar monomers (six and five carbon sugars) and they are tightly bound to lignin. Lignocellulosic biomass can be broadly classified into *virgin biomass, waste biomass* and *energy crops. Virgin biomass* includes all naturally occurring terrestrial plants such as trees, bushes and grass. *Waste biomass* is produced as a low value byproduct of various industrial sectors such as agriculture (corn stover, sugarcane bagasse, straw, etc.) and forestry (saw mill and paper mill discards). *Energy crops* that have high lignocellulosic content serve as a raw material for production of second generation biofuel; examples include switch grass (*Panicum virgatum*), Elephant grass, poplar trees, *Miscanthus giganteus*, and sugarcane which is a source of the readily fermentable sucrose and the lignocellulosic by-product bagasse. Lignocellulosic biomass, in the form of wood fuel, has a long history as a source of energy. Since the middle of the 20th century, the interest of biomass as a precursor to liquid fuels has increased, specifically the fermentation of lignocellulosic biomass to ethanol. Aside from ethanol, many other lignocellulose-derived fuels are of potential interest, including butanol, dimethylfuran, and gamma-valerolactone [80, 81].

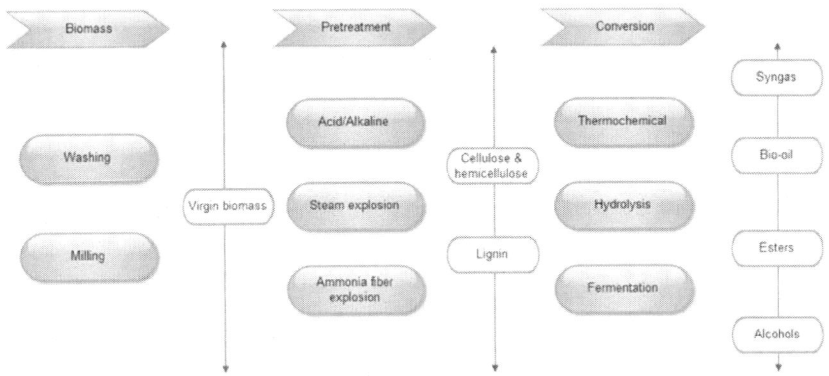

Figure 4 Substrate conversion into biofuels

Composition of lignocellulose is discussed in Table 2 and (figure 2). Lignocellulose has evolved to resist degradation and to confer hydrolytic stability and structural robustness to the cell walls of the plants. For extraction of fermentable sugars, one must first disconnect the celluloses and hemicelluloses from the lignin, and then conversion into simple monosaccharides by acid-hydrolysis or enzymatic conversion among the above hemicellulose is of meticulous interest because of its unique properties and composition that on degradation produces xylose, and their subsequent conversion into value-added products via microbial fermentation. But the recovered sugar solution after hydrolysis contains primarily pentose sugars and the fermentation of these pentoses is problematic. Only limited numbers of microorganisms that use pentose are known and the fermentation of pentose sugars at industrial scale is not established yet. And there is yet another significant pre-treatment process where the lignocelluloces and other biomass should be deoxygenated to bring efficient blending, as the presence of oxygen reduces the heat content of molecules and creates high polarity, which impairs its blending with existing fossil fuels [82-87]. Before getting deep into pre-treatment few such conversions showed the following results; detoxified hydrolysate was also used as a co-substrate for direct conversion of butyric acid to butanol in a nitrogen-free medium with butanol yield of about 0.47 C-mol butanol/C-mol carbons.

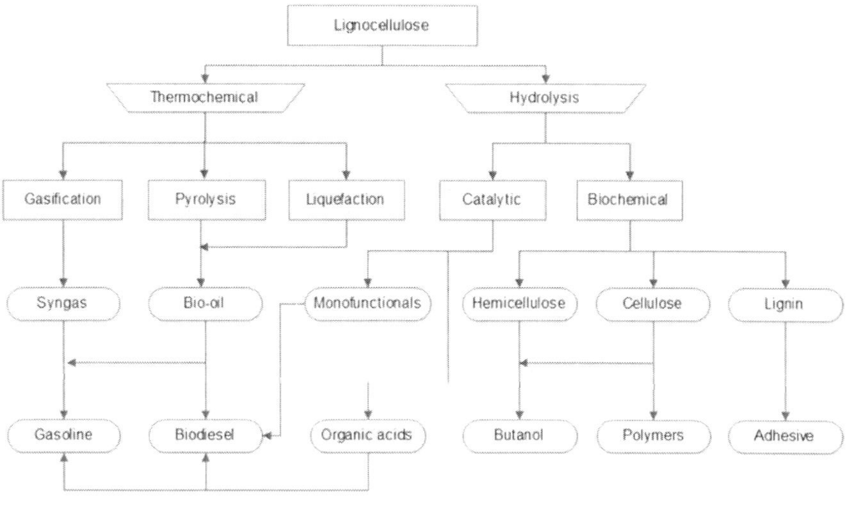

Figure 5 Lignocellulose processing using thermochemical and hydrolysis methods

This study has showed that decanter cake waste could serve as a low-cost substrate for biobutanol production [88]. Later mixed sugars containing cellobiose and xylose were studied to establish efficient butanol production process without carbon catabolite repression (CCR) that contains less glucose from lignocellulose. Cellulases produced mainly by *Trichoderma reesei* contain b-glucosidase, which results in yielding high glucose from cellulose in lignocelluloses, thus it would be possible directly to acquire a mixture of cellobiose and xylose containing no glucose from lignocelluloses since other organisms inhibits the formation of glucose [89-92]. Glycerol obtained as a byproduct of biodiesel synthesis has a significance when it is used as a major substrate and when it is used as co-substrate with glucose. The simultaneous dual-substrate obtained maximum butanol titre of 13.3 gL^{-1} and glycerol with bagasse showed maximum butanol concentration of 11.8 g L^{-1} [93, 94]. Processing of lignocelluloses by thermochemical and hydrolysis methods is shown in Figure 5.

3.4 PRE-TREATMENT AND HYDROLYSIS

Butanol produced from lignocellulosic materials, such as wood, agricultural, or forest residues, have the potential to be a valuable substitute for gasoline. Pre-treatment of the substrate is required:

(1) for destroying the lignin shell protecting cellulose and hemicelluloses,

(2) for decreasing the crystallinity of cellulose,

(3) for increasing the porosity,

(4) for breaking the shell for enzyme to access substrate (sugar), and

(5) to mitigate the concentration of inhibitors in hydrolysate that inhibits the butanol titre.

During the past few years a large number of pre-treatment methods have been developed, including chemical, physical and biological pre-treatments. Many methods have been found to result in high sugar yields, above 90% yield from lignocellulosic biomasses such as woods, grasses, corn, and so on. Study on hydrolysis investigates three major objectives: (1) optimal conditions for acid and alkali hydrolysis for the substrate, (2) optimal conditions for enzymatic hydrolysis for the substrate, and (3) comparison of butanol production by *Clostridial species* using the hydrolyzate from the hydrolysis methods used [95]. Hydrolyzate-based media are fermentable ready-by without any nutrient supplementation. The sugars in the hydrolysate are completely utilized by the microbes and produced high yields of butanol. About

90% of sugars are obtained from the enzymatic pre-treatment which is said to be advantageous than the other two methods. The common enzyme used for the conversion of sugars is xylanases and cellulases which are obtained from fungi, bacteria and actinomycetes. Enzyme cellulase, depolymerize cellulose into fermentable sugars. Cellulase synthesized by fungi and bacteria work together to degrade cellulose and this enzyme system comprises of endo-ß-glucanase, exo-ß-glucanase and ß-glucosidase [95, 96]. High production cost (~40% of total), low yield, few microorganisms capable of degrading cellulose, inhibitor formation, optimizing/understanding enzyme regulation and activity and understanding of plant cell wall structure and function would be some challenges faced while using enzymatic pretreatment. Compounds used in pre-treatment process are discussed in Tables 4, 5 & 6 with its advantages and disadvantages and conversion of biomass into biofuels is shown in Figure 4.

3.4.1 Alkaline Pre-Treatment

Using alkaline chemicals such as dilute sodium hydroxide, aqueous ammonia and lime to remove lignin has long been known to improve cellulose digestibility. Sodium hydroxide (NaOH) was widely used for pre-treatment because its alkalinity is much higher than others [95]. When 4% NaOH was employed 54.8% of lignin and 43.7% of the hemicellulose were removed. This also showed nearly 78% of the cellulose from raw cattails was converted to fermentable glucose in 48 h using a cellulase loading of 60 FPU/g glucan. Employing additional chemicals with NaOH could improve the pre-treatment performance. SEM corroborated that the NaOH/urea pre-treatment could disrupt the bonds between hemicelluloses, cellulose, and lignin, and thus changes the structure of cellulosic material. Though NaOH seemed to be efficient in pre-treatment process, it was cost expensive, where lime being another alkali source is the least expensive. But the mechanism is similar to the NaOH pre-treatment. Pre-treatment by over liming of corn stover at 25-55°C removed lignin and hemicellulose selectively and also the degree of crystallinity slightly increased from 43% to 60% with delignification, but cellulose was not affected [97]. Lime pre-treatment was also used on various biomass sources, such as sugarcane bagasse [96], switch grass [98] and poplar [99]. Over liming process is also used for detoxification which removes the inhibitors of the butanol and enhances its production [100] [101]. To reduce this potential toxicity effect, BSH was treated with lime [Ca $(OH)_2$] followed by fermentation. The treated BSH resulted in a successful fermentation and ABE concentration of 26.64 g L^{-1} was achieved. This was superior to both glucose and untreated BSH (initial sugar 60 gL^{-1}) fermentations [101]. Alkali pre-treatment methods were applied to improve fermentabilityof nano-membrane concentrated hemicellulosic hydrolysate and solvent production by *Clostridium acetobutylicum* ATCC824. Results validated that though nanofiltration could remove nearly all small molecular organic acids (acetic acid, formic acid), furfural and HMF, the resulting hydrolysate was found to be still inhibiting solvent production of *C. acetobutylicum*. At this stage overliming treatment employed significantly improve the ultimate butanol concentration to 7 g L^{-1} from 0.8 g L^{-1} [100].

Table 4 Chemical Pre-treatment

Pre-treatment type	Examples of chemical used	Conditions	Merits and demerits
Concentrated acid	H_2SO_4,HCL	Concentrated acid Low temperature	High yields Hydrolysis is mainly included Corrosion problems Material loss due to degradation High demand for chemicals Environmental issues
Diluted acid	H_2SO_4	W = 0, 5-2% T > 160°C	Corrosion problems Low yields Material loss due to degradation
ARP Ammonia recycled percolation	Ammonia	W = 15% T = ~ 170°C	Media recoverable Environmental issue due to ammonia
Lye	NaOH, Ca $(OH)_2$	W = 0, 5M T = ~ 80°C	Media not recoverable
Organic solvents	Ethanol-Water, Butanol-Water, Ethylene-Glycol	T = 150-200°C	Costs of solvent Media recoverable
Ionic liquids	PF_6^- BF_4^- $AlCl_4^-$	T = ~ 110°C	Media recoverable Low energy consumption

3.4.2 Enzymatic Hydrolysis

There was no independent enzymatic hydrolysis studied yet. This method was coupled either with alkali, acid or hot water pre-treatment methods. The ammonia fibre expansion (AFEX) process decrystallizes cellulose, depolymerises hemicelluloses, and removes lignin from cellulose/hemicelluloses that includes hydrolysis of corn stover [102], switchgrass [103], coastal Bermuda grass [104], forage and sweet sorghum bagasse [105]. The AFEX-treated hemicelluloses are in oligomeric form, which can usually be hydrolyzed to fermentable pentoses by most commercial cellulase and xylanase mixtures. In ammonia recycle percolation (ARP), ammonia being a selective reagent for lignin, noncorrosive, and a relatively less expensive chemical, is an appropriate choice for pre-treatment. SEM and FTIR spectra showed that the aqueous ammonia causes swelling and delignification of biomass at high temperatures. The ARP process could solubilise ~50% xylan from corn stover, but retains over 92% of the cellulose. Enzymatic digestibility of ARP-treated corn stover is 93% with 10 FPU/g-glucan enzyme loading [106]. The X-ray crystallography data showed that the basic cellulosic crystalline structure was not altered. By further minimizing the liquid throughput and optimizing the operation conditions for corn stover, 59–70% of lignin removal and 48–57% of xylan retention were achieved. Enzymatic digestibilities were 95, 90, and 86% with 60, 15, and 7.5 FPU/g of glucan, respectively [107]. ARP also showed great success in the delignification of sorghum [108, 109], poplar [99], switchgrass [100], etc. In the butanol production from green seaweed *Ulva lactuca*, solubilisation of over 90% of sugars was achieved by hot-water treatment followed by hydrolysis using commercial cellulases [97]. *Kluyvera species* whose strain is isolated from spent mushroom substrate has the ability to produce elevated levels of cellulase-free xylanase (5.12 U/ml).

Table 5 Biological Pre-treatment

Pre-treatment type	Examples of organisms	Merits and Demerits
Fungi	White-rot Fungi, Brown-rot Fungi, Soft-rot Fungi	Slow conversion Low energy requirement No chemicals Mild environmental conditions
Bacteria	*Sphingomonas paucimobilis*, *Bacillus circulans*	Slow conversion Low energy requirement No chemicals Mild environmental conditions

3.4.3 Hot Water Pre-Treatment

Hot water pre-treatment is ever so often called "auto hydrolysis." Advantages of this method are less expense, lower corrosion to equipment, less xylose degradation and thus fewer byproducts including inhibitory compounds in the extracts [110]. Hot water subjected to pressure can penetrate the cell structure of biomass, hydrate cellulose, and remove hemicellulose, thus efficiently improving the enzymatic digestibility of biomass cellulose. A pre-treatment at 190°C for more than 10 minutes could effectively dissolve the xylan fraction of aquatic plant as soluble oligomers. When pretreated at 190°C for 15 minutes, the highest glucose yield of 77.6% from the cellulose was achieved in 48 h using a cellulase loading of 60 FPU/g glucan [111]. Fermentation of untreated corn stover hydrolysate (CSH) exhibited no growth and no ABE production; however, upon dilution with water and wheat straw hydrolysate (WSH) in the ratio 1:1, showed 16.00 and 18.04 g L^{-1} ABE production [100].

Table 6 Physical Pre-treatments

Pre-treatment type	Conditions	Merits and Demerits
Steam Explosion	P = 2.5–7 MPa T = 180-280°C	High yields No corrosion Undesired side products possible High energy demand
LHW – Liquid Hot Water	T =170–230°C	High yields Less side products than in steam explosion No corrosion
CO_2 Explosion	P > 7.3 MPa T > 31.1°C Super critical CO_2	High cost expected Low environmental impact
AFEX Ammonia Fiber Explosion	Liquid Ammonia T = 90-100°C	Less amount of inhibitor formation Media recoverable Environmental issues due to ammonia

3.4.4 Acid Pre-Treatment

Acid hydrolysis in the conversion of cellulose to glucose is a process that has been working since many decades. Universally, there are two types of acid hydrolysis: dilute and concentrated, each having unique properties and effects on biomass and also many advantages and disadvantages. Dilute acid (0.5-1.0% sulfuric acid) pre-treatment at moderate temperatures (140-190°C) can effectively remove and recover most of the hemicellulose as dissolved sugars. By this process lignin is disrupted and partially dissolved, thus increasing cellulose susceptibility to enzymes and glucose yields from cellulose increases with hemicellulose removal to almost 100% [112]. This hydrolysis method consists of two chemical reactions: (1) converts cellulosic materials to sugar, and (2) converts sugars into other chemicals, many of which inhibit the growth of downstream fermentation microbes. [113] showed that up to 92% of the total sugars originally available in corn stover could be recovered through coupling dilute acid pre-treatment and enzymatic hydrolysis. This pre-treatment has been employed to a number of feedstocks including aquatic plants [114], switchgrass [115], hardwood [116], softwood [117], which is an effective approach that results in a high sugar yield from most biomass feedstocks that also has disadvantages such as (1) corrosivity, requiring expensive construction materials, (2) formation of inhibitory compounds, and (3) the requirement for acid neutralization. SO_2 impregnated steam-explosion has been considered as one of the most cost-effective pre-treatment processes. Coupled SO_2-catalyzed steam explosion and enzymatic hydrolysis resulted in a maximum conversion efficiency of 81% of the combined original hemicellulose and cellulose [118, 119] which increases the saccharification efficiencies of softwoods [120]. Concentrated acid hydrolysis (about 70% of acid content) uses a low temperature (100°F/38°C) and low pressure. Rate of cellulose recovery from the initial pre-treatment process and the conversion rate of cellulose to glucose under this process are much higher (90%) than with dilute acid hydrolysis. Acid after hydrolysis is removed, since large quantities of lime must be used to neutralize the sugar solution, which requires the disposal of salts increasing the cost and makes the end product more expensive [121]. By overall compositional analysis, the stem biomass contained 70% polysaccharides that were significantly higher than that of bark, which was 46%. The bark biomass was subjected to acid hydrolysis and presence of possible fermentation inhibitors led to a similar monosaccharide concentration in the stem and bark. Monosaccharide solutions from both the bark and stem biomass had problems in the transition from the acid-producing pathway to the solvent-producing pathway which was later subjected to over liming process or enzymatic pre-treatment [95].

3.4.5 Inorganic Salts Pre-Treatment

Recently, it has been discovered that Lewis acid, especially chloride salts, can react with cellulosic biomass directly, in an aqueous phase or in ionic liquid with excellent selectivity [122, 123]. Liu and Wyman (2006) [124] evaluated the effect of several inorganic salts KCl, NaCl, $CaCl_2$, $MgCl_2$, and $FeCl_3$ on xylose monomer and xylotriose degradation. $FeCl_3$ was found to significantly increase xylose monomer and xylotriose degradation, resulting in degradation ratio of 65% and 78% for xylose and xylotriose, respectively. Also, losses of xylose and xylotriose were described using first order homogeneous kinetics. Yu et al. (2011) [125] investigated the effect of the metal salts NaCl, KCl, $CaCl_2$, $MgCl_2$, $FeCl_3$, $FeCl_2$, and $CuCl_2$ solutions on the decomposition of sweet sorghum bagasse. Analysis of various substrates for method of pre-treatment is discussed in (Table 3).

3.5 ABE FERMENTATION - A ROUTE TO BUTANOL PRODUCTION

The high valued and potentially a better butanol is produced by the traditional and economical ABE fermentation. ABE fermentation was used industrially for butanol production widely and became the second largest industrial fermentation process after ethanol by using yeast as a fermenting organism. The economic importance of acetone and butanol played a significant role as a solvent in petrochemical industries. The total concentration of ABE fermentation broth was $20gl^{-1}$. Primary products obtained from ABE fermentation process are acetone, butanol and ethanol often in the ratio 3:6:1 respectively and the final concentration of butanol is about 3% [43, 45, 126, 127].

Table 7 Comparison of different methods of fermentation processes for analysing the yield and overall titre of ABE

Substrate	Strain	Fermentation process used	Yield (g/g)/ Productivity ($gl^{-1}h^{-1}$)	Maximum ABE titre (g/l)	Ref.
Barley straw	*C. beijerinckii* P260	Batch fermentation	0.43/0.39	26.64	[57]
Wheat straw	*C. beijerinckii* P260		0.41/0.31	21.42	[54]
Corn stover and switchgrass (1:1)	*C. beijerinckii* P260		0.43/0.21	21.06	[53]
Switchgrass	*C. beijerinckii* P260		0.37/0.09	14.61	[53]
Wheat straw	*C. beijerinckii* P260	Fed-batch fermentation	–/0.36	16.59	[55]
Synthetic medium with butyric acid	*C.saccharoperbutyl-acetonicum* N1– 4		0.49/0.42	16.0	[90]
Sago starch	*C.saccharobutyli-cum* DSM 13864	Continuous fermentation with free cell	0.29/0.85	9.1	[248]
Degermed corn	*C. beijerinckii* BA101		–/0.29–0.30	14.28	[187]
Starch and glucose	*C. beijerinckii* BA101		–/0.42	9.9	[249]
Whey permeate	*C. acetobutylicum* P262	Continuous fermentation with Immobi-lized cells	3.5–3.6/0.36–1.10	8.6	[250]
Lactose and yeast extract	*C. acetobutylicum* 824A		–/0.78	1.43	[244]
Corn	*C. acetobutylicum* ATCC 55025		0.42/4.6	12.50 (bu-tanol)	[246]
Defid-ered-sweet-potato slurry	*C. acetobutylicum* P262		0.20/1.0	7.73	[247]
Synthetic medium	*C. saccharoperbuty-lacetonicum* N1-4	Continuous fer-mentation with cells recycling and bleeding	–/11.0	8.58	[245]

As discussed in this chapter earlier, ABE fermentation process involves two major phases namely (1) acidogenesis, where the acids are produced together with cell growth, and (2) solventogenesis, where the growth is halted and the acids are converted into solvents. This phase is also known as solvent producing phase. This step is always carried out by aldehyde dehyrogenase (ALDH), butanol dehydrogenase (BDH), alcohol dehydrogenase (ADH) and ALDH catalyses the production of respective aldehydes from butyryl-CoA and acetyl CoA. Butyraldehyde is then converted to butanol by BDH or ADH and the functional BDH enzymes are primarily NAD(P)H-dependent. The final step of butanol production is catalysed by butanol dehydrogenase [128]. During acidogenesis, the cells are able to reproduce themselves and shift to solventogenic state with a morphological change. Under solventogenesis, the active cells become endospores, unable to reproduce themselves. Therefore the physiological states to be considered for clostridia are acidogenic phase and solventogenic phase [129-138]. Metabolic pathway of *Clostridium species* is given in (Figure 3), and the biphasic process due to the *Clostridium* life cycle is given in (Figure 6). ABE fermentation is said to be a higly sophisticated technique therefore the obstacles faced during fermentation would be formation of inhibitors such as phenolics from lignin degradation, furfural from C-5 degradation, HMF from C-6 degradation; integrated hydrolysis and fermentation into a single microbe; low conversion rates for C-5 sugars; and also the technology relied to remove inhibitors is expensive and these obstacles are solved by suitable pretreatment techniques.

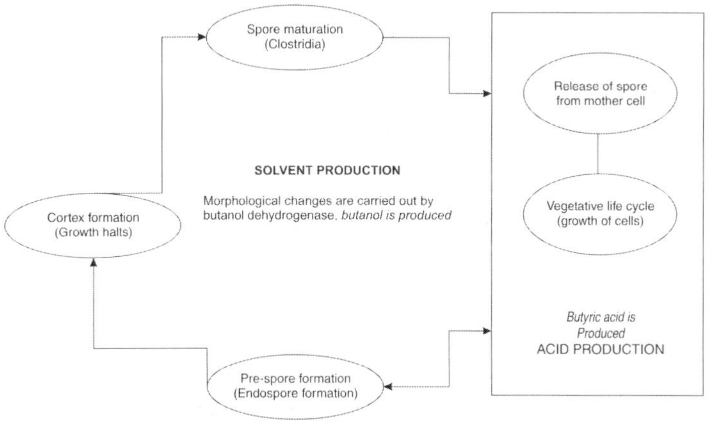

Figure 6 Microbial cycle for acid and solvent production

3.5.1 Bioreactors and Fermentation

Modern research on production process of biobutanol also focuses on finding the best reactor for fermentation since the paramount production of butanol also depends on the fermentor. Various bioreactor systems include batch, continuous, semi-continuous, cell immobilized reactors, fed batch reactors and continuous flash fermentors. An ideal reactor for butanol production would be the one which reduces the butanol toxicity on the living cells. Typically batch reactors are employed in commercial biobutanol production for its simplicity. But researchers are working on finding ideal reactors for butanol because in batch reactor as the residence time of butanol increase it will affect its utmost yield or productivity. The following excerpt discusses about the vivid reactors with its advantages and drawbacks in butanol biosynthesis.

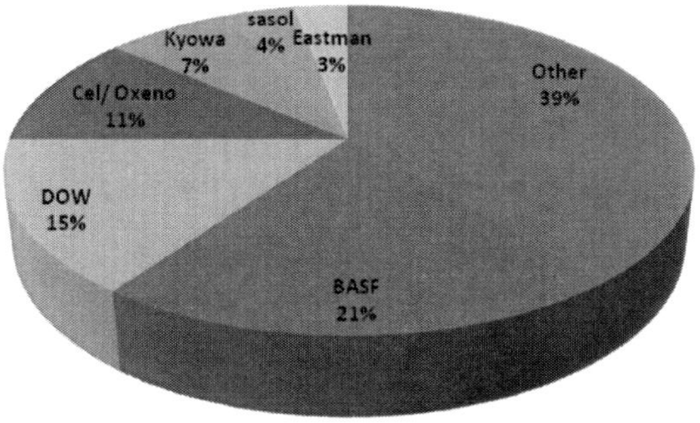

Figure 7 Current status of butanol market across the globe

3.5.1.1 ***Batch and Fed Batch Fermentation*** Batch rectors are found to be the simplest and conventional cultivator [139, 140, 141], however it has got several drawbacks due to its prolonged lag phase, preparation time and product inhibition [142]. Butyric acid clearly revealed to inhibit cellular growth with all specific growth rates declining upon the increase of butyrate concentrations. However, the presence of low and moderate levels in the medium can readily enhance the ABE-fermentation and increase butanol production through a shift induction towards the solventogenic phase controlled by the medium pH. The effect of pH on the solvent was found by Monot et al (1984). They said that when pH was 4.5 the solvent and butanol yields were 0.35 and 0.2499 (g/g-glucose), respectively. In all cases it was found that 4.0 g L^{-1} is the optimal concentration of butyrate that maximizes the yields for all ABE-solvents and butanol productivities. In batch fermentations, immobilized mutant strain *C. beijerinckii* JB 200 produced 12.7 g/L and 15.4 g/L ABE from corn fibre hydrolysate and cassava bagasse hydrolysate in a fibrous bed bioreactor, respectively. *C. beijerinckii* CC101 produced 11.35 g/L ABE from wood pulp hydrolysate, and its recombinant mutant CC101-SV6 produced 9.44 g/L ABE from sugarcane bagasse hydrolysate in free-cell batch fermentations. ABE production from wood pulp hydrolysate was further enhanced to 17.73 g/L in a gas stripping integrated ABE batch fermentation process, with a higher ABE yield of 0.44 g/g compared with 0.39 g/g from non-integrated batch process. Fermentation of dilute sulphuric acid and barley straw hydrolysate (BSH) undiluted/untreated by *Clostridium beijerinckii* P260 resulted in the production of 7.09 gL^{-1} ABE, an ABE yield of 0.33, and productivity of 10 $gl^{-1}h^{-1}$. Studies proved that BSH is toxic to the culture. This was reduced as BSH was treated with lime [Ca $(OH)_2$] followed by fermentation [143]. The treated BSH resulted in a successful fermentation and ABE concentration of 26.64 gL^{-1} was achieved. To overcome the substrate inhibition and to increase the biomass concentration, fed batch fermentation can be used [144]. Fermentation of untreated corn stover hydrolysate (CSH) exhibited no growth and no ABE production, but when diluted with water (two-fold) and wheat straw hydrolysate (WSH, ratio 1:1) the production increased to 16.00 and 18.04 g L^{-1}ABE respectively. Alternately, cultures resistant to inhibitors that are able to produce butanol at high concentrations may be another approach to improve the current process. In both the cases with BSH and CSH, a controlled experiment performed using glucose as substrate obtained a yield of 21.06 g L^{-1} total ABE [57, 58]. N-butanol fermentation by *Clostridium beijerinckii* ATCC 10132 was investigated by optimization of various physiological and nutritional parameters. Optimization of the parameters resulted in about 6-fold increase and an unprecedented titre of 20 g L^{-1} of n-butanol in 72 h incubation period. The process could be scaled up to 3 L level with similar titres. The strain was found to be tolerant to 25 g L^{-1} n-butanol under optimized conditions. This was the first report wherein high titre of n-butanol has been obtained in single batch fermentation without resorting to solvent stripping or strain improvement. Using cane molasses as substrate, total solvent of 17.88 g/L (butanol 11.86 g/L) was attained after 36h of batch fermentation in a 5 L bioreactor, the productivity and yield were 0.50 g/L/h and 0.33 g ABE/g sugar consumption, respectively by the action of *Clostridium saccharobutylicum* [145]. For fed-batch fermentation cassava bagasse hydrolysate (CBH) with additional glucose was used to initiate the fermentation. After 30 h when butanol production had reached 10 g/L, they initiated gas stripping and for extended fed batch fermentation, additional yeast extract (0.5 g, equivalent of 0.5 g/L in the 1.0 L fermentation system) was also added with the concentrated

CBH to avoid the depletion of nutrients. Results showed that a highly concentrated butanol solution of 10–16% (w/v) was obtained in the condensate collected at the end of each feeding cycle. However, acetone was less than 5% (w/v) and only a small amount of ethanol (<0.8%) and almost no acids were obtained in the condensate. Fed-batch culture also showed equally good and maximum yield from rapid consumption of both the sugars [92].

Table 8 Performance of various reactors

Process	Total solvent productivity g/l/h	Yield g/g	Solvent Concentration g/l	Productivity	Ref.
Continuous immobilized reactor	15.8	0.38	7.9	15.8 g/l	[138]
Continuous membrane cell recycle reactor	6.5	0.35	13	20 g/l of cell mass	[152]
Batch	0.34	0.42	24.2	59.8 g/l of glucose	[167]
Fed batch	0.98	0.43	165.1	384 g/l of glucose	[258]
	1.16	0.47	233	500 g/l of glucose	[199]
Continuous flash fermenter	-	0.19-0.22	130-180 g/l	7.0-8.0 g/l/h	[157]

3.5.1.2 ***Continuous and Semi-Continuous Fermentation*** The drawbacks of batch reactors like preparation time and lag phase can be circumvented by continuous bioreactors. Moreover other issues like productivity and product yield can be increased using cell recycle or immobilized bioreactors [146, 142]. Continuous reactors (alternatively referred to as flow reactors) carry material as a flowing stream. Reactants are continuously fed into the reactor and emerge as continuous stream of product [147].

Continuous fermentation processes have several advantages over batch fermentation processes as fermentation can be operated in a continuous mode. Few such advantages of continuous reactors are generally able to cope with much higher reactant concentrations due to their superior heat transfer capacities. *Plug flow reactors* have the additional advantage of greater separation between reactants and products giving a better concentration profile. *Small-sized continuous reactors* make higher mixing rates possible also the output from a continuous reactor can be altered by varying the run time by increasing the operating flexibility for manufacturers [148, 149]. Continuous fermentation was first carried out using single stage fermentor. But the existence of a true steady-state condition for A–B–E fermentation has been challenged due to the issue of solvent degeneration, in which the solvent producing ability of the bacteria was irreversibly lost after 2–3 weeks of continuous operation using a single-stage fermentor [150]. Hence the productivity was increased by the use of *Clostridium saccharobutylicum* DSM 13864 with low-value sugar-based feedstock in a two-stage semi-continuous fermentation process operated steadily for over 8 days (205 h) with an average productivity (stage II) of 1.05 g/L/h. In a 5-stage continuous fermentation, the solvent production could be steadily operated for over 200 h at dilution rate (D) of 0.1 h^{-1}, the average total solvent and productivity were 12.41 g/L and 0.25 g/L/h, respectively. Using an alkaline pre-treatment and enzyme hydrolysis procedure, corn stover hydrolysate was prepared in this study [151]. *Membrane cell recycle reactors* are another alternative to improve productivity. Thus to increase the productivity of the acetone-butanol fermentation, a hollow-fibre ultra filter is used to separate and recycle cells in a continuous fermentation of *Clostridium acetobutylicum*. Under partial cell recycling and at a dilution rate of 0.5 hr^{-1}, a cellular concentration of 20 g/l and a solvent productivity of 6.5 g/l/hr is maintained for several days at a total solvent concentration of 13 g/l [146, 152]. The highest butanol yield and productivity found in the continuous fermentation at dilution rate of 0.1 h^{-1} were 0.21 g-butanol/g-glucose and 0.81 g/L/h respectively. However, fouling of the membrane with the fermentation broth occurred revealing to be a major obstacle for this system. Lipnizki and co-workers (2000) proposed a way of overcoming this problem by allowing only the fermentation broth to undergo filtration by using the immobilized cell system [144].

Table 9 Concentration of butanol obtained in surfactant and aqueous phase after each cycle of repeated utilization of the recovered polymer (Butanol = g L^{-1}, L62 = 6%, Incubation temperature = 38 °C) [172].

Cycle	Volume ratio	Partition Coefficient	Butanol in surfactant phase g L−1	Butanol in aqueous phase g L−1
First	4	2.6	94.82	36.47
Second	4	2.7	93.3	34.55
Third	4	2.87	95.8	33.38

3.5.1.3 ***Cell Immobilization and Fibrous Bed Bioreactor*** Immobilised bioreactor involves use of localised (hence called immobilised) living cells to reduce damaging shear forces caused due to liquid flow in a bioreactor[153]and potentially increase the product yield. Immobilizing the cell enables the preservation of catalytic activity and stability of the cell. The unit can be reused repeatedly and continuously [154]. Further to improve the maximum utility of immobilized cell reactors fibrous matrices have been developed. Matrices are used as support material and it provides larger surface area, high void volume, low cost, high mechanical strength, high permeability and low pressure drop inside the vessel. FBB is successfully employed in butyric acid synthesis for its increased productivity. Generally the reactor loses its productivity over long operation periods due to continuous use or due to restricted mass transfer and build up of dead biomass and reactor or membrane deterioration may occur due to blockage and channelling effects [142]. Another advantage of FBB is that aged, latent, non-viable and non-productive cells can be immediately removed from the system and the cell density inside the bioreactor can be adjusted to prevent clogging. Cell immobilization by adsorption onto fibrous matrices involves three stages such as transport of the freely-suspended cells from the bulk liquid onto the fibre surface, cell adhesion at the surface and consequent colonization along the surface. After the cell growth, the cells get retained in the solid surface until all the available area for immobilization is completely utilised [155]. Usually a thick multi-cell layer, called film is formed [153]. Since the adhesion process depends on the balance between opposing attraction and repulsion forces in both electrostatic and hydrophobic interactions; change in the pH, temperature, ionic strength, surfactant concentration, culture age, and shearing forces may easily lead to cell leakage from the support. Cell detachment can be overcome partially by using fibres with irregular surface or tailored surfaces such as modified cotton fabric [155]. In this bioreactor the cells are immobilized in a loosely convoluted fibrous matrix and then packed inside a columnar vessel where they got the density ranging from 40–100 g l^{-1}[156]. A novel mutant strain of *C. acetobutylicum JB200* developed with high butanol tolerance and productivity through evolutionary engineering in a fibrous bed bioreactor (FBB) can produce 28.2 g/L butanol in the FBB fermentation. Fermentation process also allows the use of highly concentrated substrate, greatly reducing the water usage in the fermentation process and recovery with gas stripping and the condensate has >150 g/L of butanol.

3.5.1.4 ***Continuous Flash Fermentation*** Continuous fermentor connected cell retention system and inline product recovery equipment (flash tank) are used in order to increase the biomass concentration and to reduce the product (butanol) toxicity respectively. Cell retention system allows the fermentor to be operated at high dilution rate without cell washout and recovery of product is carried out by distillation. Final concentration of butanol was expected to be more than 20 g L^{-1} and this process could also be helpful in reduction of distillation costs and benefit environment due to lower quantity of waste water generation and it was seen that there was no effect of pressure on yield and productivity. The optimum operating ranges determined were 130g/l < S_0 < 180 g/l; 3.5 h < 4.0 <5.0h; 400m^3/h<inlet flow rate<500m^3/h; Fpu= 25m^3/h; p_{flash}=6.50 kPa [157].

(Table 7 & 8) show the efficiency of various fermentation processes based on substrate and organisms and performance of various reactors respectively.

3.6 SEPARATION TECHNIQUES

Since increase in concentration of butanol in solventogenic phase hinders the further production due to end product toxicity, there is an instilling need to separate the solvent from a continuous cultivator. But this again puts forth a challenge in terms of energy as the recovery of solvent in its low concentration (~20g/l) and high boiling point of butanol (117^{0}C) complicates the task. Traditional methods to achieve this were distillation, liquid-liquid extraction and adsorption [158] where distillation has been dismissed considering the economic viability. Yet other methods were developed recently, which includes membrane separation techniques like pervaporation and membrane solvent extraction. These novel techniques prove to be advantageous with respect to conventional one. But still there exists a heavy competition amongst them as they are advantageous over others in different ways which is explored in this excerpt.

3.6.1 Liquid-Liquid Extraction

In liquid-liquid extraction, a water insoluble organic extractant can be mixed with the broth in selectively separating the butanol. Organic solvents are preferred for separation as they are immiscible to recover butanol without disturbing rest of the system [159, 140]. This can be done with high selectivity and undertaken in the reactor itself and to achieve this, the toxicity of extraction solvent must be a prerequisite for the engineers [160]. Solvents possessing high distribution coefficient for butanol prove to be flammable, hazardous and toxic to microorganisms and humans [161, 162]. Oleyl alcohol is usually suggested due to its efficiency in extraction and low toxicity [163, 164]. Evans and Wang (1998) worked with oleyl alcohol and benzyl benzoate mixture and said that productivity has been increased by 60% when done in fed batch cultivator than in batch [140, 159, 164, 165, 166] and 20% dacanol in oleyl alcohol, enhanced butanol separation by 72% [167]. Roffler et al (1988) used the former mixture and observed that glucose conversion increased from 81 to 100 g/L and the maximum butanol productivity increased from 1.42 to g/L h in an extractive fermentation [168]. Cockren et al (1986) used 4-n-butylphenol as an extractant for the recovery of butanol from dilute solutions [169].

In common, group of esters with high molar mass were found to be compatible too. The hydrolytic products of fluorinated anions such as $[BF_4]^-$and $[PF_6]^-$ are toxic to microorganisms [170, 171]. Hence ionic liquids (ILs) possessing these anions may not be suitable for *in situ* recovery of butanol [160].The analysis of impact of different solvent properties in extraction shows that none of the ILs properties solely corresponds to streamlined extraction of butanol. However TF_2N-based ILs and [omin] $[TF_2N]$ based IL prove to be comparatively potent [160].

3.6.2 Non-Ionic Surfactants (or) Cloud Point Extraction (CPE)

The biocompatibility issue of aqueous solution organic solvent and phase system was to be overcome by non-ionic surfactants like PEG-dextrin, triton X 114, L62, L64, L61, and L62LF among which L62 proved to be the best in biocompatibility test and can be reused 3 times with no change in partition coefficient. This kind of extractive fermentation using non-ionic surfactants in a cloud point system not only seems to be biocompatible but also reduces the downstream process volume by 4-6 times [172]. Non-ionic surfactants in their aqueous solution form a coacervate (surfactant rich phase) and surfactant diluted phase at a temperature above the cloud point. This separation is due to the presence of organic compounds and the principle is known as CPE, whose advantages include mild environment, simplicity and easy scale up [172].The ampiphilic nature of surfactant, which depends on total molecular weight, relative block size and thermodynamic parameter aides the formation of two phases where hydrophilic group helps in water solubility. Above the cloud point temperature, surfactant molecule aggregates that ease the removal of targeted compounds, here butanol. This method not only eliminates the product toxicity but also brings down the production cost by reducing the process solution 4-6 times among which L62 assures to be more promising since it can be reused 3 times without affecting the partition coefficient [172]. The repeated utilization of recovered polymer non-ionic surfactants in the L62-rich phase is given in (Table 9).

Butanol capturing capacity (BCC) of a surfactant is the prerequisite in this method which can be determined using a dialysis cell and an ultra filtration membrane. After fermentation with surfactant (control), centrifugation is done to separate the cells from broth which is incubated at different temperatures (35 to 70 ^{0}C) to achieve phase separation. Wang et al (2010) in their work observed phase separation at 6% while using L62 (on incubating for 30 minutes at 70 ^{0}C). The lower phase seems to be rich in surfactant with butanol and upper phase being aqueous. The samples from each phase is assayed for butanol concentration and surfactant rich phase appears to be comparatively rich in butanol which is further enriched by incubating at 38 ^{0}C [173]. Through evaporation between 120 to 130 ^{0}C butanol is separated from surfactant rich phase. The BCC of some selected surfactants has been given in (Table 10).

Table 10 Butanol capturing capacity (BCC) of certain surfactants at 30°C [172]

Surfactant	Butanol capturing capacity (kg kg^{-1})	Cloud point temperature (°C)
Triton X 114	0.6	25
L62LF	0.52	28
L62	0.32	32
L64	0.06	58
L61	Insoluble in water	24

3.6.3 Pervaporation

Pervaporation is a membrane separation technique that uses a selective membrane driven by chemical potential difference in which the feed (retentate) remains in the liquid phase. Generally it is carried out to dehydrate organic solution. This seemed to be a more cost effective, commercially competitive separation technique and is the only process where membrane phase transition occurs. Advantages include simple processing, single step separation, energy efficient, cleaner end product, cost effective multipurpose unit operation, commercially competitive separation technique and is the only process where membrane phase transition occurs whereas the disadvantages are lack of information, increase in capital, risk associated with reliability of the process, biocompatibility and membrane fouling. The steps involved in mechanism of transport are selective sorption, selective diffusion through membrane and desorption into vapour phase on permeate. Important factors influencing the separation are permeate membrane interaction (determines solubility), molecular size, shape and mass (determines diffusivity), flux (low fluxes reduces the efficiency), membrane swelling and concentration polarisation [174]. In general for a given membrane, separation factor and flux determines the effectiveness of a pervaporation system [163, 164, 175]. Selectivity and flux of pervaporation depends on membrane material, thickness of the membrane, vacuum or sweep gas partial pressure, feed temperature and feed composition [138, 176, 177]. It has been identified that by reducing the membrane thickness, the flux increased, but the selectivity decreased. Pore size, porosity and roughness of membrane surface aides in reduction of the membrane thickness [178]. The operating conditions for pervaporation and membrane evaporation are given in (Table 11).

Table 11 Optimum operating conditions of membrane separation techniques like pervaporation and membrane evaporation experiments

Condition	Value	Range
Feed solution flow rate	1.2×10^{-3} - 3.0×10^{-3}	L min^{-1}
Concentration of solute	500	Ppm
Volume of feed solution	2.3×10^{-1}	L
Sweep gas flow rate	0.6-1.3	L min^{-1}
Temperature	30	°C

Pervaporation is done for vivid solvent separation; in this part the works done towards finding the best membrane for butanol separation has been discussed. Among the membranes found so far, silicates prove to be promising for its high selectivity towards mass production of butanol [138, 165] [179]. Studies showed the effect of ultrasound irradiation on butanol pervaporation using silicone tubing as membrane to check on its improvement and was found that by running pervaporation unit with ultrasound irradiation for 8 hours at 60 ^{0}C, concentration of butanol and flux increase in permeate side by 18 g/L and 13% respectively, i.e., from 203 to 221 g/L (ABE fermentation). PDMS (polydimethylsiloxane) membranes generally prove to be good among polymeric membranes. Fadeeve et al (2000) examined PDMS and PTMSP dense films in butanol pervaporation and found that latter showed high selectivity and permeate flux. But after 20 hours, due to membrane fouling inhibition [180] flux-decreased [181-182] and works were done to improve PDMS membrane performance using ceramic as a support which likely increased the flux. Even silicates PDMS membranes were combined to check on its performance and it was found that separation factor increased by 35%. Qureshi et al (1999) investigated silicate in silicone membrane and found that there was an increase in the flux and separation factor [183]. Vankelecom et al (1995) said that zeolite Y and ZSM-5 embedded separately in a PDMS membrane resulted in increase in water flux and organic flux respectively. PDMS membrane filled with surface-functioned silica nano particles were used by Beltrane et al (2013) [185] and they found that this membrane performed better at 60 ^{0}C and lower temperature. Liu et al (2013) did similar work on PEBA. They used zeolitic imidazole frame works (ZIF-71) particles as the filler for PEBA, which resulted in increased flux and separation factor compared to conventional PEBA membrane [186]. Various membranes and its efficiency in terms of its flux and separation factor are discussed in (Table 13).

3.6.4 Gas Stripping

Gas stripping is a mass transfer operation in which the solute is removed from a liquid by contacting it with a gas. This is a physical separation process where the gas stream must be added as a separating agent. This proves to be a reliable separation technique for butanol because of its versatility [187, 188, 189] since combination of membrane vapour separation [158], two stage fermentation, fluidized bed reactor, two stage gas stripping [175] or usage of separate packed bed stripper [190] is possible. Even separation of volatile compounds can be achieved by bringing down the pressure, heat or by using an inert gas. On average the solution to be separated is fed counter current to the gas (Inert) in a column and separation of components (Butanol, Ethanol and Acetone) can be achieved [191]. The exit gas containing ABE and water is always around its equilibrium partial pressure. After which the stripping gas is condensed to separate the inert gas and the ABE. It is again fed for another extraction cycle [174, 192]. This is an *in situ* process where fermentation and stripping (recovery) are done simultaneously [192–

195, 163, 159, 164, 140] without creating any possible damage to the cell though it results in foaming which brings down the performance. Thus, the gas flow rate, anti foam and presence of other components has to be put under surveillance here. Park et al (1991) identified an increase in gas flow rate and temperature leads to an increase in glucose conversion and butanol separation [196]. Highest butanol selectivity is obtained when gas stripping is performed at 67 ^{0}C [197]. Ezeji et al (2005) said that stripping rate is proportional to the gas recycle rate [198], stated that for 1 L fermentor, 4.8L/min is an optimized gas recycle flow rate to keep butanol in a concentration less than its toxic level and also found that the presence of acetone does not affect butanol separation by gas stripping. They observed from different fermentation systems coupled with gas stripping that the yield and productivity in fed-batch and continuous fermentations are the same and higher than batch fermentation. But fed batch fermentation failed due to the accumulation of inhibitors [199-201]. Mollah and Stuckey (1993) also used gas stripping in a 2.5L fluidized bed reactor of calcium alginate beads immobilizing *C.acetobutylicum* using glucose as substrate [202]. The carrier gas was a mixture of 20% hydrogen, 10% carbon dioxide and 70% nitrogen. They found that increasing the gas flow rate up to 1.67 L/min led to an increase in productivity but further increase resulted in reduction in productivity [198] [203]. Ennis et al (1986) used 1.2 L batch reactor where a 2.7L/min nitrogen gas stream was used to strip the butanol produced [204] that showed an increase in the productivity and sugar utilization, though the ABE yield decreased from 0.39 to 0.27 g solvent/g sugar. This may be due to the incomplete solvent capture in the condenser. Groot et al (1989) used batch and continuous fermentation using a separate packed bed stripper [190] and found that substrate consumption increased from 37 to 126g/L in batch fermentation, whereas the product yield remains fairly constant. In continuous fermentation, an increase in biomass concentration, a threefold increase in productivity, a 10% increase in substrate utilization but a low productivity and selectivity were witnessed due to gas stripping. Qureshi et al (2001) obtained a substrate utilization of 30 g/L in batch fermentation, while the sugar consumption was 199g/L in fed-batch fermentation coupled with gas stripping for butanol separation. These methods though increase the sugar utilization and productivity, the selectivity for butanol separation still staggers. Yet another study was done using carbon dioxide (stripping gas), cold trap and NRTL (unisim thermodynamic fluid design) R400 (Honeywell) [175] to determine the final concentration of the butanol recovered. In this procedure a simulated fermentation broth consisting ABE mass percentage of 0.6 : 1.2 : 0.1 at 37 ^{0}C gave mass fraction of butanol in final recovered product between 0.24 to 0.32 at cold trap temperature between -40 ^{0}C to 10 ^{0}C when a gas stream is in thermodynamic equilibrium with the fermentation broth at the exit of the fermentor. Thus it is evident now that when cold trap temperature is increased from -40 ^{0}C to 10 ^{0}C the butanol recovered decreases by 20%. This is because of the decrease in amount of acetone being condensed as temperature increases [175]. Since butanol has low partial pressure than acetone at a given temperature, the percentage recovery of butanol is much higher than acetone. When the results were studied under the same condition containing 0.3: 0.6: 0.1 the final mass fraction decreases as the mass percentage of solvents were halved. The decrease was found to be because of increased proportion of water in gas stream exiting the fermentor.

Table 12 Results for butanol separation from fermentation broths by pervaporation using various membranes [175]

Membrane	Butanol conc. (g/L)	Flux(g/m^{2}h)	Membrane thickness (μm)	Separation factor	Feed Temp (°C)	Ref
Silicone	-	34.8	600	18.8	41	[138]
Silicone tube	-		250	20	37	[258]
Silicone tubes	17.5	13.5*	1000	46	37	[259]
	16	10*	1000	52	37	[259]
	14.8	8*	1000	57	37	[259]
	14	7*	1000	58	37	[259]
	12	6.2*	1000	64		[259]
Silicone tubes	2	1.5*	240	-	32	[260]
	8.5	6*	240	-	32	[260]
	12.5	8.9*	240	-	32	[260]
	15.5	11.5*	240	-	32	[260]
Silicone	3.3	27	-	7	36	[138]

Membrane	Butanol conc. (g/L)	Flux(g/m²h)	Membrane thickness (μm)	Separation factor	Feed Temp (°C)	Ref
	4.4	31	-	11	36	[138]
	5	31	-	13	36	[138]
	6	29	-	19	36	[138]
Silicone membrane	10	84	170	44	78	[184]
Silicone membrane	6	13	-	30	35	[261]
Silicone membrane	7	122	306	70	78	[261]
Silicone-silicalite composite membrane[1]	12	89	306	97	78	[179]
Silicone-silicalite composite membrane[2]	10	119	226	70	78	[179]
Silicone membrane	9.1	84	170	44	78	[179]
Thin film silicalite filled silicon composite membrane	15-16	850-900	13-15	50	70	[262]
PDMS	-	621	201	19.8	37	[263]
PDMS	-	10.4	250	11	30	[264]
PDMS	-	418	201	15.7	35	[265]
PDMS/PE/perforated alloy meal	-	57	-	18-20	37	[266]
PDMS/ceramic	-	741	-	21.4	37	[267]
PDMS/ceramic composite membrane	11	1000	-	15	37	[182]
PDMS/ceramic composite membrane	-	338-847	-	5.1-27.1	37	[181]
NA	NA	NA	NA	30	37	[268]
40 membrane fibres made of PP	-	7.1*	400	5	-	[269]
PP hollow fibre	1	1.5	-	3.2	35	[270]

Membrane	Butanol conc. (g/L)	Flux(g/m²h)	Membrane thickness (μm)	Separation factor	Feed Temp (°C)	Ref
	4	-	-	3	35	[270]
	6.5	-	-	2.8	35	[270]
Hydroxyterminatedplybutadiene-based poly ethaneurea(HTPB-PU)	11	9.7	140	13.7	40	[271]

*Butanol flow rate (g/m^2h)
[1]Silicalite/silicone ratio (1.5:1)
[2]Silicalite/silicone ratio (1:1)

3.6.5 PERSTRACTION

Perstraction is also a membrane separation technique which involves usage of a suitable membrane that separates the broth containing butanol from extractant and that could provide enough surface area for the immiscible phases to exchange butanol. Since no direct contact of solvent is involved in this it overcomes the drawbacks of liquid–liquid extraction like toxicity to cells, emulsion formation, rag layer formation and phase dispersion. Another notable trait of this system is its ability to support butanol diffusion alone and other components like acetic or butyric acid will be retained in the aqueous phase itself [205]. Butanol transfer rate or diffusion rate depends entirely on the membrane which may act as a physical barrier. Since total mass transfer depends on rate of diffusion of butanol, the membrane should be chosen in such a way that it facilitates butanol movement to the organic side [140, 175]. But the high cost of the membrane and the clogging and fouling property strangles the evolution of the method [159, 165, 160, 206]. Some of the commonly used or standard extractants are oleyl alcohol, polypropylene glycol, tributyrin, 1-octanol, 1-dodecanol and 2-ethyl-1-hexanol. Silicon membrane, hydrophobic and hydrophilic membranes are widely used so far. Jeon and Lee (1987) used oleyl alcohol flowing in a silicon tubing to recover butanol and observed that the substrate utilization increased from 60 to 601g/l and butanol production increased from 11.5 to 147.4 g/l when fed batch fermentation was used instead of simple batch fermentation [207] and found that the average extraction efficiency was 87% and the average product yield was 0.224

g (product)/ g (glucose consumed) when they integrated the reactor separator system. Here continuous feeding and withdrawal was started when the butanol concentration reached about 10g/l [208]. Shukla et al (1989) also studied butanol perstraction using hydrophobic hollow fibre tubular fermentor-extractor and 2-ethyl-1-hexanol as the extractant [209] and stated that butanol distribution coefficient in the extractant was 6.09 and separation factor was 276.7 but the drawback was the gas build up in the shell side of the fermentor during perstraction. This led to recommend that increasing the number of fibres can solve this problem. Groot et al (1990) investigated the butanol separation using silicon tubes and different polar and non-polar extractants such as hexane, oleyl-alcohol and isopropyl myristate for perstraction. And when the results were compared with pervaporation it was found that, though butanol selectivity in perstraction is high, permeate flux in pervaporation is much higher than that of perstraction, which may be associated to higher mass transfer resistance in solvent phase [206]. Grobben et al (1993) studied perstraction using a hydrophobic membrane and oleyl-alcohol or decane mixture as an extractant [210] and found that in their system an increase in yield was seen after 50 h of operation and the drawback caused from membrane fouling was solved by using hydrophilic membrane and micro filtration was used to remove solids and cells. This integrated process which uses MFA (methylated fatty acids) as an extractant gave higher yield but lower productivity which may be due to a slow extraction rate, a detrimental shear effect on cells due to rotation in micro filtration membrane or maybe due to the undesirable effect of polysulphone microfiltration on the cells. Qureshi and Maddox (2005) did perstraction for butanol from a broth using oleyl alcohol as an extractant and silicone tubing as a membrane. There seemed to be a problem due to low ABE flux through the membrane and stated that, this problem may be solved using a better membrane having higher flux for butanol such as silicate-based membranes [211].

Table 13 The economics of current and future biofuels

45 MM gal/yr	Corn ethanol	Sugarcane ethanol	Soyabean biodiesel	Corn butanol	Corn stover ethanol (biochem)	Corn stover ethanol (thermo-chem)
Total production cost ($M/yr)	$74	$58	$121	$88	$64	$58
Total project investment ($M)	$138	$88	$23	$276	$183	$241
Production cost ($/gal)	$1.53	$1.29	$2.55	$1.96	$1.48	$1.32
Energy density (BTU/gal)	76,330	76,330	119,550	99,837	76,330	76,330
Production cost with energy equivalent to gasoline ($/gal)	$2.33	$1.95	$2.48 equivalent to gasoline; $2.74 equivalent to diesel	$2.28	$2.25	$2.00

3.6.6 Reverse Osmosis (RO)

The issue of dilute solvent concentration swallowing more energy for distillation in ABE fermentation can be eliminated by using RO to dewater the fermentation liquor. Here high pressure is used to remove water via a semi permeable membrane [212, 175]. This can be incorporated to even separate butanol, provided the suitable membrane is known [175]. Garcia et al (1986) worked with polyamide membranes in a continuous separation system for butanol extraction and found that the permeate flux was only one-third of the flux obtained with model solutions and when hydrolysed to separate the cells and suspend the solids using ultra-filtration unit, the results showed that RO unit prevented the membrane fouling. In the RO system it was found that 90% of the butanol could be removed but the flux through the membrane for the fermentation broth was significantly low due to the presence of other components [213]. Their work enlightened the scope of membrane separation technology to battle with the problems of low yield and low solvent concentration in butanol production by fermentation broth. The above overview of various separation techniques brought the advantages of the membrane techniques like pervaporation, perstraction and reverse osmosis over non-membrane techniques like gas stripping, liquid-liquid extraction and sorption to the forefront.

Table 14 Production description of butanol in various industries

COMPANIES	PRODUCTION DESCRIPTION
GEVO	Colorado-based company. Genetically modified yeast that produces only isobutanol from glucose. Lab results of 105 g/L of isobutanol (Glassner, 2009).
COBALT TECNOLOGIES	California-based biofuel company has proprietary technologies in microbial physiology, strain development, fermentation. They use non-modified *Clostrium* (ABE fermentation in a continuous process) to break down components of plant matter including cellulosic, hemicellulosicand starch.
BUTAMAX ADVANCED BIOFUELS	They have recently announced commencement of commercialization of butanol with the name Butamax Advanced Biofuels which is planned to be operational in 2010 and the first commercial plant is expected to be operational by 2013.
DU PONT	Their process is based on fermentation route of producing butanol using variety of feedstocks.
BUTYL FUEL, LLC	Production of n-butanol with modified *Clostridium*. This patented technology claims to be able to produce 1.3 to 1.9 times more butanol (per weight of raw material).
GREEN BIOLOGICS	UK-based company has developed their own thermophillic bacteria for rapid enzymatically release of fermentable sugar from biomass, this bacterial strain can also digest hemicellulose (5-carbon sugar) feedstocks. GBL using genetic engineering has developed a strain to produce high yields of butanol and can tolerate close to 4% butanol concentration. Recently, GBL has announced signing agreement between GBL and Laxmi Organics, India for construction of commercial demonstrator for bio-butanol.
SYNTEC BIOFUEL	Syntec is a Canada-based renewable energy company working on processes of converting biomass to alcohols (B2A). Right now they are working on development of catalyst for conversion of cellulosic biomass to pure isobutanol using thermochemical process. Syntec biofuel is focusing on utilizing wide range of biomass like, MSW, forestry and agriculture waste as a feedstock.

COMPANIES	PRODUCTION DESCRIPTION
TETRAVIATE BIOSCIENCE	Illinois-based company. Uses a patented mutant strain of *C. beijerinckii* for ABE fermentation with higher yields.
BUTALACO *(bio-based innovations)*	Yields with headquarters in Zug, Switzerland, genetically optimized yeasts which in turn enables increased yields in the production of biobutanol, using C5/C6 sugars for fermentation.
METABOLIC EXPLORER	Based in Clermont-Ferrand, France, the company has designed high-performance microorganisms that can transform plant-derived raw materials into a flexible and competitive fermentation process for butanol production.

3.6.7 IMPROVING BUTANOL TITRE VIA GENETIC ENGINEERING

Genetic engineering plays a vital role in the biobutanol production. It proves its eminence in modifying microorganisms, substrates that the organisms depend for growth and substrates that are used as a source for biofuel production. Conversion of cellulosic biomass to biofuel, which is both abundant and renewable, is a promising alternative. However, the cellulases and pre-treatment processes involved are very expensive. Genetically engineered plants helps to produce cellulases and hemicellulases, and to reduce the need for pre-treatment processes through lignin modification, are promising paths to solving this problem, together with other strategies, such as increasing plant polysaccharide content and overall biomass [214]. Bacteria were the first organisms to be modified in the laboratory, due to their simple genetics by incorporating genes that code for efficient butanol production by cloning techniques [215] [216]. Metabolic engineering provided the targeted gene for manipulation and it came to light that organisms were incapable to produce significant amounts of fermentative products. Metabolic engineering specifically seeks to mathematically model these networks, calculate a yield of useful products, and pin point parts of the network that constrain the production of these products. Genetic engineering techniques can then be used to modify the network in order to relieve these constraints. Genetically engineered *Clostridia* species shows its excellence in the high productivity of butanol from ABE fermentation process [217].

Mutation process involves few genes from various host organisms producing AB with greater tolerance and high yield. Some of those genes include—*adc* gene with the help of acetoacetate decaboxylase enzyme produces acetone by deletion from the host organism *C. acetobutylicum* where the yield has been improved by 57–70.8% [218]; *thiL, hbd, crt, bcd-etfB-etfA,* and *adhe1* (*adhe*) genes in the presence of acetyl-CoA acetyl transferase, b-hydroxybutyryl-CoA dehydrogenase, crotonase, butyryl-CoA dehydrogenase, butyraldehydedehydrogenase, and butanol dehydrogenase produced butanol through insertion from *E. coli* as a host organism inferred that expression of all genes were detected except *bcd-etfB-etfA* [219] and also these genes from the host organisms *E. coli, P. putida,* and *B. subtilis* showed that the butanol titre was greatest in *E. coli* from the pathway that was constructed in three organisms [220]. Insertion of *thiL, hbd, crt, bcd-etfB-etfA,* and *adhe2* and deletion of *adhE, ldhA, frdBC, fnr, pflB* from *E. coli* with the action of Acetyl-CoA acetyltransferase, b-hydroxybutyryl-CoA dehydrogenase, crotonase, butyryl-CoA dehydrogenase, butyraldehyde dehydrogenase, and butanol dehydrogenase produces butanol [221]. Genes such as *ctfA, ctfB, adc,* and *aad* through Acetyl-CoA acetyl transferase, b-hydroxybutyryl-CoA dehydrogenase, crotonase, butyryl-CoA dehydrogenase, butyraldehydedehydrogenase, butanol dehydrogenase, and acetoacetate decaboxylase titres butanol and acetone via insertion from the host cell of *C. acetobutylicum* [222]. Genes like *aldA, ctfA, bdhA, bdhB, hbdA, crt A, bcd-etfB-etfA, adhA, adh-1, adcA,* and *aad* from *E. coli* with the help of Crotonase, butyryl-CoA dehydrogenase, electrontransport protein subunits A and B, 3- hydroxybutyryl-CoA dehydrogenase, alcohol dehydrogenase, CoA-transferase (subunits A and B), aldehyde dehydrogenase, and acetoacetate decaboxylase produces butanol, acetone, and ethanol by insertion process [223]. Genes from *S. cerevisiae* such as *hbd, crt, bcd, etfA, etfB,* and *adhe2* has encoding enzymes like 3-Hydroxybutyryl-CoA dehydrogenase, crotonase, butyryl CoA, electron-transport protein subunits A and B dehydrogenase titres Bbutanol through insertion [224]. The TargeTron technology was utilized to disrupt acetoacetate decarboxylase gene *(adc),* which was responsible for acetone production in hyper-butanol producing industrial strain EA 2018. As a result, butanol ratio enhanced from 70% to 80.05%, while acetone production was reduced to 0.21 g L^{-1}. Recently, whole genome of two butanol producing microorganisms has been sequenced. In ABE fermentation *C. acetobutylicum* is characterized by the unique shift from acid to solvent production. The mechanism of the solventogenic switch involves the induction of several enzymes, including NADH-dependent butanol dehydrogenase (BDH) at the onset of solventogenesis. To characterize the genetic control of this gene, we have cloned and expressed it in *E. coli* [131, 225]. A catabolite control protein (*CcpA*) that shows pleiotropism key regulation was functionally identified in *C. acetobutylicum*. It was found that deletion of *CcpA* gene could eliminate glucose repression on xylose metabolism, where (1) *CcpA* mediates carbon catabolite regulation; (2) *CcpA* is involved in positive regulation of the key genes responsible for acidogenesis and solventogenesis; (3) *CcpA* is necessary for sporulation of *C. acetobutylicum*, achieving the simultaneous fermentation of glucose/xylose mixture and will thus benefit the utilization of lignocellulosic biomass in *C. acetobutylicum* [226]. Coenzyme A (CoA)-transferase (acetoacetyl-CoA: acetate/butyrate; CoA-transferase [butyrate-acetoacetate CoA-transferase] [EC 2.8.3.9] of *C. acetobutylicum* ATCC 824 is an important enzyme in the metabolic shift between the acid-producing and solvent-forming states of this organism. The genes encoding the two subunits of this enzyme have been cloned by using synthetic oligodeoxynucleotide probes designed from amino-terminal sequencing data from each subunit of the CoA-transferase [21, 227]. Compared with *C. beijerinckii* 8052, *C. beijerinckii* BA101 showed increased expression of butyryl-CoA and butanol formation genes, including *bcd* and *bdh*, during the solventogenic phase, suggesting differences in the regulation of primary metabolic genes in the two strains. Maximal levels of induction of sporulation genes in *C. beijerinckii* BA101 were significantly lower than those in *C. beijerinckii* 8052, consistent with the observed lower level of endospore formation. The *spo0A* protein is a master regulator of gene expression in the sporulating cells of clostridia and bacilli [228, 229]. Phosphorylated *spo0A* protein has been found to activate a variety of targets, such as the solventogenic operon *adhE1-ctfA-ctfB* in *C. acetobutylicum* and *adc* in *C. acetobutylicum* and *C. beijerinckii*, as well as multiple sporulation factor genes including the *sigF* operon and *sigG* in *C. acetobutylicum* and *B. subtilis* [228, 230-233]. In *C. beijerinckii* BA101, reduced expression of *spo0A* may lead to weakened induction of the *sigF* operon and the *sigE* and *sigG* regulons and thus an overall reduction in sporulation.

Spo0A has also been shown to negatively regulate the expression of chemotaxis/motility operons [234-236]. Similar to the solventogenesis in *C. acetobutylicum* ATCC 824, the solvent formation in *C. beijerinckii* 8052 coincided with the appearance of *Clostridia* forms and high-level expression of *spo0A*, the *sigF* operon, and the solventogenic genes *ald*, *ctfA-ctfB*, and *adc*, as well as the downregulation of the chemotaxis/motility genes [237, 238]. Extracellular electron shuttles to alter the metabolic pathways in *Clostridium beijerinckii* NCIMB 8052 fermentation increase butanol yield with glucose as a primary substrate. When anthraquinone-2,6-disulfonate (AQDS), the oxidized form of the electron shuttle, was introduced into batch cultures of *C. beijerinckii* it increased butanol yields by 2.5 to 5 times relative to glucose alone fermentation batches. Adding Fe(III) (as ferrihydrite or ferric citrate), which also acts as an electron sink, increased butanol yield by 5 times relative to the glucose-only control [239]. *C. acetobutylicum* ATCC 824 when mutated by butanol it gets mutated to SA-1 gene which results in enhanced butanol tolerance (121% higher than parent organism) [240]. Likewise when *C. acetobutylicum* PTCC-23 is exposed to UV, N-methyl-N-nitro-N nitrosoguanidine and ethyl methane sulphonate it gets mutated to MEMS-7 gene that result in enhancing the yield by 20% [241]. In *C. acetobutylicum* ATCC 824 (or *C. beijerinckii* NCIMB 8052), when N-methyl-N0-nitro-N-nitrosoguanidine together with selective enrichment on the glucose analog 2-deoxyglucose acts as mutagens, mutation takes place and gets mutated to BA101, BA-105 genes; this increases the amylolytic activity increased by 82% (BA-101), 25% (BA-105), and increase to nearly 2% final concentration of butanol [242, 243]. *E. coli* JCL16 when mutated by 4-aza-D,L-leucine, a structural analogue to L-leucine it changes to AL-1, AL-2, as a result the yield has reached 33% of theoretical maximum (0.33 g/g) [244].

3.7 ECONOMICS OF BUTANOL ACROSS THE WORLD

In recent years economic studies have been done on the production of butanol from corn whey permeate, wheat straw and molasses [194, 195, 256]. In these studies it was discovered that distillation as separation technique is not feasible economically when compared to the petrochemical route. Nevertheless works involving *C.beijerinki* BA101, *C.acetobutylicum* P260, hydrolysed DDGS and wheat straw states that commercial production of biobutanol from agricultural wastes is drawing closer. Recently DuPont (US) and BP (UK) confessed their idea that the first plant would operate on sugar or corn starch, however it is likely that agricultural waste would become a very potential substrate in future [140]. Further recent advances in microbial culture, process technologies and waste substrates requires developed fermentation based biobutanol industry that can compete with petrochemical routed butanol. It is foreseen that further research may include works on GM microbes as second generation cultures and on recovery techniques involving membranes and usage of agricultural-byproducts as substrates [140].

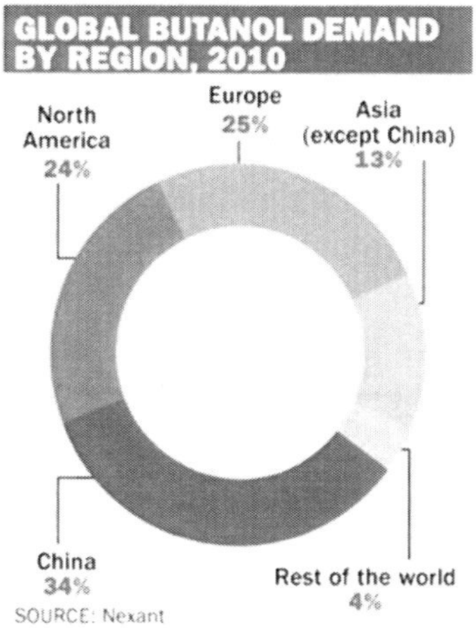

Figure 8 Demand for butnol across the globe in the year 2010 Adopted [272]

Later on several economic studies have been performed on the production of butanol (traditional ABE fermentation) from corn, whey permeate, and molasses. It is not economical when compared with butanol derived from the current petrochemical route. New developments in process technology for butanol production could allow for a significant reduction in the price of butanol. Petrochemical industries gave capacity to reduce the price of butanol. For a fuel to be economically viable the main parameter accountable is feedstock availability. In 2008, the global market for 1-butanol was 2.8 million tons [257], estimated to be worth approximately $5 billion. But in the mere future the average growth is expected to be 3.2% pa with demand concentrated in North America (28%), Western Europe (23%) and North East Asia (35%). The current butanol market in U.S. alone is 2.9 billion lbs per annum; this was recorded in the 2009 [46]. With the current demand for fuels, butanol is produced globally exceeding 1.2 billion gallons per year and valued market will be an early target for lower cost butanol. The market expands 3% per year which accounts for $7-8.4 billon at current price. And the market estimated by 2020 is $247 billion due to its advantages over ethanol and biodiesel [258]. China took a remarkable step in re-commercializing the ABE fermentation process. China has invested over $200 billion to install 0.21 million tons per annum of solvent volume with the expansion plans for 1 million tons per annum. About 30,000 t of butanol is produced per annum from corn starch by 61 major plants set up in China. Further due to its increasing demand, the market will expand significantly and will become a gateway to other chemical derivatives. On the whole it is simply known as the next generation fuel. Table shows the following observation of economics of butanol based on feedstock availability; Corn to ethanol: US$ 1.53/gal (0.40/L); Corn to butanol: US$ 1.96/gal (0.52/L); Sugarcane to ethanol (Brazil): US$ 1.29/gal (0.34/L). Refer (Table 13) for the current and future economics of bio-fuels and (Table 14) & (Fig. 7, 8 & 9) shows the production of butanol per annum and demand for butanol worldwide.

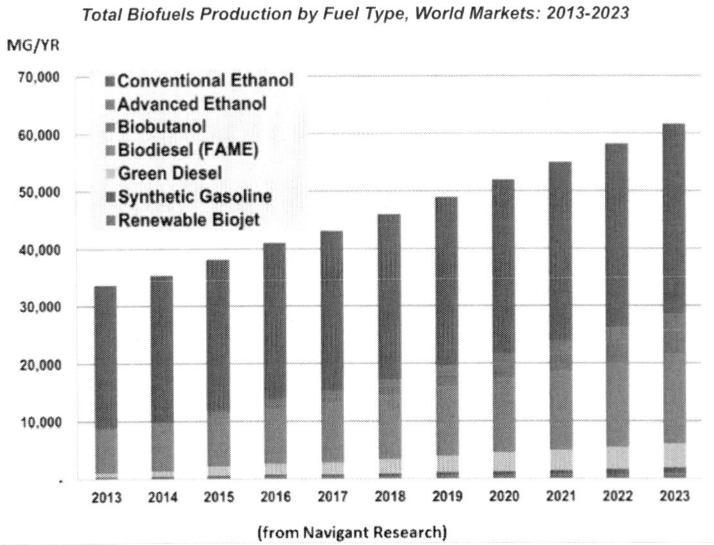

Figure 9 Total biofuel production by fuel type. Adopted [273]

3.8 CONCLUSION

Due to increasing demand for energy source and rapid depletion of fossil fuels and petroleum products the world is going in search of alternative fuels. Butanol as an alternative energy source acts as an efficient extender; it blends with gasoline up to 95% which proves its uniqueness over other fuels. Agricultural waste and lignocellulosic materials play a vital role in the production of butanol through ABE fermentation with the help of the kingmaker *Clostridium* species through acidogenesis and solventogenesis. The major challenge faced during the process is the production of toxic inhibitor which inhibits the productivity of butanol. In order to overcome this obstacle pre-treatment methods are implemented so as to reduce the toxicity. Pre-treatment is not only used for reducing the toxicity but it also helps in the breaking down of lignocelluloses into cellulose, hemicelluloses and lignin and the above overview of various separation techniques brought the advantages of the membrane techniques like pervaporation, perstraction and reverse osmosis over non-membrane techniques like gas stripping, liquid-liquid extraction and sorption to the forefront. Several *in situ* techniques have also come up but these cannot be manipulated by the industries for large scale production of butanol. More research and work is required to be initiated in this field for the manipulation of these techniques for the large scale production of butanol. By foreseeing its advantages and increasing demand the market is expected to expand by 2020.

3.9 REFERENCES

1. Agrawal AK. Biofuels (alcohols and biodiesel) applications as fuels for internal combustion engines. Prog Energy Combustion Sci 2007;33:233-71.

2. Escobar JC, Lora ES, Venturini OJ, Yanez EE, Castillo EF, Almazan O. Biofuels: environment, technology and food security. Renew. Sustain. Energy.Rev. 2009;13:1275-87.

3. Zhao R, Bean SR, Wang D, Park SH, Schober TJ, Wilson JD. Small-scale mashing procedure for predicting ethanol yield of sorghum grain. J Cereal Sci 2009;49(2):230-8.

4. Singh A, Pant D, Korres NE, Nizami AS, Prasad S, Murphy JD. Key issues in life cycle assessment of ethanol production from lignocellulosic biomass: challenges and perspectives. Bioresour Technol 2010;101(13):5003-12.

5. Prasad S, Singh A, Jain N, Joshi HC. Ethanol production from sweet sorghum syrup for utilization as automotive fuel in India. Energy Fuel 2007;21(4):2415-20.

6. Gullison RE, Frumhoff PC, Canadell JG, Field CB, Nepstad DC, Hayhoe K, et al. Tropical forests and climate policy. Science 2007;316:985-6.

7. He Y, Wang S, Lai KK. Global economic activity and crude oil prices: a cointegration analysis. Energy Econ; 2010;. doi:10.1016/j.eneco.2009.12.005.

8. Singh A, Smyth BM, Murphy JD. A biofuel strategy for Ireland with an emphasis on production of biomethane and minimization of land-take.Renew. Sustain. Energy. Rev. 2010;14(1):277-88.

9. Prasad S, Singh A, Joshi HC. Ethanol as an alternative fuel from agricultural, industrial and urban residues. Resour Conserv Recycling 2007;50:1-39.

10. Bhatti HN, Hanif MA, Qasim M, Ata-ur-Rehman. Biodiesel production from waste tallow. Fuel 2008;87:2961-6.

11. Dennis JS, Scott SA, Stephenson AL. Improving the sustainability of the production of biodiesel from oilseed rape in the UK. Process Saf Environ Protect 2008;86:427-40.

12. Poonam Singh Nigam, Anoop Singh. Production of liquid biofuels from renewable resources. Prog Energy Combustion Sci 37 (2011) 52-68.

13. Lin L, Cunshan, Z, Vittayapadung, S, Xiangqian, S, Mingdong D. Opportunities and challenges for biodiesel fuel. Appl. Energy.; 2010.doi:10.1016/ j.apenergy.2010.09.029.

14. Hoekman SK. Biofuels in the US-challenges and opportunities. Renew. Energy.2009;34:14–22.

15. Yang QZ, Song B. Sustainability assessment of biofuels as alternative energy resources. ICSET; 2008.

16. Huang H, Liu H, Gan YR. Genetic modification of critical enzymes and involved genes in butanol biosynthesis from biomass. Biotechnol Adv; 2010.doi:10.1016/j.biotechadv.2010.05.015.

17. Ma H, Oxley L, Gibson J, Li W. A survey of China's renewable energy economy. Renew. Sustain. Energy. Rev. 2010;14:438–45.

18. Demirbas A. Biofuels securing the planet's future energy needs. Energy. Convers. Manage. 2009;50:2239–49.

19. Demirdas A. Political, economic and environmental impacts of biofuels: a review. Appl. Energy. 2009;86:108–17.

20. Demirbas MF. Biorefineries for biofuel upgrading: a critical review. Appl Energy 2009;86:s151–61.

21. Kent SK. Biofuels in the US – challenges and opportunities. Renew. Energy. 2009;34:14–22.

22. Borjesson P. Good or bad bioethanol from a greenhouse gas perspective – what determines this? Appl. Energy. 2009;86:589–94.

23. He J, Zhang W. Techno-economic evaluation of thermo-chemical biomass-toethanol. Appl. Energy.; 2010. doi:10.1016/j.apenergy.2010.10.022.

24. Balat M, Balat H. Progress in biodiesel processing. Appl. Energy. 2010;87:1815–35.

25. Naik SN, Goud VV, Rout PK, Dalai AK. Production of first and second generation biofuels: a comprehensive review. Renew. Sustain. Energy.Rev.2010;14:578–97.

26. Gressel J. Transgenics are imperative for biofuel crops. Plant Sci 2008;174:246–63.

27. Chattopadhyay S, Karemore A, Das S, Deysarkar A, Sen R. Biocatalytic production of biodiesel from cottonseed oil:

standardization of process parameters and comparison of fuel characteristics. Appl. Energy.; 2010. doi:10.1016/j.apenergy.2010.10.007.

28. Huang G, Chen F, Wei D, Zhang X, Chen G. Biodiesel production by microalgal biotechnology. Appl. Energy. 2010;87:38–46.

29. Demirbas A. Competitive liquid biofuels from biomass.Appl. Energy. 2011;88:17–28.

30. Trent J. OMEGA: The future of biofuels? In: Conference. The ASEAN algae biofuels conference, Singapore; 2012.

31. Ramey D. Butanol advances in biofuels. The Light Party, http://www.lightparty.com /Energy/Butanol.html; 2004.

32. Dürre P.: Fermentative butanol production bulk chemical and biofuel. Ann. N. Y. Acad. Sci., 2008, 1125, 353-362.

33. Antoni D, Zverlov VV, Schwarz WH. Biofuels from microbes. Appl. Microbiol. Biotechnol.2007;77:23-35.

34. Fortman J.L., Chhabra S., Mukhopadhyay A., Chou H., Lee T.S., Steen E. and Keasling J.D.: Biofuel alternatives to ethanol: pumping the microbial well. Trends in Biotechnol., 2008, 26(7), 375-381.

35. Wackett L.P.: Biomass to fuels via microbial transformations. Curr. Opinion Chem. Biol., 2008, 12,187-193.

36. Zhou A, Thomson E. The development of biofuels in Asia.Appl. Energy.2009;86:11–20.

37. Balat M. BH: recent trends in global production and utilization of bio-ethanol fuel. Appl. Energy. 2009;86:2273–82.

38. Yan J, Lin T. Biofuels in Asia. Appl. Energy. 2009;86:1–10.

39. Durre P. Biobutanol: an attractive biofuel. Biotechnol J 2007(2):1525–34.

40. Pfromm PH, Boadu VA, Nelson R, Vadlani P, Madl R. Bio-butanol vs. bioethanol: a technical and economic assessment for corn and

switchgrass fermented by yeast or Clostridium acetobutylicum. Biomass Bioenrgy; 2010. doi:10.1016/j.biombioe.2009.12.017.

41. Swana J, Yang Y, Behnam M, Thompson R. An analysis of net energy production and feedstock availability for biobutanol and bioethanol. Bioresour Technol; 2010. doi:10.10.16/j.biotech.2010.08.051.

42. Brekke K. Butanol an energy alternative? Ethanol Today; March 2007:36e9.

43. EBTP. Biobutanol. European biofuels technology platform (EBTP), http://www.biofuelstp.eu/butanol.html; 2009.

44. Wu M, Wang M, Liu J, Huo H. Life-cycle assessment of corn-based butanol as a potential transportation fuel. Argonne National Laboratory; 2007.ANL/ ESD/07e10.

45. Jones DT, Woods DR. Acetone-butanol fermentation revisited: Microbiol. Rev 1986;50(4):484–524.

46. Ezeji TC, Qureshi N, Blaschek HP. Bioproduction of butanol from biomass:from genes to bioreactors. Curr.Opin.Biotechnol. 2007;18:220-7.

47. Durre P. New insights an novel developments in clostridial acetone/butanol/isopropane fermentation. Appl. Microbiol. Biotechno.1998;49:639–48.

48. Keis S, Shaheen R, Jones TD. Emended descriptions of Clostridium acetobutylicum and Clostridium beijerinckii, and descriptions of Clostridium saccharoperbutylacetonicum sp. nov. and Clostridium saccharobutylicum sp. nov. Int. J. Syst. Evol.Microbiol. 2001;51:2095–103.

49. Comyns AE. Bio-futures. Focus Catal 2007;2007(8):1.

50. Comyns AE. Biobutanol: a European jv for BP and DuPont. Focus Catal.2009;2007(8):3.

51. Comyns AE. BP and Du plan 'biobutanol'. Focus Catal. 2006;2006(8):7.

52. Zhang Z, Lohr L, Escalante C, Wetzstein M. Food versus fuel: what do prices tell us? Energy Policy 2010;38:445–51.

53. Qureshi N, Saha BC, Hector RE, Dien B, Hughes S, Liu S, et al. Production of butanol (a biofuel) from agricultural residues: part II – use of corn stover and switchgrass hydrolysates. Biomass Bioenrgy; 2010.doi:10.1016/j.biombioe.2009.12.023.

54. Qureshi N, Saha BC, Hector RE, Hughes SR, Cotta MA. Butanol production from wheat straw by simultaneous saccharification and fermentation usingClostridium beijerinckii: part I – batch fermentation. Biomass Bioenrgy; 2008;32:168–75.

55. Qureshi N, Saha BC, Cotta MA. Butanol production from wheat strawby simultaneous saccharification and fermentation using Clostridiumbeijerinckii: part II – fed-batch fermentation. Biomass Bioenrgy; 2008;32:176–183.

56. Qureshi N, Ezeji TC, Ebener J, Dien BS, Cotta MA, Blaschek HP. Butanolproduction by Clostridium beijerinckii. Part I: use of acid and enzymehydrolyzed corn fiber. Bioresour Technol 2008;99:5915–22.

57. Qureshi N, Saha BC, Dien B, Hector RE, Cotta MA. Production of butanol (a biofuel) from agricultural residues: part I – use of barley straw hydrolysate.Biomass Bioenrgy 2010;34:559–65.

58. Sanchez S, Lozano LJ, Godinez C, Juan D, Perez A, Hernandez FJ. Carbo pod as a feedstock for the production of bioethanol in mediterranean areas. Appl. Energy. 2010;87:3417-24.

59. Larson DE. Biofuel production technologies: status, prospects andimplications for trade and development. In: United Nations conference ontrade and development, New York and Geneva; 2008.

60. Chauhan S. Biomass resources assessment for power generation: a case study from haryana state, India. Biomass Bioenrgy; 2010. doi:10.1016/j.biombioe.2010.04.003:1-9.

61. Christi Y.: Biodiesel from microalgae beats bioethanol. Trends Biotechnol., 2008, 26(3), 126-131.

62. Christi Y.: Biodiesel from microalgae. Biotechnol.Advan., 2007, 25, 294-306.

63. Fengxue Xin, Jianzhong He: Characterization of a thermostable xylanase from a newly isolated Kluyvera species and its application for biobutanol production;Bioresource Technol 135 (2013) 309–315.

64. Hetty van der Wal, Bram L.H.M. Sperber, Bwee Houweling-Tan, et all: Production of acetone, butanol, and ethanol from biomass of the green seaweed Ulva Lactuca; Bioresource Technol 128 (2013) 431–437.

65. Mohamed H.A., Abdel-Wahab E. E.: Production of acetone-butanol-ethanol from spoilage date palm (Phoenix dactylifera L.) fruits by mixed culture of Clostridium acetobutylicum and Bacillus subtilis; Biomass Bioenrgy 42 (2012) 17 2 -178.

66. Hanh Thi My Tran, Benjamas Cheirsilp, et all: Potential use of Bacillus subtilis in a co-culture with Clostridium butylicum for acetone–butanol–ethanol production from cassava starch; Biochem. Eng. J. 48 (2010) 260–267.

67. Jasmine Isar, Vidhya Rangaswamy: Improved n-butanol production by solvent tolerant Clostridium beijerinckii; Biomass Bioenrgyy 37 (2012) 9-15.

68. Qureshi N, Blaschek HP. Economics of butanol fermentation using hyperbutanol producing Clostridium beijerinckii BA101. Trans IChemE 2000;78: 139–44.

69. Lenz TG, Morelra AR. Economic evaluation of the acetone–butanol fermantation. Ind Eng Chem Prod Res Dev 1980;19:478–83.

70. Qureshi N, Blaschek HP. ABE production from corn: a recent economic evaluation. J Ind Microbiol Biotechnol 2001;27:292–7.

71. Gapes JR. The economics of acetone–butanol fermentation: theoretical and market considerations. J Microbiol Biotechnol 2000;2(1):27–32.

72. Serrano-Ruiz JC, West RM, Dumesic J. Catalytic Conversion of Renewable Biomass Resources to Fuels and Chemicals. Chemical and Biomolecular Engineering-Annual Review 2010; 1: 79-100.

73. Sivakumar G, Vali DR, Xu J, Burner DM, Jr JOL, Ge X, Weathers PJ. Bioethanol and Biodiesel: Alternative Liquid Fuels for Future Generations. Engineering and Life Science 2010; 10(1): 8-18.

74. Visser EM, Filho DO, Martins MA, Steward BL. Bioethanol Production Potential from Brazilian Biodiesel Co-products. Biomass Bioenrgy2011; 35: 489-494.

75. Alvandi S, Agrawal A. Experimental Study of Combustion of Hydrogen Syngas/Methane Fuel Mixtures in a Porous Burner. Int. J. Hydrogen Energy 2008; 33(4): 1407-1415.

76. Kaparaju P, Serrano M, Thomsen AB, Kongjan P, Angelidaki I. Bioethanol, Biohydrogen and Biogas Production from Wheat Straw in a Biorefinery Concept. Bioresour. Technol 2009; 100: 2562-2568.

77. Fischer CR, Klein-Marcuschamer D, Stephanopoulos G. Selection and Optimization of Microbial Hosts for Biofuels Production. Metab.Eng. 2008; 10: 295-304.

78. Johnson JMF, Coleman MD, Gesch R, Jaradat A, Mitchell R, Reicosky D, Wilhelm WW. Biomass-bioenergy Crops in the United States: A Changing Paradigm. American J of Plant Sci and Biotechnol 2007; 1: 1-28.

79. Weber C, Farwick A, Benisch F, Brat D, Dietz H, Subtil T, Boles E. Trends and Challenges in the Microbial Production of Lignocellulosic Bioalcohol Fuels. Appl. Microbiol. Biotechnol. 2010; 87: 1303-1315.

80. Andrew Carroll and Chris Somerville "Cellulosic Biofuels" Annual Review of Plant Biology 2009, Vol. 60, pp. 165-182. doi:10.1146/annurev.arplant.043008.092125

81. Barbara A. Tokay "Biomass Chemicals" in Ullmann's Encyclopedia of Industrial Chemistry 2002, Wiley-VCH, Weinheim. doi:10.1002/14356007.a04_099

82. A Research Roadmap Resulting from the biomass to biofuels workshop; U.S. Department of energy; Dec 7-9, 2005

83. Cherubini F, Stromman AH. Production of Biofuels and Biochemicals from Lignocellulosic Biomass: Estimation of Maximum Theoretical Yield and Efficiencies using Matrix Algebra. Energy Fuels 2010; 24: 2657-2666.

84. Saha BC. Hemicellulosic Bioconversion. J. Ind Microbial Biotechnol 2003; 30 279-291.

85. Carvalheiro F, Duarte LC, Gírio FM Hemicellulose Biorefineries: A Review on Biomass Pretreatments. J of Sci & Ind Res 2008; 67 849-864.

86. Girio FM, Fonseca C, Carvalheiro F, Duarte LC, Marques S, Bogel-Lukasik R. Hemicellulosesfor Fuel Ethanol: A Review. Bioresour Technol 2010; 101 4775-4800.

87. Kuhad RC, Gupta R, Khasa YP, Singh A, Zhang Y.-H. P Bioethanol Production from Pentose Sugars: Current Status and Future Prospects. Renew. Sustain. Energy.Rev.2011; 15 4950-4962.

88. Sawang L., Benjamas C., Kamontam U.: Decanter cake waste as a renewable substrate for biobutanolproduction by Clostridium beijerinckii; Proc Biochem 48 (2013) 1933–1941.

89. Mach, L. R., Seiboth, B., Myasnikov, A., et all: The bgl1 gene of Trichoderma reesei QM 9414 encodes an extracellular, cellulose-inducible b-glucosidase involved in cellulase induction by sophorose, Mol. Microbiol., 16, 687e697 (1995).

90. Sun, Y. and Cheng, J.: Hydrolysis of lignocellulosic materials for ethanol production: a review, Bioresour. Technol.,83, 1e11 (2002).

91. Abdel-Rahman, M. A., Tashiro, Y., and Sonomoto, K.: Lactic acid production from lignocellulose-derived sugars using lactic

acid bacteria: overview and limits, J. Biotechnol., 156, 286e301 (2010).

92. Takuya Noguchi,Yukihiro Tashiro,Tsuyoshi Yoshida, et all: Efficient butanol production without carbon catabolite repression from mixed sugars with Clostridium saccharoperbutylacetonicum N1-4; J of Biosci.Bioeng. VOL. 116 No. 6, 716-721, 2013.

93. Ma, F.R., Hanna, M.A., 1999. Biodiesel production: a review. Bioresour. Technol. 70, 1-15.

96. Wei-Chen Kao, De-Shun lin, Chieh-lun Cheng, et all: Enhancing butanol production with Colstridium pasterurianum CH4 using sequential glucose-glycerol addition and simultaneous dual-substrate cultivation strategies; Bioresour. Technol. 135 (2013) 324-330.

95. Piyoungkoon Noomtim, Benjamas Cheirsilp: Production of butanol from Palm Empty Fruit Bunches Hydrolysate by Clostridium acetobutylicum; Energy Procedia 9 (2011) 140-146.

[96] Rabelo SC, Filho RM, Costa AC (2009) Lime Pretreatment of Sugarcane Bagasse for Bioethanol Production. App.l Biochem. Biotechnol. 153: 139-150.

97. Shahbazi A, Wang L, Diallo O, Whitmore A (2010) Alkali Pretreatment and Enzymatic Hydrolysis of Cattails from Constructed Wetlands. Am J Eng Applied Sci 3: 328-332.

98. Kim S, Holtzapple M (2006) Effect of structural features on enzyme digestibility of corn stover. Bioresour Technol 97: 583-591.

99. Xu J, Cheng JJ, Sharma-Shivappa RR, Burns JC (2010) Lime pretreatment of switchgrass at mild temperatures for ethanol production. Bioresour Technol 101: 2900-2903.

100. Sierra R, Garcia LA, Holtzapple MT (2010) Selectivity and delignification kinetics for oxidative and nonoxidative lime pretreatment of poplar wood, part III: Long-term.Biotechnol. Prog. 26: 1685-1694

101. Wyman C, Dale B, Elander RTH, Ladisch MR, Lee YY, et al. (2009) Glucose and xylose yields from switchgrass for ammonia fiber expansion, ammonia recycle percolation, dilute sulfuric acid, hot water, lime, and sulfur dioxide pretreatments followed by enzymatic hydrolysis.In the 31st Symposium on Biotechnology for Fuels and Chemicals SanFranciscoCA.

102. Teymouri F, Laureanoperez L, Alizadeh H, Dale B (2005) Optimization of the ammonia fiber explosion (AFEX) treatment parameters for enzymatic hydrolysis of corn stover. Bioresour Technol 96: 2014-2018.

103. Alizadeh H, Teymouri F, Gilbert T, Dale B (2005) Pretreatment of switchgrass by ammonia fiber explosion (AFEX). Appl. Biochem.Biotechnol. 124: 1133-1141.

104. Lee JM, Jameel H, Venditti RA (2010) A comparison of the autohydrolysis and ammonia fiber explosion (AFEX) pretreatments on the subsequent enzymatic hydrolysis of coastal Bermuda grass. Bioresour Technol 101: 5449-5458.

105. Li B-Z, Balan V, Yuan Y-J, Dale BE (2010) Process optimization to convert forage and sweet sorghum bagasse to ethanol based on ammonia fiber expansion (AFEX) pretreatment. Bioresour Technol 101: 1285-1292.

106. Kim T, Lee Y, Sunwoo C, Kim J (2006) Pretreatment of corn stover by low-liquid ammonia recycle percolation process. Appl. Biochem.Biotechnol. 133: 41-57.

107. Kim T, Lee Y (2005) Pretreatment and fractionation of corn stover by ammonia recycle percolation process. Bioresour Technol 96: 2007-2013.

108. Salvi D, Aita G, Robert D, Bazan V (2010) Ethanol production from sorghum by a dilute ammonia pretreatment. J. Ind Microbiol Biotechnol 37: 27-34.

109. Gupta R, Lee YY (2009) Pretreatment of hybrid poplar by aqueous ammonia.Biotechnol.Prog.25: 357-364. [110] Huang H, Ramaswamy S, Tschirner U, Ramarao B (2008)

A review of separation technologies in current and future biorefineries.Sep. Purif. Technol. 62: 1-21.

111. Zhang B, Shahbazi A, Wang L, Diallo O, Whitmore A (2010) Hot-water pretreatment of cattails for extraction of cellulose. J. Ind Microbiol.Biotechnol. 38: 819-824.

112. Knappert DR, Grethlein HE, Converse AO (1981) Partial acid hydrolysis of poplar wood as a pretreatment for enzymatic hydrolysis.Biotechnol.Bioeng.Symp 11: 67-77. [113] Lloyd T, Wyman C (2005) Combined sugar yields for dilute sulfuric acid pretreatment of corn stover followed by enzymatic hydrolysis of the remaining solids. Bioresour Technol 96: 1967-1977. [114] Zhang B, Shahbazi A, Wang L, Diallo O, Whitmore A (2011) Dilute-sulfuric acid pretreatment and cellulose conversion of cattails. In ASABE Annual Meeting, Louisville, Kentucky. [115] Yang Y, Sharma-Shivappa R, Burns JC, Cheng JJ (2009) Dilute Acid Pretreatment of Oven-dried Switchgrass Germplasms for Bioethanol Production†. Energy Fuels 23: 3759-3766. [116] Torget R, Himmel ME, Grohmann K (1991) Dilute sulfuric acid pretreatment of hardwood bark. Bioresour Technol 35: 239-246.

117. Nguyen Q, Tucker M, Keller F, Eddy F (2000) Two-stage dilute-acid pretreatment of softwoods.Appl. Biochem.Biotechnol. 84-86: 561-576.

118. Bura R, Mansfield S, Saddler J, Bothast R (2002) SO_2-catalyzed steam explosion of corn fiber for ethanol production. Appl. Biochem.Biotechnol. 98-100: 59-72.1

119. Bura R, Bothast R, Mansfield S, Saddler J (2003) Optimization of SO2- catalyzed steam pretreatment of corn fiber for ethanol production. Appl. Biochem.Biotechnol. 106: 319-335.

120. Ewanick SM, Bura R, Saddler JN (2007) Acid-catalyzed steam pretreatment of lodgepole pine and subsequent enzymatic hydrolysis and fermentation to ethanol.Biotechnol.Bioeng. 98: 737-746.

121. U.S. Department of Energy - Energy Efficiency and Renewable Energy Biomass Program, Concentrated Acid Hydrolysis.

122. Wan Y, Chen P, Zhang B, Yang C, Liu Y, et al. (2009) Microwave-assisted pyrolysis of biomass: Catalysts to improve product selectivity.J of Analy. Appl. Pyrolysis. 86: 161-167. [123] Zhao H, Holladay JE, Brown H, Zhang ZC (2007) Metal Chlorides in Ionic Liquid Solvents Convert Sugars to 5-Hydroxymethylfurfural. Science 316: 1597-1600. [124] Liu C, Wyman C (2006) The enhancement of xylose monomer and xylotriose degradation by inorganic salts in aqueous solutions at 180°C.Carbohydr. Res.341: 2550-2556.

125. Yu Q, Zhuang X, Yuan Z, Qi W, Wang Q, et al. (2011) The effect of metal salts on the decomposition of sweet sorghum bagasse in flow-through liquid hot water.Bioresour Technol 102: 3445-3450.

126. Peter H. Pfromm, Vincent Amanor-Boadu , Richard Nelson: Bio-butanol vs. bio-ethanol: A technical and economic assessment for corn and switchgrass fermented by yeast or Clostridium acetobutylicum; Biomass Bioenrgy34 (2010) 515 – 524.

127. Qureshi, N., Li, X.L., Hughes, S., Saha, B.C., Cotta, M.A., 2006. Butanol production from corn fiber xylan using Clostridium acetobutylicum. Biotechnol.Progr. 22, 673– 680

128. Welch, R.W., Rudolph, F.B., Papoutsakis, E.T., 1989. Purification and characterization of the NADH-dependent butanol dehydrogenase from Clostridium acetobutylicum (ATCC 824). Arch. Biochem. Biophys. 273 (2), 309–318.

129. Adsul, M.G., Singhvi, M.S., Gaikaiwari, S.A., Gokhale, D.V., 2011. Development of biocatalysts for production of commodity chemicals from lignocellulosic biomass. Bioresour. Technol. 102 (6), 4304–4312.

130. Cheng, C.L., Che, P.Y., Chen, B.Y., Lee, W.J., Chien, L.J., Chang, J.S., 2012. High yield bio-butanol production by solvent-producing bacterial microflora. Bioresour. Technol. 113, 58–64.

131. Dürre, P., Fischer, R.J., Kuhn, A., Lorenz, K., Schreiber, W., Stürzenhofecker, B., Ullmann, S., Winzer, K., Sauer, U., 1995. Solventogenic enzymes of Clostridium acetobutylicum: catalytic properties, genetic organization, and transcriptional regulation. FEMS Microbiol. Rev. 17 (3), 251–262.

132. Nair, R.V., Bennett, G.N., Papoutsakis, E.T., 1994. Molecular characterization of an aldehyde/alcohol dehydrogenase gene from Clostridium acetobutylicum ATCC 824. J. Bacteriol. 176 (3), 871–885.

133. Shi, Z., Blaschek, H.P., 2008. Transcriptional analysis of Clostridium beijerinckii NCIMB 8052 and the hyper-butanol-producing mutant BA101 during the shiftfrom acidogenesis to solventogenesis. Appl. Environ. Microbiol. 74 (24), 7709–7714.

134. Thormann, K., Feustel, L., Lorenz, K., Nakotte, S., Dürre, P., 2002. Control of butanol formation in Clostridium acetobutylicum by transcriptional activation. J.Bacteriol. 184 (7), 1966–1967.

135. Wilkinson, S.R., Young, D.I., Morris, J.G., Young, M., 1995. Molecular genetics and the initiation of solventogenesis in Clostridium beijerinckii (formerly Clostridium acetobutylicum) NCIMB 8052. FEMS Microbiol. Rev. 17 (3), 275–285.

136. Wilkinson, S.R., Young, M., 1995. Physical map of the Clostridium beijerinckii (formerly Clostridium acetobutylicum) NCIMB 8052 chromosome. J. Bacteriol. 177 (2), 439–448.

137. Mutschlechner O, Swoboda H, Gapes JR. Continuous twostage ABE-fermentation using Clostridium beijerinckii NRLL B592 operating with a growth rate in the first stage vessel close to its maximal value. J. Mol. Microbiol.Biotechnol. 2000; 2(1):101-5.

138. Qureshi N, Blaschek HP. Production of acetone butanol ethanol (ABE) by a hyper-producing mutant strain of Clostridium beijierinckii BA101 and recovery by pervaporation. Biotechnol. Prog. 1999;15(4):594-602.

139. Peter Durre: Fermentative production of butanol- academic perspective; Peter Durre: Fermentative production of

butanol- academic perspective; Curr. Opin.Biotechnol.22 (2011) 331-336.22 (2011) 331-336.

140. Ezeji, Qureshi, Blaschek et al: Bioproduction of butanol from biomass: from genes to bioreactors: Curr. Opin. Biotechnol.18 (2007) 220-227.

141. Gracia,Pakkila,Ojamo et al: challenges in biobutanol production: how to improve the efficiency?:Renew. Sustain. Energy. Rev.: 15 (2011) 964-980.

142. Laili Gholizadeh Baroghi: Enhanced Butanol Production by Free and Immobilized Clostridium sp. Cells using Butyric Acid as Co-Substrate: Biotechnol (Bioprocess Eng – Biofuels): (2009).

143. Congcong Lu, M.S.: Butanol Production from Lignocellulosic Feedstocks by Acetone-Butanol-Ethanol Fermentation with Integrated Product Recovery; The Ohio State University 2011; 1-135

144. Lipnizki, F. et al: Use of Pervaporation-bioreactor Hybrid Process in Biotechnology: Chem. Eng. Technol.; 23 (2000) 569-577.

145. Congcong Lu, Jingbo Zhao, Shang-Tian Yang et al: Fed-batch fermentation for n-butanol production from cassava bagasse hydrolysate in a fibrous bed bioreactor with continuous gas stripping: Bioresour. Technol. 104 (2012) 380–387.

146. Lee, Park, Jang et al: Fermentative Butanol Production by Clostridia: Biotechnol. Bioengg; 101(2008) 209–228.

147. Ezeji TQN, Blaschek HP. Production of acetone–butanol–ethan ol (ABE) in a continuous flow bioreactor using degermed corn and Clostridium beijerinckii. Process Biochem 2007;42:34–9. {47}

148. Nimc evic DaG JR.The acetone–butanol fermentation in pilot plant and preindustrial scale. J. Mol. Microbiol.Biotechnol. 2000;2(1):15–20.{42}

149. Qureshi N, Lai LL, Blaschek HP. Scale-up of a high productivity continuous biofilm reactor to produce butanol by adsorbed cells

of Clostridium beijerinckii. Food Bioprod Process 2004;82:164–73.

150. Stephens, G.M., Holt, R., Gottschal, J., Morris, J.,: Studies on the stability of solvent production by Clostridium acetobutylicum in continuous culture: J. Appl. Microbiol; 58 (1985) 597–605.

151. Ye Ni, Zhi-Hao Sun: Continuous Butanol Fermentation from Low-Value Sugar-Based Feedstocks by Clostridium saccharobutylicum DSM 13864; The Key Laboratory of Industrial Biotechnology, Ministry of Education, Schoolof Biotechnology, Jiangnan University, Wuxi 214122, Jiangsu, China

152. Pierrot,P. et al:Continuous Acetone-Butanol Fermentation with High Productivity by Cell Ultrafiltration and Recycling. Biotechnol Letters: 8(4) (1986) 253-256.

153. Kourkoutas, Y. et al: Immobilization technologies and support materials suitable in alcohol beverages production: a review: Food Microbiol; 21 (2004) 377-3.

154. Lima, N., Mota et al: Biotecnologia – Fundamentos e aplicações. Biocatálise aplicada Lidel-Edições Técnicas, Lda.(2003).

155. Ramey, D., Yang et al : Production of Butyric Acid and Butanol from Biomass – Final Report: U.S. Department of Energy, Morgantown, WV: (2004)

156. Silva, M.N, S-T Yang : Kinetics and Stability of a Fibrous-bed Bioreactor for Continuous Production of Lactic Acid from Unsupplemented Acid Whey: J. Biotechnol; 41 (1995) 59-70.

157. Mariano, Costa, Angelis et al: Optimisation of a continuous flash fermentation for butanol:Chem. Eng. Res. Des.; 88 (2010) 562 -571.

158. Yakovlev, Shalygin, Matson et al: Separation of diluted butanol-water solutions via vapour phase by organophilic membranes based on high permeable polacetylenes:J. Membr. Sci.; 434 (2013) 99- 105.

159. Durre.P: biobuatnol: an attractive biofuel: biotecnol J; 2 (2007) 1525-1534.

160. Sung, Mai, Yoon et al: Butanol recovery from aqueous solution into ionic liquids by liquid- liquid extraction: Process Biochem; 45 (2010) 1899-1903.

161. Rofler, Blanch, Wilke: In situ recovery of fermentation products: Trends.Biotechnol. ; 2 (1984) 129-136.

162. Cull, Holbrey, Vargas et al: Room temperature ionic liquids as replacements for organic solvents in multiphase bioprocess operations: Biotechnol. Bioeng; 69 (2000) 227-233.

163. Ezeji, Qureshi, Blanschek: Butanol fermentation research: upstream and downstream manipulations: Chem Rec; 4 (2004) 305-314.

164. Zhenji, Li LZ, Xian et al: Problems with the microbial production of butanol; J Ind Microbiol.Biotechnol.: 36 (2009) 1127-1138.

165. Ezeji, Milne, Price et al: Achievements and perspectives to overcome the poor solvent resistance in acetone and butanol-producing microorganisms: Appl. Microbiol.Biotechnol.: 85 (2010) 1697-1712.

166. Roffler, Blanch, Wilke et al: In-situ recovery of butanol during fermentation:Part1. batch extractive fermentation:Bioprocess Eng.: 2(1987) -12.

167. Evans, Wang et al: Enhancement of butanol formation by Clostridium acetobutylicum in the precence of decanol-oleyl alcohol mixed extractants: Appl Environ Microbiol: 54(7) (1988) 1662-1667.

168. Roffler, Blanch, Wilke et al: In situ extractive butanol fermentation of acetone and butanol: Biotechnol Bioeng: 31(1988) 135-143.

169. Cockrem, Meyer, Lightfoot et al: Recovery of n-butanol from dilute solution by extraction: Sep. Sci. Technol.:21(10) (1986) 1059-1074.

170. Najdanovic-Visak V, Esperanca JMSS, Rebelo LPN et al: Phase behaviour of room temperature ionic liquid solutions: an unusally large co-solvent effect in (water plus ethanol): Phys. Chem. Chem. Phys.: 4 (2003) 1701-3.

171. Swatloski RP, Holbrey JD, Rogers RD: Ionic liquids are not always green: hydrolysis of 1-butyl-3-methylimidiazolium hexafluorophosphate: Green. Chem.; 5 (2003) 361-363.

172. Dhamole, Wang, Liu et al: Extractive fermentation with non-ionic surfactants to enhance butanol production: Biomass Bioenrgyy; 40 (2012) 112-119.

173. WangZ, Feng H: Double cloud point of ethylene oxide propylene oxide triblock copolymer in an aqueous solution; colloid surf A; 362(1-3) (2010) 110-116

174. Abdehagh,Tezel, Thinault: Separation techniques in butanol production: challenges and developments: Biomass Bioenrgyy: xxx (2013) 1-25.

175. Ezeji, Qureshi, Blaschek et al: Production of acetone butanol (AB) from liquefied corn starch, a commercial substrate, using Closidrium beijerinckii coupled with product recovery by gas stripping: J. Ind. Microbiol.Biotechnol.: 34(2007) 771-777.

176. Qureshi, Blaschek et al: Butanol recovery from model solution/ fermentation broth by pervapouration: evaluation of membrane performance: Biomass Bioenrgy: 17(1999) 175-184.

177. Lipnizki, Hausmanns, Laufenberg et al: Review: use of pervapouration-bioreactor hybrid processes in biotechnology: Chem. Eng. Technol.: 23(2000) 569-577.

178. Peters,, Poeth, Benes et al: Ceramic-supported thin PVA pervapouration membranes combining the flux-selectivity paradigm: J. Membr. Sci.: 276(2006) 42-50.

179. Qureshi, Meagher, Huang et al: Acetone butanol ethanol (ABE) recovery by pervapouration using silicalite-silicone

composite membrane from fed-batch reactor of Clostridium acetobutylicum: J. Membr. Sci.: 187(2001) 93-102.

180. Fadeev, Meagher, Kelley et al: Fouling of poly(-1-(trimethylsilyl)-1-propyne) membranes in pervapourative recovery of butanol from aqueous solutions and ABE fermentation broth: J. Membr. Sci.: 173(2000) 133-144.

181. Wei, Zhu et al: Acetone-butanol-ethanol (ABE) fermentation using Clostridium acetobutylicum XY16 and in situ recovery by PDMS/ceramic composite membrane: Bioprocess Biosyst. Eng.: 35(2012) 1057-1065.

182. Li G, Wie, Wu H et al:Pervapouration performance of PDMS/ceramic composite: 373(2011) 121-129.

183. Jonquieres, Fane et al: Filled and unfilled composite GFT PDMS membranes for the recovery of butanols from dilute aqueous solutions:influence of alcohol polarity: J. Membr. Sci.: 125(1997) 245-255.

184. Qureshi, Meagher, Hutkins et al: Recovery of butanol from model solutions and fermentation broth using a silicaliate/silicone membrane: J. Membr. Sci.: 158(1999) 115-125.

185. Beltran AB, Nisola GM, Choi SS, Kim Y, Chung WJ. Surface functionalized silica nanoparticles as fillers in polydimethylsiloxane membrane for the pervaporative recovery of 1-butanol from aqueous solution. J. Chem. Technol. Biotechnol. (2013).

186. Liu, Zhao, Jin et al: Hydrophobic-ZIF-71 filled PEBA mixed matrix membranes for recovery of biobutanol via pervapouration: J. Membr. Sci.: (2013).

187. Qureshi N., Maddox: continuous production of acetone- butanol-ethanol using immobilized cells of clostridiumacetobutilicum and integration with product removal by liq-liq extraction: J.Ferment. bioeng: 80(2) (1995) 185-189.

188. Liu F, liu L, Feng: separation of acetone- butanol –ethanol (ABE) from dilute aqueous solution by pervaporation: Sep. Purifi. Technol.: 42 (2005) 273-282.

190. Groot, Van der lans, Luyben: Batch and Continuous butanol fermentations with free cells: Integration with product recovery by gas- stripping: Appl. Microbiol.Biotechnol.: 32 (1989) 305-308.

192. Ezeji, Qureshi, Blaschek: Industrially relevant fermentation: In handbook on clostridia: Taylor and Francis group: (2005) 797-812.

193. Ezeji, Qureshi, Blaschek, Karcher: Butanol production from corn: In alcoholic fuels: fuels for today and tomorrow: Taylor and Francis: (2006) 99-122.

194. Ezeji, Qureshi, Blaschek: Process for continuous solvent production: US Provisional Patent: (2005) no.60/ 504,280.

195. Qureshi N, Ezeji TC: Butanol, 'a superior biofuel' production from agricultural residues (renewable biomass): recent progress in technology: Biofuels.Bioprod. Bioref.;2 (2008)319-330.

196. Park CH, Okos MR, Wankat PC. Acetone-butanol-ethanol (ABE) fermentation and simultaneous separation in a trickle bed reactor. Biotechnol.Prog. (1991);7:185-94.

197. Qureshi N, Maddox IS. Integration of continuous production and recovery of solvents from whey permeate: use of immobilized cells of Clostridium acetobutylicum in a fluidized bed reactor coupled with gas stripping. Bioprocess Eng. (1991);6:63-9.

198. Ezeji, Karcher, Qureshi, Blaschek: Improving performance of a gas stripping-based recovery system to remove butanol from Clostridium beijerinckii fermentation:Bioprocess Biosyst Eng: 27(2005) 207-214.

199. Ezeji, Qureshi, Blaschek: Acetone butanol ethanol (ABE) production from concentrated substrate: reduction in inhibition by fed-batch technique and product inhibition by gas strippinig: Appl. Microbial.Biotechnol.: 63 (2004) 653-658.

200. Ezeji TC, Qureshi N, Blaschek HP: Microbial production of a biofuel (acetone - butanol-ethanol) in a continuous bioreactor: impact of bleed and simultaneous product removal. Bioprocess. Biosyst. Eng.;36(2013) 109-16.

201. Ezeji TC, Qureshi N, Blaschek HP: Production of acetone, butanol and ethanol by Clostridium beijerinckii BA101 and in situ recovery by gas stripping: World J. Microbiol. Biotechnol.;19(2003) 595-603.

202. Mollah AH, Stuckey DC: Feasibility of in situ gas stripping for continuous acetone-butanol fermentation by Clostridium acetobutylicum: Enzyme.Microb. Technol.;15(1993)200-7.

203. Park CH, Okos MR, Wankat PC: Acetone-butanol-ethanol (ABE) fermentation and simultaneous separation in a trickle bed reactor: Biotechnol. Prog.;7(1991)185-94.

204. Ennis BBM, Marshall CT, Maddox IS, Paterson AHH: Continuous product recovery by in-situ gas stripping/ condensation during solvent production from whey permeate using Clostridium acetobutylicum. Biotechnol.Lett.;8(10) (1986)725-30

205. Qureshi, Maddox: Reducton in butanol inhibition by perstraction: Utilization of concentrated lactose/ whey permeate by clostridium acetobutilicum to enhance butanol fermentation economics: J Eur. Fed. Chem Eng:Food Bioprod Process. Part C 83 (2005) 43-52.

206. Groot WJ, Soedjak HS, Donck PB et al: Butanol recovery from fermentations by liquid-liquid extraction and membrane solvent extraction. Bioprocess Eng.;5(1990)203-16.

207. Jeon YJ, Lee YY: Membrane-assisted extractive butanol Fermentation: Ann N Y Acad Sci;506(1987) 536-42.

208. Jeon YJ, Lee YY.In situ product separation in butanol fermentation by membrane-assisted extraction.Enzyme.Microb. Technol. (1989);11:575-82.

209. Shukla R, Kang W, Sirkar KK: Acetone-butanol-ethanol (ABE) production in a novel hollow fiber fermentor-extractor: Biotechnol.Bioeng.;34(1989)1158-66.

210. Grobben NG, Eggink G, Cuperus FP, Huizing HJ. Production of acetone, butanol and ethanol (ABE) from potato wastes: fermentation with integrated membrane extraction: Appl. Microbiol.Biotechnol.; 39(1993) 494-8.

211. Qureshi N, Maddox IS. Reduction in butanol inhibition by perstraction: utilization of concentrated lactose/whey permeate by Clostridium acetobutylicum to enhance butanol fermentation economics. Food.Bioprod.Process. (2005);83(C1):43-52.

212. Plaza, Merlet et al: separation of butanol from ABE mixtures by sweep gas pervaporation using a supported gelled ionic liquid membrane: Analysis of transport phenomena and selectivity: J. Memb. Sci.: 444(2013) 201-212.

213. Garcia A, Lannotti EL, Fischer JL: Butanol fermentation liquor production and separation by reverse osmosis: Biotechnol. Bioeng.;18(1986)785-91.

214. Mariam B. Sticklen, Dpt of Crop and Soil Sciences, Michigan State University East Lasing, Michigan.

215. Panesar, Pamit et all (2010): Enzyme in food Pocessing fundamentals and potential Applications chapter 10, I C International Publishing house; IISBN 978-9380026336.

216. Walter H. Anneberg (Genetically Modified Bacteria) Anneberfoundation.

217. Yang, Y.T., Bennel, G.N., San, K.Y., (1998) Genetic and Metabolic Enginneering; Electronic: J. Biotechnol.; ISSN 07117-3458.

218. Jiang Y, Xu C, Dong F, Yang Y, Jiang W, Yang S. Disruption of the acetoacetate decarboxylase gene insolvent-producing Clostridium acetobutylicum increases the butanol ratio.: Metabol.Engg. 2009;11: 284–91.

219. Inui M, Suda M, Kimura S, Yasuda K, Suzuki H, Toda H, et al. Expression of Clostridium acetobutylicum butanol synthetic genes in Escherichia coli. Appl. Microbiol. Biotechno.2008;77:1305–16.

220. Nielsen DR, Leonard E, Yoon SH, Tseng HC, Yuan C, Prather KLJ.Engineering alternative butanol production platforms in heterologous bacteria.Metab. Eng. 2009;11:262–73.

221. Atsumi S, Cann AF, Connor MR, Shen CR, Smith KM, Bryniidsen MP, et al. Metab. Eng. of Escherichia coli for 1-butanolproduction:Metab. Eng.2008;10:305–11.

222. Cornillot E, Nair RV, Papoutsakis ET, Soucaille P. The gene for butanol and acetone formation in Clostridium acetobutylicum ATCC 824 reside on a large plasmid whose loss leads to degeneration of the strain: J Bacteriol 1997;179(17):5442–7.

223. Berezina OV, Brandt A, Yarotsky S, Schwarz WH, Zverlov VV. Isolation of a new butanol-producing Clostridium strain: high level of hemicellulosic acidity and structure of solventogenesis genes of a new Clostridium saccharobutylicum isolate: Syst. Appl. Microbiol 2009;32:449–59.

224. Metab.Eng. of Saccharomyces cerevisiae for the production of nbutanol.<http://wwwmicrobialcellfactoriescomcontent/7/1/36> [accessed 08.09.10].

225. Petersen DJ, Welch RW, Walter KA, Mermelstein LD, Papoutsakis ET, Rudolph FB, Bennett GN.Rice University, Department of Biochemistry and Cell Biology, Houston, Texas 77251.

226. Cong Ren, Yang Gu, Yan Wu, Sheng Yang, Weihong Jiang: CcpA, a Pleiotropic Key Regulator in Butanol-producingClostridium acetobutylicum, China

227. (D. P. Weisenborn, F. B. Rudolph, and E. T. Papoutsakis, Appl. Environ. Microbiol. 55:323-329, 1989)

228. Cloning and Expression of Clostridium acetobutylicum ATCC 824 Acetoacetyl-Coenzyme A:Acetate/Butyrate:Coenzyme ATransferase in Escherichia coli.

229. Alsaker, K. V., T. R. Spitzer, and E. T. Papoutsakis. 2004. Transcriptional analysis of spo0A overexpression in Clostridium acetobutylicum and its effect on the cell's response to butanol stress. J. Bacteriol. 186:1959-1971.

230. Dürre, P., and C. Hollergschwandner. 2004. Initiation of endospore formation in Clostridium acetobutylicum. Anaerobe 10:69-74.

231. Harris, L. M., N. E. Welker, and E. T. Papoutsakis.2002. Northern, morphological, and fermentation analysis of spo0A inactivation and overexpression in Clostridium acetobutylicum ATCC 824. J. Bacteriol.184:3586-3597.

232. Ravagnani, A., K. C. B. Jennert, E. Steiner, R. Grunberg, J. R. Jefferies, S. R. Wilkinson, D. I. Young, E. C. Tidswell, D. P. Brown, P. Youngman, J. G. Morris, and M. Young. 2000. spo0A directly controls the switch from acid to solvent production in solvent-forming clostridia. Mol. Microbiol. 37:1172-1185.

233. Tomas, C. A., K. V. Alsaker, H. P. J. Bonarius, W. T. Hendriksen, H. Yang, J. A. Beamish, C. J. Paredes, and E. T. Papoutsakis.2003. DNA array-based transcriptional analysis of asporogenous, nonsolventogenic Clostridium acetobutylicum strains SKO1 and M5. J. Bacteriol.185:4539-4547.

[234] Wu, J. J., M. G. Howard, and P. J. Piggot.1989. Regulation of transcription of the Bacillus subtilis spoIIA locus. J. Bacteriol.171:692-698.

235. Molle, V., M. Fujita, S. T. Jensen, P. Eichenberger, J. E. González-Pastor, J. S. Liu, and R. Losick. 2003. The Spo0A regulon of Bacillus subtilis: Mol. Microbiol. 50:1683-1701.

236. Paredes, C. J., K. V. Alsaker, and E. T. Papoutsakis. 2005. A comparative genomic view of clostridial sporulation and physiology: Nat. Rev. Microbiol. 3:969-978.

237. Alsaker, K. V., and E. T. Papoutsakis. 2005. Transcriptional program of early sporulation and stationary-phase events in Clostridium acetobutylicum: J. Bacteriol. 187:7103-7118.

238. Jones, S., C. Paredes, B. Tracy, N. Cheng, R. Sillers, R. Senger, and E. T. Papoutsakis. 2008. The transcriptional program underlying the physiology of clostridial sporulation. Genome Biol. 9:R114.

239. Anne Haluska: Increasing Fermentative Butanol Production In Clostridium Beijerinckii Using Oxidized Extracellular Electron Shuttling Molecules.

240. Lin YL, Blaschek HP. Butanol production by a butanol-tolerant strain of Clostridium acetobutyricum in extruded corn broth. Appl. Environ. Microbiol 1983;45(3):966–73.

241. Syed Q, Nadeem M, Nelofer R. Enhanced butanol production by mutant strains of Clostridium acetobutylicum in molasses medium. Turk J Biochem 2008;33(1):25–30.

242. Annous BA, Blaschek HP. Isolation and characterization of Clostridium acetobutylicum mutants with enhanced amylolytic activity. Appl. Environ.Microbiol. 1991;57(9):2544–8.

243. Formanek J, Mackie R, Blaschek HP. Enhanced butanol production by Clostridium beijerinkii BA101 grown in semidefined P2 medium containing 6 percent maltodextrin or glucose. Appl. Environ.Microbiol. 1997;63(6):2306–10.

244. Connor MR, Cann AF, Lio JC. 3-Mithyl-1-butanol production in Escherichia coli: random mutagenesis and two-phase fermentation. Appl. Microbiol. Biotechno.2010;86:1155–64.

245. Napoli F, Olivieri G, Marzocchella A, Russo ME, Salatino P. Production of butanol in a continuous packed bed reactor of Clostridium acetobutylicum: Chem. Eng. Trans. 2010;20:193–8.

246. Tashiro Y, Takeda K, Kobaayashi G. Sonomoto: high production of acetone–butanol–ethanol with high cell density culture by cell-recycling andbleeding: J. Biotechnol. 2005;120:197–206.

247. Huang WC, Ramey DE, Yang ST. Continuous production of butanol byClostridium acetobutylicum immobilized in a fibrous bed bioreactor. Appl.Biochem.Biotechnol. 2004;115:887–98.

248. Badr HR, Toledo R, Hamdy MK. Continuous acetone ethanol butanolfermentation by immobilized cells of Clostridium acetobutylicum.Biomass Bioenrgy 2001;20:119–32.

249. Liew ST, Arbakariya A, Rosfarizan M, Raha AR. Production of solvent (acetone–butanol–ethanol) in continuous fermentation by Clostridium saccharobutylicum DSM 13864 using gelatinised Sago starch as a carbon source.Malaysian J. Microbiol. 2005;2(2):42–5.

250. Ezeji TC, Qureshi N, Blaschek HP. Continuous butanol fermentation and feed starch retrogradation: butanol fermentation sustainability using Clostridium beijerinckii BA101: J. Biotechnol. 2005;115:179–87.

251. Ezeji T, Qureshi N, Blaschek HP. Production of acetone–butanol–ethanol (ABE) in a continuous flow bioreactor using degermed corn and Clostridium beijerinckii. Process Biochem. 2007;42: 34–9.

252. Joshua T. Ellis, Neal N. Hengge, et al: Acetone, butanol, and ethanol production from wastewater algae; Bioresour. Technol. 111 (2012) 491–495.

253. S.-H. Han, D.H. Cho, Y.H. Kim , S.-J. Shin: Biobutanol production from 2-year-old willow biomass by acid hydrolysis and acetone-butanol-ethanol fermentation; Energy 61 (2013) 13-17

254. Zhijie Sun, Shijie Liu: Production of n-butanol from concentrated sugar maple hemicellulosic hydrolysate by Clostridia acetobutylicum ATCC824; Biomass Bioenrgy39 (2012) 39-47

255. Ni Y, Sun Z: Recent progress on industrial fermentative production of acetone–butanol–ethanol by Clostridium acetobutylicum in China:Appl. Microbiol. Biotechno.2009, 83:415-423.

256. Cascone R: Biobutanol—a replacement for biocthanol? Chem. Eng. Prog. 2008:S4-S9

257. N. Qureshi a, B.C. Saha , M.A. Cotta , V. Singh : An economic evaluation of biological conversion of wheat straw to butanol: A biofuel: Energy Convers. Manage.65 (2013) 456–462.

258. Groot WJ, Van den Oever CE, Kossen NWF.Pervaporation for simultaneous product recovery in the butanol/ isopropanol batch fermentation.Biotechnol.Lett. 1984;6(11):709-14.

259. Larrayoz MA, Puigjaner L. Study of butanol extraction through pervaporation in acetobutylic fermentation.Biotechnol.Bioeng. 1987;30:692-6

260. Geng Q, Park CH. Pervaporative butanol fermentation by Clostridium acetobutylicum B18.Biotechnol.Bioeng. 1994;43:978-86

261. Meagher MM, Ghureshi N, Hutkins R. Silicalite membrane and method for the selective recovery and concentration of acetone and butanol from model ABE solutions and fermentation broth. US Patent 5755967, 1998 May 26.

262. Huang J, Meagher MM. Pervaporative recovery of n-butanol from aqueous solutions and ABE fermentation broth using thin-film silicalite-filled silicone composite membranes: J. Membr. Sci. 2001;192:231-42.

263. Hecke WV, Hofmann T, De Wever H. Pervaporative recovery of ABE during continuous cultivation: enhancement of performance. Bioresour. Technol. 2013;129:421-9.

264. Groot WJ, Luyben KChAM. Continuous production of butanol from a glucose/xylose mixture with an immobilized cell system coupled to pervaporation. Biotechnol.Lett. 1987;9(12):867-70.

265. Hecke WV, Vandezande P, Claes S, Vangeel S, Beckers H, Diels L, et al. Integrated bioprocess for long-term continuous cultivation of Clostridium acetobutylicum coupled to pervaporation with PDMS composite membranes. Bioresour. Technol. 2012;111:368-77.

266. Li SY, Srivastava R, Parnas RS. Study of in situ 1-Butanol pervaporation from A-B-E fermentation using a PDMS composite membrane: validity of solution-diffusion model for pervaporative A-B-E fermentation. Biotechnol.Prog. 2011;27(1):111-20.

267. Liu G, Hou D, Wei W, Xiangli F, Jin W. Pervaporation separation of butanol-water mixtures using polydimethylsiloxane/ceramic composite membrane. China J. Chem. Eng. 2011;19(1):40-4.

268. Groot WJ, Schoutens GH, Van Beelen PN, Van den Oever CE, Kossen NWF. Increase of substrate conversion by pervaporation in the continuous butanol fermentation. Biotechnol.Lett. 1984;6(12):789-92.

269. Friedl A, Qureshi N, Maddox IS.Continuous acetonebutanol-ethanol (ABE) fermentation using immobilized cells of Clostridium acetobutylicum in a packed bed reactor and integration with product removal by pervaporation.Biotechnol. Bioeng.1991;38:518e27.

270. Qureshi N, Maddox IS, Friedl A. Application of continuous substrate feeding to the ABE fermentation: relief of product inhibition using extraction, perstraction, stripping, and pervaporation. Biotechnol.Prog. 1992;8:382-90.

271. Chen Y, Zhou T, Liu D, Li A, Xu S, Liu Q, et al. Production of butanol from glucose and xylose with immobilized cells of Clostridium acetobutylicum. Biotechnol Bioprocess Eng 2013;18:234-41

272. Nexant; Green Chemicals: Growing number of chemical firms enter bio-butanol space. http://www.icis.com/resources/news/2011/12/05/9513406/green-chemicals-growing-number-of-chemical-firms-enter-bio-butanol-space/.

273. Herman K. Trabish; The Expanding Biofuels Market: http://nenmore.blogspot.in/2013_04_01_archive.html.

4

Biohydrogen – Feedstocks, Enhancement Process and Economics

4.1 INTRODUCTION

Gaseous fuels are hydrocarbons, hydrogen and carbon monoxide mixtures present in gaseous state which forms the basis of potential heat energy or light energy that can be readily disseminated by means of pipes from the origin to the place of consumption. Gaseous fuels are contrasted from liquid fuels and solid fuels, though some gaseous fuels are liquefied for storage or transport. Their nature has advantages over liquid and solid fuels in transporting and the dangers of spillage. Gaseous fuels are classified into two broad classes of fuel gases, based not on their chemical composition, but their source and the way they are produced differs widely: those found naturally and those manufactured from other materials. Manufactured gaseous fuel are those produced via artificial process, usually gasification, these gaseous fuels include: Coal gas, water gas, producer gas, syngas, wood gas, uncompressed hydrogen or compressed hydrogen, biogas and blast furnace gas. Natural gas, composed primarily of methane, became the dominant source of fuel gas, as instead of having to be manufactured in various processes, it could be extracted from deposits in the earth. Natural gas may be combined with hydrogen to form a mixture known as HCNG. This chapter and the next focus on the imminent production of manufactured gaseous fuels biohydrogen and biogas respectively.

4.2 BIOHYDROGEN

Hydrogen has been recognized from the evolution of wood, animal fat, coal, petroleum, to natural gas as one of the most potential and clean fuel for the sheer future, where it acts as a promising and eco-friendly alternative (releases only water vapor and energy on combustion) to conventional fossil fuels which is fast depleting as it is used indiscriminately and is detrimental to the surrounding environment due to the harmful gases released on its combustion. Hydrogen (H_2) has reached the succession in fuel evolution with promising technical and socio-economic levels apart from its environmental benefits (does not contribute to greenhouse emission, acid rain, ozone depletion and energy-linked environmental impacts). Also when studied, hydrogen had the highest energy content of 142 kJ/g or 61,000 Btu/lb of any known fuel. Today 40% hydrogen is produced from natural gases, 30% from heavy oil and naptha, 18% from coal, 4% from electrolysis and 1% from biomass. Hydrogen is used as either a direct combustion fuel for an internal combustion engine or as fuel for fuel cells. However, the largest hydrogen users are fertilizers and petroleum industries with 50% and 37% respectively. But, due to the utilization of renewable energy resources, sustainability, considering the crisis of energy and environmental distress, and its simple operation at contiguous temperature and atmospheric pressure, biological hydrogen production is receiving climbing attention in the current era [1-9]. Disparate to chemical and electrochemical foils, biological processes of producing hydrogen are catalyzed by microorganisms in an aqueous environment at optimal temperature and atmospheric pressure. The characteristics of these microorganisms widely differ from one another with respect to substrate and process conditions. Additionally, these techniques are much fitting for decentralized energy production in pilot-plant in locations where biomass and wastes are easily available, dodging energy expenditure and transport costs [10, 11]. Biological systems for the production of biohydrogen include photobiological process (direct biophotolysis, indirect biophotolysis), photo-fermentation (light dependent, where small-chain organic acids are used by photosynthetic bacteria)) and dark-fermentation (light independent, uses both facultative and obligate anaerobes). Further the biosystems for the production of biohydrogen are classified widely as:

(1) Photodecomposition of organic compounds by photosynthetic bacteria,

(2) Dark-fermentation of organic compounds and,

(3) Hybrid systems,

- Use of fermentative and photosynthetic bacteria,
- Use of bioelectrochemical-assisted bioreactor [3, 10, 12-15].

4.2.1 Production Methods of Biohydrogen

The catalyzing chemical reaction of hydrogen production is $2H^+ + 2e^- \rightarrow H_2$. Enzymes play the significant role in catalyzing the biohydrogen production. Recent studies of all known enzyme potentials of hydrogen fruition show that they hold complex metallo-clusters as active sites. In this case the three prevalent enzymes involved in this reaction are: nitrogenase, Fe-hydrogenase and Ni-hydrogenase in which Fe-hydrogenase is widely utilized in biophotolysis and nitrogenase in photo-fermentation process [16]. As discussed earlier biohydrogen production methods are classified as photobiological process and dark fermentation process shown in (Figure 1).

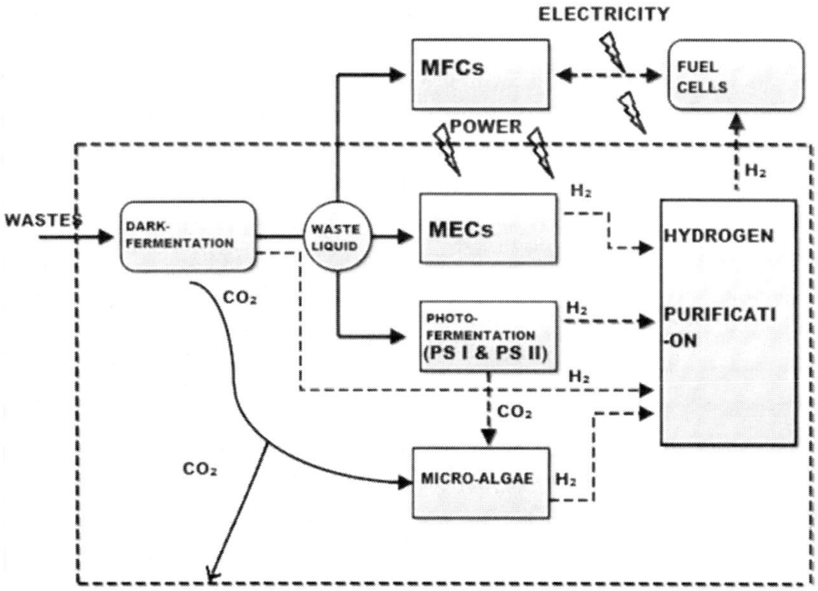

Figure 1 Production pathway of biohydrogen by various methods

4.2.1.1 *Photobiological Process*

4.2.1.1.1 *Direct Biophotolysis* Direct biophotolysis is analogous to the photosynthesis process that transpires in plants and algae. In this process, the readily available substrate and water in the presence of solar energy, light energy absorbed from Photo System I (PSI) and Photo System II (PSII) is converted into oxygen and hydrogen respectively, $2H_2O$ + light energy→ $2H_2 + O_2$ (demonstrated with cell free chloroplast-ferridoxin-hydrogenase system) [16,17]. However, only under special conditions enzyme Fe-hydrogenase plays its activity in hydrogen production which is exceedingly oxygen sensitive. In this case inhibition of the enzyme activity due to oxygen release will slug down the hydrogen evolution process. Photosynthetic energy conversion efficiency can be as high as 10% under low light intensities and under very low partial pressures of O_2 corresponds to the conversion efficiency of 22% of visible light energy into hydrogen but was dissipated as heat or fluroscence whose pathway is shown in (Figure 2). This was found in some green algae (*Chlamydomonas reinhardtii*) that naturally possess an inducible reversible hydrogenase. And for the simultaneous oxygen and hydrogen production the O_2 partial pressure must be below 0.1% which is less than one micromolar O_2 in liquid phase. It is not possible to maintain such low partial pressure in direct biophotolysis, due to the huge amount of diluents gas and power inputs required for transfer.

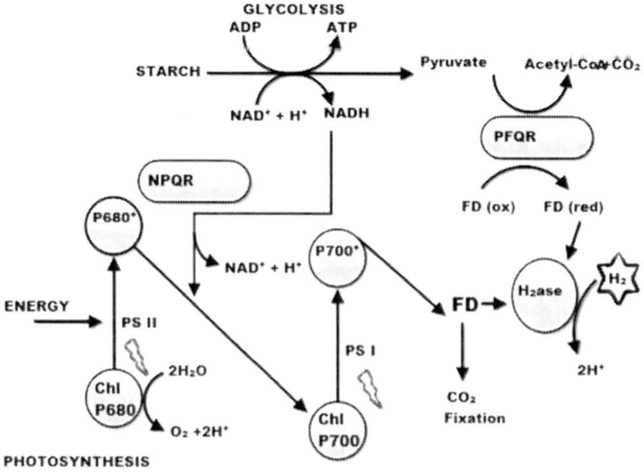

Figure 2 Photosynthetic pathway of biohydrogen evolution.

A direct biophotolysis process must perforce to operate at a partial pressure of near one atmosphere of O_2 which is thousand more than the maximum to be tolerated. Thus, O_2 sensitivity of the hydrogenase enzyme and the supporting reductant generating pathway was the pertaining problem since years [18-20]. Approaches to overcome this problem includes the use of O_2 scavengers, however, the involvement of irreversible and reversible O_2 absorbers reduces the net result. Furthermore an increase of the respiratory activity and effort to increase oxygen tolerance to the key metalloenzymes are underway. Protein engineering and mutagenesis of the enzymes did not produce any significant result on reducing the O_2 sensitivity. Hence, a better found solution for this pertaining problem was the use of heterocystous cyanobacteria. Heterocystous cyanobacteria produces hydrogen in the anerobic conditions by using nitrogenase enzyme complex. This shows similar output to that of the hydrogenase, however the process requires more chemical energy leading it to be economically unfeasible even though ATP generated via photophosphorylation may not always be a limiting factor and due to the increase in the enzyme copy number, hydrogen evolution can reach significant levels [21-32]. In direct biophotolysis, hydrogen production rates of the order of 0.07 m mol/h per litre have been reported [33, 34].

4.2.1.1.2 ***Indirect biophotolysis*** Hydrogen is produced by biochemical reduction of organic compounds during Calvin-cycle, however, indirect biophotolysis have issues of the sensitivity of the hydrogen generation process which are potentially outwitted by separating O_2 and hydrogen evolution whose separation takes place in two stages which is coupled through CO_2 fixation/evolution.

$$12H_2O + 6CO_2 + \text{'light energy'} \rightarrow C_6H_{12}O_6 + 6O_2$$

$$C_6H_{12}O_6 + 12H_2O + \text{'light energy'} \rightarrow 12H_2 + 6CO_2$$

The above concept is elaborated into four conspicuous steps:

(1) Production in open ponds at 10% solar efficiency of a biomass high in storage carbohydrates,

(2) Concentration of the biomass from the ponds in a settling pond,

(3) Anaerobic dark fermentation to yield $4H_2$ and glucose stored in the algal cells, plus 2 acetates, and

(4) A photobioreactor in which the algal cells would convert the two acetates to 8 moles of H_2. At last the algal biomass would be returned to the ponds, to repeat the cycle.

Support systems include the anaerobic digestion (methane fermentation) of any wasted biomass (assumed at 10% for each cycle), an inoculum production system to provide make-up biomass and a gas handling and separation system (to recycle the CO_2 from the H_2 back to the ponds) [35]. Cyanobacteria (also known as blue-green algae, cyanophyceae, or cyanophytes) are a large and diverse group of photoautotrophic microorganisms [36]. Cyanobacteria contain photosynthetic pigments, such as chl *a*, carotenoids, and phycobiliproteins and can perform oxygenic photosynthesis. They are a morphologically diverse group that includes unicellular, filamentous, and colonial species. Within the filamentous cyanobacteria, vegetative cells may develop into structurally modified and functionally specialized cells, such as the akinetes (resting cells) or heterocysts [32]. The nutritional requirements of cyanobacteria are simple: air (N_2 and O_2), water, mineral salts, and light [37]. The rates of hydrogen evolution are as wide-ranging as the species and conditions used in these studies. Rates of hydrogen production by non-nitrogen-fixing cyanobacteria range from 0:02 µmol H_2/mg chl *a*/h (*Synechococcus* PCC 6307) to 0:40 µmol H_2/mg chl *a*/h (*Aphanocapsa montana*) [38]. These rates are very low compared with those of heterocystous cyanobacteria, which range from 0:17 µmol H_2/mg chl a/h (*Nostoc linckia* IAM M-14) to 4:2 µmol H_2/mg chl a/h (*Anabaena variablilis* IAM M-58) [39]. Because of the higher rates of H_2 production by *Anabaena* species and strains, these have been subject to intense study [40]. In indirect biophotolysis, mutant strains of *A. variabilis* have demonstrated hydrogen production rate in the order of 0.355 m mol/h per litre [41].

4.2.1.1.3 *Photo-fermentation* Photosynthetic bacteria produces molecular hydrogen catalyzed by nitrogenase under nitrogenleficient conditions using light energy and reduced organic acids [40]. Calculations on efficiency of conversion of light energy into hydrogen often show values approaching 100%, but these estimates generally gnore the energy content of the organic substrate. These bacteria hemselves are not powerful enough to split water. However, under anaerobic conditions, these bacteria are able to use simple organic acids, ike acetic acid as electron donors. These electrons are transported to he nitrogenase by ferredoxin using energy in the form of ATP. When nitrogen is not present, this nitrogenase enzyme can reduce proton nto hydrogen gas again using extra energy in the form of ATP [42]. The overall reaction of hydrogen production can be given as

$$C_6H_{12}O_6 + 6H_2O + \text{light energy} \rightarrow 12H_2 + 6CO_2$$

Photoheterotrophic bacteria have been investigated for their potential to convert light energy into hydrogen using waste organic compounds as substrate [43-53] and hydrogen production rates of he order of 145–160 mmol/h per litre have been reported [40] True photosynthetic efficiencies are much lower under ideal (low) light conditions. Despite reports of impressive hydrogen production yield, with essentially stoichiometric conversion of substrates to hydrogen, here are several drawbacks to this type of system: (1) the use of the nitrogenase enzyme with its inherent high energy demand, (2) the ow solar energy conversion efficiencies, and (3) the requirement for elaborate anaerobic photobioreactors covering large areas. After extensive R&D on this system in Japan under the RITE (Research on Innovative Technologies for the Earth) program, it was dropped from further study.

4.2.1.2 ***Dark-Fermentation*** In dark-fermentation hydrogen is produced by anaerobic bacteria, grown in the dark (absence of light) on carbohydrate-rich substrates. Fermentation reactions can be operated at mesophilic, thermophilic, extreme thermophilic, or hyperthermophilic temperatures based on the bacteria used. Depending on the fermentation pathway and end-products, hydrogen production differed greatly with respect to bacterial species such as Enterobacter, Bacillus, and Clostridium and various form of carbohydrates (glucose, isomers of hexoses, or polymers in the form of starch or cellulose, preferred as substrates. When acetic acid is the end-product, a theoretical maximum of 4 mole H_2 per mole of glucose is obtained:

$$C_6H_{12}O_6 + 2H_2O \rightarrow 2CH_3COOH + 4H_2 + 2CO_2$$

When butyrate is the end-product, a theoretical maximum of 2 moles H_2 per mole of glucose is obtained:

$$C_6H_{12}O_6 + 2H_2O \rightarrow CH_2CH_2CH_2COOH + 2H_2 + 2CO_2 \quad [54]$$

Hydrogen production by these bacteria is highly dependent on the process parameters such as pH, hydraulic retention time (HRT), and gas partial pressure that affects the metabolic balance. Thus fermentation end-products produced by a bacterium depend on the environmental conditions in which it grows. Reduced fermentation end-products like ethanol, butanol and lactate contain hydrogen that has not been liberated as gas and in order to maximize the yield of H_2, the metabolism of the bacterium must be directed away from alcohols and reduced acids towards volatile fatty acids (VFA). *C. pasteurianum* is a classic H_2 and VFA producer, but its metabolism can be directed away from hydrogen production and towards solvent production by high glucose concentrations (12.5% w/v), by CO (which inhibits Fe-hydrogenase), and by limiting Fe concentrations[55]. The majority of microbial hydrogen production is determined by the anaerobic metabolism of pyruvate. The degradation of pyruvate is catalyzed by one of two enzyme systems:

(1) Pyruvate:formate lyase (PFL)

Pyruvate + CoA → acetyl-CoA + formate
PFL

(2) Pyruvate:ferredoxin (>avodoxin) oxidoreductase (PFOR)

Pyruvate + CoA + 2Fd (ox) → acetyl-CoA + CO_2 + 2Fd (red)
PFORL

Hence, in both these biological systems, the pyruvate that was generated by glycolysis is used in the absence of oxygen to produce acetyl CoA from which ATP is derived. The enteric bacteria derive hydrogen from formate and strict anaerobes derive hydrogen from Fd (red). The overall yield in these metabolisms is relatively low; 1 to 2 hydrogen produced per molecule of pyruvate. Also, in many organisms the actual yield of hydrogen is reduced by hydrogen recycling due to the presence of one or more uptake hydrogenases, which consume a portion of the hydrogen produced. It is unknown to what extent hydrogen production could be increased through metabolic engineering and manipulation of culture conditions [20]. The partial pressure of hydrogen (pH_2) is an extremely important factor for continuous hydrogen synthesis. Hydrogen synthesis pathways are sensitive to H_2 concentrations that lead to end-product inhibition. As hydrogen concentration increases, H_2 synthesis decreases and metabolic pathways shift to production of more reduced substrates such as lactate, ethanol, acetone, butanol, or alanine. But when the temperature increases, the condition that favours the reaction is less affected by H_2 concentration. Continuous H_2 synthesis requires pH_2 of < 50 kPa at 60 C [56], < 20 kPa at 70 C [57], and <2 kPa at 90 C [58]. In conclusion, the rates and efficiency of hydrogen production involving photobiological processes fall far short of even plausible economic feasibility as organic substrates are the ultimate source of hydrogen in photofermentations or indirect biophotolysis processes, therefore it can be argued that it should be simpler and more efficient to extract the hydrogen from such substrates using a dark fermentation process.

4.2.2 Feedstock Available for Biohydrogen Production

Potential renewable feedstocks are brought to being in order to produce renewable hydrogen energy. Apart from water, resources such as agricultural crops and their waste by-products, lignocellulosic products such as wood and wood waste, waste from food processing and aquatic plants and algae, and effluents produced in human habitats are used as feedstocks. Also energy produced from water-containing biomass such as sewage sludge, agricultural and livestock effluents, and animal excreta is recovered mainly by microbial fermentation. Biomass being a potential resource has become a significant source of renewable hydrogen. Thus, countries with large agricultural economies have the potential for significant economic development through incorporation of bioenergy into bio industry.

$$\text{Biomass} + O_2 \rightarrow CO + H_2 + CO_2 + \text{Energy}$$

Studies showed that, for the biohydrogen fermentative processes, carbohydrates are the main source therefore; wastes and biomass rich in sugars and/or complex carbohydrates appear to be suitable feedstock for biohydrogen evolution [59]. The major criteria that have to be met for the selection of substrates suitable for fermentative biohydrogen evolution includes: (1) availability, (2) cost, (3) carbohydrate content, and (4) biodegradability. Hexoses such as glucose, sucrose and lactose are biodegradable and thus preferred as key substrate models for hydrogen production. However, pure carbohydrate sources are expensive raw materials for real-scale hydrogen production, which is viable provided the sources are renewable and economical [10].

4.2.2.1 ***Carbohydrates*** As discussed above, biohydrogen can be produced from a broad spectrum of carbohydrates as they act as a key substrate, but most studies have been limited to the usage of food products such as monosaccharides [60, 61], disaccharides [62, 63], and polysaccharides [64-67]. Simple sugars such as glucose and sucrose are easily degraded and thus a short period of fermentation would suffice but it takes relatively longer fermentation time to degrade complex sugars such as starch and cellulose, thus they are hydrolyzed into monosaccharides using pretreatment before being used for hydrogen production. The maximum hydrogen yield obtained from these pure carbohydrates vary from 2.40 moles H_2/mole of hexose using cellulose [68] to 3.33 moles H_2/mole of hexose from starch [69] and glucose or sucrose [57], indicating that these carbohydrates are suitable as feedstock for dark fermentation. However, hydrogen conversion potential from various substrates is associated with the microbial species as hydrogen yield from similar cultures obtained under same conditions are rather inconsistent [70-72].

4.2.2.2 *Municipal Wastes* The renewable biomasses that have easy accessibility are found to be Municipal Solid Wastes (MSWs). MSW shares the high composition of wastes that are derived from food-related wastes, food industry wastes/wastewater and food residues. The food manufacturing industry produces high-strength organic wastes or wastewater being a significant feedstock for biological hydrogen production. The biotransformation of wastes and wastewater to hydrogen can be considered quite appealing from both the environmental and the economic view [59]. Various studies have been conducted on the recovery of biohydrogen from variety of substrates to investigate its conversion. These substrates include rice and wheat bran [73], bean curd manufacturing waste, noodle manufacturing waste [74], apple processing and potato processing wastewater [75], cheese whey [76], sugar factory wastewater [77], rice winery wastewater [78,79], molasses [80], sugar beet wastewater [81], dehydrated brewery mixture [82], starch manufacturing wastes [83,84] and organic wastewater [85]. Hydrogen yield obtained from food processing wastes or wastewater ranges from 0.68 to 2.70 moles H_2/mole of hexose, which was comparable to those obtained from pure carbohydrates. Of the yield reported, the highest yield was established with fermenting starch residue in a mixed continuous culture, whereas the lowest one was demonstrated in mixed batch culture fed with dehydrated brewery solid waste. Complex solid wastes, such as wastes from kitchen, food processing, mixed wastes, and municipal wastes, have also been tested as feedstock for fermentative hydrogen production [59]. Exclusive of carbohydrates, such wastes usually have quite high contents of protein and fat and thus their biotransformation to hydrogen is comparatively lower than those obtained from carbohydrate-based wastewater. The hydrogen production potential of carbohydrate-based wastes was reported as 20 times higher than that of fat-based and protein-based wastes [86].

Also Organic Fraction of Municipal Solid Wastes (OFMSW) proves to be an important feedstock for hydrogen evolution as they are easily degradable among food waste due to their composition (carbohydrate and protein) [87]. Among the OFMSW, it was reported that hydrogen production potentials of individual carbohydrates varied from 26.3-61.7 (cabbage) and 44.9-70.7 (carrot) to 19.3-96.0 mL/g VS (rice), and carbohydrates produced the most hydrogen compared with proteins (egg: 2.60-7.07 mL/g VS; lean meat: 2.47-7.68 mL/g VS) or lipids (chicken skin: 3.56-10.2 mL/g VS; fat: 4.41-11.14 mL/g VS) [88]. Later, the above result was investigated and a similar observation was reported supporting the fact that the hydrogen production potential of carbohydrate-rich High-Solid Organic Wastes (HSOW) (rice and potato) is approximately 20 times larger than that of fat-rich HSOW (fat meat and chicken skin) and protein-rich HSOW (egg and lean meat) stating that carbohydrates have the maximum potential to produce high yield of hydrogen than lipids and proteins [86].

Furthermore investigations were made to find the addition of sewage sludge to food waste and found that up to 13-19% increase in culture alkalinity enhanced overall hydrogen production potential due to the high protein content in sewage sludge [88]. This is largely attributed to the fact of ammonia production from the proteinaceous substrates such as peptone can avoid culture pH drops by neutralizing fatty acids produced from glucose fermentation [89].

In addition to food waste, wasted sludge generated during wastewater treatment process mainly consists of microorganisms whose organic substance composition differs greatly from carbohydrate-rich substrates such as glucose or starch. Additional studies suggested that proteins are the largest constituent of the waste-activated sludge, accounting for about 32-41% of total dry solid [90], while higher proportion of proteins is present in sewage sludge (0.20 g carbohydrate-COD/g VS and 0.73 g protein-COD/g VS). Lower hydrogen yield from such wastewater sludge would be expected because of the large protein proportion. In addition to alkaline treatment, pre-treatment of the sludge through freezing and thawing and sterilization markedly increased the hydrogen yield from 20.2 to 49.3-71.7 mL H_2/g DS [91]. However, the yield obtained was not comparable to hydrogen evolved from carbohydrate stating protein-rich substrate to be uneconomical. Various types of waste and wastewater feedstocks with respect to mode of operations are discussed in (Table 1).

Table 1 Various types of wastes with respect to type of fermentation process

Industry Category	Type of Fermentation	Type of Feedstocks
Food processing industry	Dark	Food processing wastewater Citric acid wastewater Slaughterhouse waste Sweet sorghum syrup/extract Liquid swine manure
Dairy based industry		Dairy processing waste Dairy waste permeate/waste lactose Cattle wastewater
Alcohol based industry		Brewery wastewater Wine processing wastewater Vinasse
Plant/Agricultural based wastewater		Vegetable based market waste Lignocelluloses derived organic acids Cassava stillage Citrus peelings
Organic based industries		Phenol containing wastewater Organic wastewater
Oil based industries		Glycerine from biodiesel production Palm oil mill effluent (POME) Landfill leachate Probiotic wastewater
General	Photo	Olive mill waste Potato stem peel hydrolysate Tofu wastewater Ground wheat solution Starch fermentation

4.2.2.3 Future Aspects for Feedstocks However, the continuously rising food prices, the sustainability doubts and the energy equation challenges have led to a hostile response against the use of energy crops as feedstock for biofuel generation. This response was the lead to food-vs.-fuel debate, as many country's huge agricultural areas have been turned into feedstock farms for the production of biofuels. The exploitation of crops as feedstock for energy production and biotransformation to biofuels has been escalating in recent years. In general, the sustainability of such processes can be assured only if (a) the crops are produced at low cost, with minimum nutrient and water requirements; (b) they are resistant to environmental stresses, and (c) they are high biomass yielding. Even in the case of non-food crop cultivation, the sustainability issue is under question. As an answer to those issues, the production of second-generation biofuels is proposed that is, biofuels produced by feedstock such as wastes and residues that are not a competition to edible crops [59]. Moreover plants, suitable for hydrogen production via dark fermentation, should also have high sugar and/or carbohydrates content and low lignin content [54]. According to their chemical composition, energy crops used for fermentative hydrogen production can be divided into sugar-based crops, starch-based crops, and lignocellulose-based crops including herbaceous and woody.

4.2.3 Bioreactors in Hydrogen Production

Bioreactor together with operation modes is associated with the high production rates of hydrogen that governs performance of the engineering system and influences (1) the microenvironment in the reactor, (2) hydrodynamic behaviour, (3) wastewater-biocatalyst contact, and (4) survivability of microbial population [92, 93]. Recent advances in the reactor configuration have significantly improved the Hydrogen Yield (HY) and Hydrogen Production Rates (HPR). Broad spectrum of reactor configuration includes suspended growth, biofilm/packed-bed/fixed bed where biofilm/attached-growth systems are generally robust to shock loads and provide resilience and resistance to changes in the process parameters [94, 95], fluidized bed, expanded bed, Up flow Anaerobic Sludge Blanket (UASB), granular sludge, membrane-based systems, immobilized systems, etc., in biohydrogen research. These reactors facilitate improved reaction potential leading to stable and robust system for treating wastewater. Studies made on different supporting materials viz., SBA-15 (mesoporous) and activated carbon (granular; GAC and powder; PAC)] in comparison with suspended growth of culture showed the effect of self-immobilization of enriched acidogenic mixed consortia on hydrogen production [96]. Attached growth showed higher hydrogen yield (up to nine-folds), lower substrate degradation efficiency particularly at higher loading rates. Consequently, different adaptive bioreactors have been developed including Continuously Stirred Tank Reactors (CSTRs), Anaerobic Fluidized-Bed Reactors (AFBRs), Packed-Bed Reactors

PBRs), and UASB reactors for dark fermentation, as well as specific photobioreactors and MECs. Batch, fed-batch, semi batch/continuous, periodic discontinuous batch (sequencing batch operation), and continuous modes of reactor operations have been studied on hydrogen production. Due to fill-draw mode operation, fed-batch operation reduces the accumulation of soluble metabolic intermediates [95, 97-99]. And in batch-mode operations poor biomass retention/cell washout could be reduced to some extent [100]. To maintain stable and robust cultures suitable for treating highly variable wastewater, batch mode operation was coupled with biofilm configuration where the combination and the dual operational advantages of both the systems helps [94-96, 98, 101-104]. Amid the various options of dark-fermentation reactors, CSTRs have been conventionally adopted attributed to their simplicity and ease of operation, but they generally suffer from a low HPR and poor operation stability due to biomass washout [105]. Furthermore, high energy input is usually needed to achieve a vigorous mechanical stirring. Comparatively, the AFBRs, PBRs, and UASBs enable effective biomass retention and allow for operation under higher Organic Loading Rates (OLRs), which lead to better hydrogen production performance [106]. Moreover, several modifications of CSTRs have also been proposed to enhance biomass retention by incorporating into biomass-liquid separation techniques. In the intervening time, efforts have also been devoted to the development of efficient photobioreactors and microbial electrolysis cells (MECs) to ensure efficient hydrogen production. However, it is widely accepted that the reactor configuration and operating mode can pose significant influences on the system biohydrogen production performance by directly affecting the hydrodynamics and bacterial properties. Both volumetric HPR and HY have been considered as important index to evaluate the performance of a biohydrogen production system. And, it is noted that the HY is defined by the amount of hydrogen produced per substrate consumed, which is more dependent upon the microbial properties, substrate type and environmental conditions rather than the reactor configurations. In fact, HYs obtained for dark fermentation with different systems are inconsistent and mostly no greater than those are achieved with CSTRs [107, 108]. Therefore, there is no reason to think that the reactor type would essentially influence the HY in dark-fermentation systems, and a comparison of these reactors here shall mainly be based on their HRPs. However for the phototrophic and MEC hydrogen production processes, the relatively low HPR value of such systems, the HY is usually used for reactor evaluation because of the fewer microbial species and substrates involved as well as. Further, the operating stability is also a critical factor to be considered for the biological hydrogen production systems. Distinct bioreactors for hydrogen production processes have been developed with different degrees of success. Comparisons of performances of various reactors are discussed in (table 2). Reactors including dark-fermentation photobioreactors, and MECs, have their own unique advantages and inherent limitations.

Table 2 Comparison of performances of various types of reactors

Reactor	Substrate	Culture	Maximum HPR (LL-1h-1)	Maximum HY (mol-H_2 mol-1-hexose)	VSS (g/L)	Ref.
Integrated CSTR	Glucose	HPF	0.4	2.8	2.3	[109]
Integrated CSTR	Glucose	HPF	0.32	–	5.8	[110]
Integrated CSTR	Sucrose	HPG	15.4	3.5	35.4	[111]
AFBR	Sucrose	HPG	1.32	0.57	15.0	[112]
AFBR	Glucose	HPG	2.36	1.19	21.5	[61]
AFBR	Glucose	HPG	0.25	2.6	29.3	[113]
PBR	Sucrose	HPG	7.3	3.03	26	[114]
PBR	Glucose	HPG	0.98	1.11	18-24	[115]
UASB	Winery wastewater	HPG	0.16	2.14	9.4-11.7	[78]
UASB	Sucrose	HPG	0.71	3.47	19.4	[116]
UASB	Starch	HPG	1.64	0.94	8.26	[117]
Photobioreactor	DL malic acid	PSB	0.007	6.75	–	[118]
Photobioreactor	Acetate	PSB	0.038	9.45	–	[119]
Photobioreactor	Lactate	PSB	0.021	–	–	[120]
MEC	Acetate	ARB	0.011	8.7	–	[121]
MEC	Glycerol	ARB	0.025	10.8	–	[122]

The application of broad spectrum immobilization and mixing techniques adds up to the complexity and versatility of reactors. A systematic comparison of different biohydrogen producing reactor systems is given in (Table 2). Commonly, considerable higher HPRs (ranging from 0.11 to 15.4 $LL^{-1}h^{-1}$) were found in dark-fermentation reactors compared to the biophotoreactors (0.007-0.038 $LL^{-1}h^{-1}$) and MECs (0.0006-0.26 $LL^{-1}h^{-1}$) [123]. On the contrary, much higher HYs are commonly seen in the photobioreactors (6.75-9.45 mol-H2 mol^{-1}-hexose) and MECs (4.8-16.56 mol-H2 mol^{-1}-hexose). Thus in conclusion, the dark-fermentation reactors may be a superior choice for degradation of readily degradable sugar-based substrate, while the photobioreactor and MEC systems are more favourable for bioconversion. Among the various dark-fermentation reactors, the granule-based reactors display distinctly better performances than suspended-growth HPF systems in terms of biomass retention and HPR. Although the highest HPR (15.0 $LL^{-1}h^{-1}$) reported so far was achieved in a CSTR with silicone-immobilized granules [111], the other granule-based configurations, especially UASB reactors, demonstrate higher stability and economical efficiency in long-term operation. Furthermore, the biohydrogen production processes can be influenced by a number of other factors such as operation conditions, substrate type, and microbial species. Therefore, it is not easy to draw a conclusion with regard to what configuration is better, even under a specific set of conditions notwithstanding the differentiated performances of different reactor systems.

4.2.4 Factors Influencing Biohydrogen Production

4.2.4.1 ***Redox Condition*** Based on the organisms and their growth conditions, changes in external pH can bring about alterations in several primary physiological parameters, including the internal pH, concentration of other ions, membrane potential, and proton-motive force. Redox condition is an important factor for fermentative H_2 production process where the activity of AB is considered to be crucial and rate limiting [98, 124-126]. This also influences the efficiency of substrate metabolism, protein and storage materials synthesis, and release of metabolic byproducts. Under acidic conditions, pyruvate converts into VFA together with H_2 by AB. Neutral operation leads to the generation of CH_4 and CO_2 by MB, while basic operation leads to solventogenesis. Repression of MB indirectly promotes H_2 production within the system [102, 127, and 128]. AB functions well below pH 6, while for MB optimum range is between 6.0 and 7.5 [125, 126]. To avoid both methanogenesis and solventogenesis, the pH range of 5.5-6.0 is reported to be ideal [102, 125, 126 and 129]. Good hydrogen yield was observed when the pH was maintained in and around 6 compared to near-neutral pH [102, 126 and 130]. However, highly acidic pH (<4.5) is detrimental to H_2 production as it inactivates AB [126,128]. Both dehydrogenase and hydrogenase were the other important enzymes that maintain H^+ concentration in the cell through H^+ shuttling leading to increased chances of H_2 generation in acidophilic and alkaline operations [129]. However, hydrogenase enzyme activity gets inhibited by maintaining low or high pH beyond optimum range [124, 126].

4.2.4.2 ***Temperature*** Temperature is another important factor that affects the hydrogen production, metabolites distribution, substrate degradation, and bacterial growth. Hydrogen production was studied under ambient (15-27 °C), mesophilic (30-45 °C), and moderate thermophilic (50-60 °C) temperatures, and a few studies were reported even under extreme thermophilic conditions (over 60 °C) [126, 131]. The optimal temperature for H_2 production widely varies based on the nature of the biocatalyst and the type of substrate used. For mixed microflora diverse optimum temperatures were reported [132], whereas for pure cultures, the optimal temperatures were in the range of 37-45°C. Both mesophilic and thermophilic temperatures were observed to be optimal for fermentative hydrogen production processes. However thermophilic conditions were reportedly advantageous due to its thermodynamics [132] that gives higher reaction rate with better process performance and decrease problems with contaminating H_2-consuming microorganisms. Although higher temperatures favour reaction kinetics, rapid changes in the system pH may inhibit H_2-producing bacteria [126]. The changes in soluble metabolite composition were also observed with change in the operating temperature which resulted in metabolic shifts to dominant bacterial function at that particular temperature [133]. Therefore, temperature control might not be a feasible option for process control at all instances [126].

4.2.4.3 *Fermentation Time* Fermentation time or hydraulic retention time (HRT) is one of the important process control variables which significantly influences the H_2 generation by way of restricting methanogenic activity. Under batch mode using dairy, chemical, and distillery wastewaters as substrate, maximum H_2 yield was observed between 0 and 14 h in various experiments performed [97, 99, 101, 134]. Metabolic shift from the acidogenic to the methanogenic process which is unfavourable for H_2 production was observed under induced and longer HRT. Therefore, it was reported that use of short HRT helps to restrict the Methanogenic Bacteria (MB) growth [54]. Optimum HRT between 8.0 and 14 h was reported for effective H_2 production [54]. However, HRT mostly depends on the nature and composition of the substrate, nature of biocatalyst, loading rate, and redox condition [126].

4.2.4.4 *Reactor Configuration and Operation Modes* Bioreactor together with operation modes is associated with the high production rates of hydrogen and is responsible for performance of the engineering system and influences the following:

(1) the microenvironment in the reactor,
(2) hydrodynamic behaviour,
(3) wastewater-biocatalyst contact, and
(4) survivability of microbial population [92, 93].

4.2.5 Challenges and Future Prospects

The main challenge in fermentative production of hydrogen is that less than 15% of the energy from the organic source can typically be obtained in the form of hydrogen [135]. Therefore, it is not surprising that major efforts are directed to substantially increase the hydrogen yield. The U.S. Department of Energy (2007) program goal for fermentation technology is to realize yield of 4 and 6 mol H_2/mol of glucose by 2013 and 2018, respectively, as well as to achieve 3 and 6 months of continuous operation for the same years. Moreover, through coupled processes it is found that, more hydrogen or energy per mole of substrate can be achieved in second stage. Some integrated strategies are now under development, such as the two-step fermentation process (acidogenic and photobiological or acidogenic and methanogenic processes) or the use of modified microbial fuel cells whose efficiency can be improved using pre-treatment discussed in (table 3) [136-138] Further enhancement in hydrogen yield may be possible by using (1) a suitable microbial strain, (2) process modification, (3) efficient bioreactor design (optimization and scale up), and (4) genetic and molecular engineering techniques, to redirect the metabolic pathway. Although, the acceptability of genetically modified microorganisms is a challenge, because of the possible risk of horizontal transference of genetic material, this could be ruled out by chromosomal integration and the elimination of plasmids containing antibiotic markers with the available molecular tools [139].

Table 3 Various pretreatment methods to improve the efficiency of bacterial strain

Conditions	Pretreatment Method	Function
Extreme acidic microenvironment (pH < 4)	Acid	Enrich spore-forming bacteria by suppressing the MB specifically.
Alkaline microenvironment (pH > 9)	Alkaline	Inhibition of MB (Non-specific)
In the presence of higher substrate concentration	Load shock	Leads to the accumulation of high organic acids which prevents growth of MB
Extreme temperature (>80 °C)	Heat shock	Suppress non-spore-forming bacteria and helps in harvesting spore-forming bacteria.
2-bromoethanesulfonic acid (BESA,100 mmol)	Chemical	Inhibits coenzyme-M reductase complex (chief component for methanogenesis)
Iodopropane		Iodopropane is a corrinoid antagonist that prevents the functioning of the B12 enzymes as a methyl group carrier
Acetylene In the presence of oxygen/air (0.5 mg/l)	Oxygen shock	Nonspecific inhibition of MB MB are obligative anaerobes, exposure to oxygen lowers their adenylate charge and leads to death
Combination of two or more pretreatment methods	Combined	To achieve more specific enrichment of AB

New pathways must be discovered to directly take full advantage of the 12 mol H_2 available in a mole of hexose. Unstable production is another challenge in hydrogen production and is possibly attributed to the metabolic shift of hydrogen-producing bacteria, and this could be minimized. A major technical prerequisite for stable and efficient hydrogen fermentations is the maintenance of low hydrogen partial pressures through continuous removal of hydrogen from the fermentation broth as discussed earlier. Hydrogen is reported currently more expensive than other fuels. Biohydrogen production employing renewable biomass may be a potential answer to overcome some of the economic constraints to fulfill many of our energy needs, hence there is a hunt for alternatives such as agricultural wastes possibly molasses and sugarcane baggase that can be used as they possess high carbohydrate content in them which in turn helps in the reducing the cost of production.

4.2.6 ECONOMICS OF BIOHYDROGEN PRODUCTION

The world hydrogen production is not monitored, but estimated at 45 million tons (500 million cubic metres) per year. And overall, total hydrogen production costs were estimated at USD 10/GJ. The capital costs were almost 90% of total costs at a 25% annual capital charge. The economic analyses discussed above indicates that photobiohydrogen could be produced at a cost between USD 10/GJ and USD 20/GJ [140]. This is a reasonable maximal cost target for renewable hydrogen fuel [141]. An indirect micro algal biophotolysis system consists of open ponds (140 ha) and photobioreactors (14 ha) where the plant was assumed to generate 1.2 million GJ/yr at 90% plant capacity, with estimated total capital costs for the system estimated at USD 43 million and annual operating costs at USD 12 million. And the algal ponds estimated at a cost of USD 6 per m^2, while the photobioreactors with assumed costs of USD 100 per m^2 were the major capital and operating cost factors [142]. The analysis to determine the proposed photobioreactor design could meet the cost requirements for hydrogen production through single-stage biophotolysis. An initial cost for a large-scale (>100 ha) single-stage algal or cyanobacterial biophotolysis process in a near-horizontal tubular reactor system was analyzed [143]. The tubular photobioreactor offers superior features for biohydrogen production due to the internal gas exchange and the effective wate spray cooling. The costs of the tubular photobioreactor were estimatec at USD 50 per m^2 based on 10% solar energy conversion efficiency. Th capital fixed costs were estimated at 80% of total costs, with the tubula material for the photobioreactor as the major cost. The hydroger production costs were estimated at USD 15/GJ, which are comparabl to the costs projected for hydrogen produced in a two-stage proces from biomass residues projected at Euro 19/GJ [143]. It is clear from th economic analyses that the development of low-cost photobioreactor and the optimization of photosynthetic efficiency are the major R&I challenges. And it is reported that the sales of hydrogen has climbec by 6% annually in the last five years which is presumed that there wil sure be a rise in market globally in 2020.

4.3 CONCLUSION

In conclusion gaseous fuels are the basis of heat energy which is usec directly as fuels for various engines. Both biohydrogen and biogas car be produced from easily available sources such as MSWs, industria effluents, agricultural wastes and domestic wastes respectively. Alsc these fuel gases are versatile and cleaner energy source with non polluting characteristics purging the tribulations caused by conventiona fossil fuels. Various technologies have been studied to find the easie and cheaper way to produce these fuels gases. In case of biohydrogen production of this versatile gas is enhanced by improving the reactor efficiency, and use of suitable genetic, metabolic manipulation anc process modifications.

4.4 REFERENCES

1. Pallavi Sinha, Anjana Pandey. An evaluative report anc challenges for fermentative biohydrogen production. Int J Hydrogen Energy 2011:36:7460-7478.

2. Das D. Advances in biohydrogen production process: an approach towards commercialization. Int J Hydrogen Energy 2009;34:7349-57.

3. Debabrata Das, T. Nejat Veziroglu Advances in biological hydrogen production processes Int J Hydrogen Energy 2008:33: 6046–6057.

4. Suzuki Y. On hydrogen as fuel gas. Int J Hydrogen Energy 1982;7:227–30.

5. Nath K, Das D. Hydrogen from biomass. Curr Sci 2003;85:265–71.

6. Shireen Meher Kotay, Debabrata Das. Biohydrogen as a renewable energy resource-Prospects and potentials. Int J Hydrogen Energy 33 (2008) 258 – 263.

7. Boyles D. Bioenergy technology-thermodynamics and costs. NewYork: Wiley; 1984.

8. Benemann JR. Hydrogen biotechnology: progress and prospects. Nat Biotechnol1996;14:1101–3.

9. Momirlan M, Veziroglu TN. Current status of hydrogen energy. Renew Sustain Energy Rev 2002;6:141–79.

10. Das D, Veziroglu TN. Hydrogen production by biological processes: a survey of literature. Int J Hydrogen Energy 2001;26:13–28.

11. Nandi R, Sengupta S. Microbial production of hydrogen: an overview. Crit Rev Microbiol 1998;24:61–84.

12. GaGron H, Rubin J. Fermentative and photochemical production of hydrogen in algae.J Gen Physiol 1942;26: 219–40.

13. Pierik AJ, Roseboom W, Happe RP, Bagley KA, Albracht SPJ. Carbon monoxide and cyanide as intrinsic ligands to iron in the active site of [NiFe]-hydrogenase NiFe(CN)2CO, Biology's way to activate H2. J Biol Chem 1999;274:3331–7.

14. Rousset M, Montet Y, Guigliarelli B, Forget N, Asso M, Bertrand P, Fonteccilla- Camps JC, Hatchikian EC. [3Fe–4S] to [4Fe–4S] cluster conversion in Desulfovibrio fructosovorans [NiFe] hydrogenase by site-directed mutagenesis. Proc Natl Acad Sci USA 1998;95:1625–11630.

15. Basak N, Das D. The prospect of purple non-sulfur photosyntheti bacteria for hydrogen production: the present state of art. Worl J Microbiol Biotechnol 2007;98:1183–90.

16. Hallenbeck PC, Benemann JR. Biological hydrogen production fundamentals and limiting processes. Int J Hydrogen Energ 2002;27:1185–93.

17. Benemann JR, Berenson JA, Kaplan NO, Kamen MD. Hydroger evolution by chloroplast-ferredoxin-hydrogenase system. Pro Natl Acad Sci USA 1973;70:2317–20.

18. Greenbaum E. Energetic e2ciency of hydrogen photoevolutior by algal water splitting. Biophys J 1988;54:365–8.

19. Spruit CJP. Simultaneous photoproduction of hydrogen anc oxygen by Chlorella. Meded Landbouwhogesch Wageninger 1958;58:1–17.

20. Patrick C. Hallenbecka, John R. Benemannb Biologica hydrogen production; fundamentals and limiting processes In J of Hydrogen Energy 27 (2002) 1185 – 1193

21. Rosenkranz A, Krasna AJ. Stimulation of hydroger photoproduction in algae by removal of oxygen by reagents tha combine reversibly with oxygen. Biotech Bioeng 1984;26:1334-42.

22. Melis A, Zhang L, Foestier M, Ghirardi ML, Seibert M Sustained photobiological hydrogen gas production upor reversible inactivation of oxygen evolution in the green alga Chlamydomonas reinhardtii. Plant Physiol 2000;122:127–33.

23. McTavish H, Sayavedra-Soto LA, Arp DJ. Substitution o Azotobacter vinelandii hydrogenase small subunit cysteines by serines can create insensitivity to inhibition by O2 and preferentially damages H2 oxidation over H2 evolution. J Bacteriol 1995;177:3960–4.

24. Ghirardi ML, Togasaki RK, Seibert M. Oxygen sensitivity of algal H2-production. Appl Biochem Biotechnol 1997;63–65:141–51.

25. Seibert M, Flynn TY, Ghirardi ML. Strategies for improving oxygen tolerance of algal hydrogen production. In: Miyake J, Matsunaga T, San Pietro A, editors. Biohydrogen II. Amsterdam: Pergamon, Elsevier Science, 2001. p. 67–77.

26. Benemann JR, Weare NM. Hydrogen evolution by nitrogen-1xing Anabaena cylindrica cultures. Science 1974;184:174–5.

27. Weissman JC, Benemann JR. Hydrogen production by nitrogen-starved cultures of Anabaena cylindrica. Appl Environ Microbiol 1978;33:123–31.

28. Hallenbeck PC, Kochian LV, Weissman JC, Benemann JR. Solar energy conversion with hydrogen producing cultures of the blue-green alga, Anabaena cylindrica. Biotech Bioeng Symp 1978;8:283–98.

29. Miyamoto K, Hallenbeck PC, Benemann JR. EGects of nitrogen supply on hydrogen production by cultures of Anabaena cylindrica. Biotech Bioeng 1979;21:1855–60.

30. Miyamoto K, Hallenbeck PC, Benemann JR. Solar energy conversion by nitrogen-limited cultures of Anabaena cylindrica. J Ferment Technol 1979;57:287–93.

31. Benemann JR, Miyamoto K, Hallenbeck PC. Bioengineering aspects of biophotolysis. Enzyme Microb Technol 1980;2:103–11.

32. Tamagnini P, Axelsson R, Lindberg P, Oxelfelt F, Wunschiers R, Lindblad P. Hydrogenases and hydrogen metabolism of cyanobacteria. Microbiol Mol Biol Rev 2002;66:1–20.

33. Kosourov S, Tsygankov A, Seibert M, Ghirardi ML. Sustained hydrogen photoproduction by *Chlamydomonas reinhardtii*: effects of culture parameters. Biotechnol Bioeng 2002;78:731–40.

34. Melis A, Zhang L, Forestier M, Ghirardi ML, Seibert M. Sustained photobiological hydrogen gas production upon reversible inactivation of oxygen evolution in the green alga *Chlamydomonas reinhardtii*. Plant Physiol 2000;122:127–35.

35. Benemann JR. Processes analysis and economics of biophotolysis of water. A preliminary assessment. Report to the International Energy Agency Hydrogen Program, Subtask B, Annex 10 Photoproduction of Hydrogen, IEA/H2/10/TR2-98, 1998.

36. Schopf JW. The fossil record. Tracing the roots of the cyanobacterial lineage. In: Whitton BA, PottsM, editors. The ecology of cyanobacteria. Dordrecht, The Netherlands: Kluwer Academic Publishers, 2000.

37. Hansel A, Lindblad P. Mini-review: toward optimization of cyanobacteria as biotechnologically relevant producers of molecular hydrogen, a clean energy source. Appl Environ Microbiol 1998;50:153–60.

38. Howarth DC, Codd GA. The uptake and production of molecular hydrogen by unicellular cyanobacteria. J Gen Microbiol 1985;131:1561–9.

39. Masukawa H, Nakamura K, Mochimaru M, Sakurai H Photohydrogen production and nitrogenase activity in some heterocystous cyanobacteria. BioHydrogen 2001;II:63–6.

40. Levin DB, Pitt L, Love M. Biohydrogen production: prospects and limitations to practical application. Int J Hydrogen Energy 2004;29: 173–85.

41. Sveshnikov DA, Sveshnikov NV, Rao KK, Hall DO. Hydrogen metabolism of *Anabaena variabilis* in continuous cultures and under nutritional stress. FEBS Lett 1997;147:297–301.

42. Akkerman I, Janssen M, Rocha J, Wijffels RH. Photobiological hydrogen production: photochemical efficiency and bioreactor design. Int J Hydrogen Energy 2002;27:1195–208.

43. Arik T, Gunduz U, Yucel M, Turker L, Sediroglu V, Eroglu I. Photoproduction of hydrogen by Rhodobacter sphaeroides OU001. In: Viroglu TN, Winter CJ, Baselt JP, Kreysa G, editors. Proceedings of the 11th World Hydrogen Energy Conference, Stuttgart, Germany. Frankfurt: Scon & Wetzel GmbH, 1996. p. 2417–26.

44. Bolton JR. Solar photoproduction of hydrogen. Sol Energy 1996;57:37–50.

45. Fedorov AS, Tsygankov AA, Rao KK, Hall DO. Hydrogen photoproduction by Rhodobacter sphaeroides immobilised on polyurethane foam. Biotechnol Lett 1998;20:1007–9.

46. Tsygankov AA, Hirata Y, Miyake M, Asada Y, Miyake J. Photobioreactor with photosynthetic bacteria immobilized on porousglas sfor hydrogen photoproduction. J Ferment Bioeng 1994;77:575–8.

47. Tsygankov AA, Fedorov AS, Laurinavichene TV, Gogotov IN, Rao KK, Hall DO. Actual and potential ratesof hydrogen photoproduction by continuousculture of the purple non-sulphur bacteria Rhodobacter capsulatus. Appl Microbiol Biotechnol 1998;49:102–7.

48. ZurrerH, BachofenR. Hydrogenproductionbythephotosynthetic bacterium, Rhodospirillum rubrum. Appl Environ Microbiol 1979;37:789–93.

49. Fascetti E, Todini O. Rhodobacter sphaeroides RV cultivation and hydrogen production in a one- and two-stage chemostat. Appl Microbiol Biotechnol 1995;22:300–5.

50. Zurrer H, Bachofen R. Aspects of growth and hydrogen production of the photosynthetic bacterium Rhodospirillum rubrum in continuousculture. Biomass1982;2:165 –74.

51. Francou N, VignaisPM. Hydrogen production by Rhodopseudomonas capsulata cellsentrapped in carrageenan beads. Biotechnol Lett 1984;6:639–44.

52. Vincenzini M, Materassi R, Sili C, Florenzano G. Hydrogen production by immobilized Cells III. Prolonged and stable H_2 photoevolution by Rhodopseudomonas palaustris in light-dark-cycles. Int J Hydrogen Energy 1986;11:623–6.

53. Tsygankov AA, Fedorov AS, Talipova IV, Laurinavichene TV, Miyake J, Gogotov IN, Rao KK, Hall DO. Application of immobilized phototrophic bacteria for simultaneous waste

water treatment and hydrogen photoproduction. Appl Biochem Microbiol 1998;34:1–5 (in Russian).

54. Hawkes FR, Dinsdale R, Hawkes DL, Hussy I. Sustainable fermentative biohydrogen: challenges for process optimization. Int J Hydrogen Energy 2002;27:1339–47.

55. Dabrock B, Bahl H, Gottschalk G. Paramenters affecting solvent production by Clostridium pasteurianum. Appl Environ Microbiol 1992;58:1233–9.

56. Lee MJ, Zinder SH. Hydrogen partial pressures in a thermophilic acetate-oxidizing methanogenic co-culture. Appl Environ Microbiol 1988;54:1457–61.

57. van Niel EWJ, Claassen PAM, Stams AJM. Substrate and product inhibition of hydrogen production by the extreme thermophile Caldicellulosiruptor saccharolyticus. Biotechnol Bioeng 2002;81:255–62.

58. AdamsMWW. The metabolism of hydrogen by extremely thermophilic sulphur-dependent bacteria. FEMS Microbiol Rev 1990;75:219–38.

59. Ntaikou, I., Antonopoulou, G., Lyberatos, G., 2010. Biohydrogen production from biomass and wastes via dark fermentation: a review. Waste Biomass Valorization 1 (1), 21–39.

60. Horiuchi, J., Kikuchi, S., Kobayashi, M., Kanno, T., Shimizu, T., 2001. Modeling of pH response in continuous anaerobic acidogenesis by an artificial neural network. Biochem. Eng. J. 9 199–204.

61. Zhang, Z.P., Tay, J.H., Show, K.Y., Yan, R., Liang, D.T., Lee, D.J. et al., 2007. Biohydrogen production in a granular activated carbon anaerobic fluidized bed reactor. Int. J. Hydrogen Energy 32, 185–191.

62. Lee, K.S., Lo, Y.S., Lo, Y.C., Lin, P.J., Chang, J.S., 2003. H-2 production with anaerobic sludge using activated-carbon supported packed-bed bioreactors. Biotechnol. Lett. 25, 133-138.

63. Yu, H.Q., Mu, Y., 2006. Biological hydrogen production in a UASB reactor with granules. II: reactor performance in 3-year operation. Biotechnol. Bioeng. 94, 988–995.

64. Czernik, S., French, R., Feik, C., Chornet, E., 1999. Production of hydrogen from hemicellulose-rich fractions. Abstr. Pap. Am. Chem. Soc. 217, U808–U809

65. Van Ginkel, S.W., Oh, S.E., Logan, B.E., 2005. Biohydrogen gas production from food processing and domestic wastewaters. Int. J. Hydrogen Energy 30, 1535–1542.

66. Yokoi, H., Tokushige, T., Hirose, J., Hayashi, S., Takasaki, Y., 1998. H2 production from starch by a mixed culture of Clostridium butyricum and Enterobacter aerogenes. Biotechnol. Lett. 20, 143–147.

67. Zhang, T., Liu, H., Fang, H.H.P., 2003. Biohydrogen production from starch in wastewater under thermophilic condition. J. Environ. Manage. 69, 149–156.

68. Ueno, Y., Kawai, T., Sato, S., Otsuka, S., Morimoto, M., 1995. Biological production of hydrogen from cellulose by natural anaerobic microflora. J. Ferment. Bioeng. 79, 395–397.

69. Kanai, T., Imanaka, H., Nakajima, A., Uwamori, K., Omori, Y., Fukui, T., et al., 2005. Continuous hydrogen production by the hyperthermophilic archaeon, Thermococcus kodakaraensis KOD1. J. Biotechnol. 116, 271–282.

70. Ferchichi, M., Crabbe, E., Hintz, W., Gil, G.H., Almadidy, A., 2005b. Influence of culture parameters on biological hydrogen production by Clostridium saccharoperbutylacetonicum ATCC 27021. World. J. Microbiol. Biotechnol.21, 855–862.

71. Kumar, N., Das, D., 2000. Enhancement of hydrogen production by Enterobacter cloacae IIT-BT 08. Process Biochem. 35,589–593.

72. Logan, B.E., Oh, S.E., Kim, I.S., Van Ginkel, S., 2002. Biological hydrogen production measured in batch anaerobic respirometers. Environ. Sci. Technol. 37, 2530–2535.

73. Noike, T., Mizuno, O., 2000. Hydrogen fermentation of organi municipal wastes. Water Sci. Technol. 42 (12), 155–162.

74. Mizuno, O., Ohara, T., Shinya, M., Noike, T., 2000. Characteristic of hydrogen production from bean curd manufacturing wast by anaerobic microflora. Water Sci. Technol. 42, 345–350.

75. Van Ginkel, S.W., Oh, S.E., Logan, B.E., 2005. Biohydrogen ga production from food processing and domestic wastewaters Int. J. Hydrogen Energy 30, 1535–1542.

76. Ferchichi, M., Crabbe, E., Gil, G.H., Hintz, W., Almadidy, A. 2005a. Influence of initial \ pH on hydrogen production fron cheese whey. J. Biotechnol. 120, 402–409.

77. Ueno, Y., Otsuka, S., Morimoto, M., 1996. Hydrogen productior from industrial wastewater by anaerobic microflora in chemosta culture. J. Ferment. Bioeng. 82, 194–197.

78. Yu, H., Zhu, Z., Hu, W., Zhang, H., 2002. Hydrogen productior from rice winery wastewater in an upflow anaerobic reactor by using mixed anaerobic cultures. Int. J. Hydrogen Energy 27 1359–1365.

79. u, H.Q., Hu, Z.H., Hong, T.Q., 2003. Hydrogen production fron rice winery wastewater by using a continuouslystirred reactor. J Chem. Eng. Jpn. 36, 1147–1151.

80. Tanisho, S., Ishiwata, Y., 1995. Continuous hydrogen productior from molasses by fermentation using urethane foam as a suppor of flocks. Int. J. Hydrogen Energy 20, 541– 545.

81. Hussy, I., Hawkes, F.R., Dinsdale, R., Hawkes, D.L., 2005 Continuous fermentative hydrogen production from sucros and sugarbeet. Int. J. Hydrogen Energy 30, 471–483.

82. Fan, K.S., Chen, Y.Y., 2004. H-2 production through anaerobic mixed culture: effect of batch S-0/X-0 and shock loading in CSTR. Chemosphere 57, 1059–1068.

83. Yokoi, H., Saitsu, A., Uchida, H., Hirose, J., Hayashi, S., Takasaki Y., 2001. Microbial hydrogen production from sweet potatc starch residue. J. Biosci. Bioeng. 91, 58–63.

84. Yokoi, H., Maki, R., Hirose, J., Hayashi, S., 2002. Microbial production of hydrogen from starch-manufacturing wastes. Biomass Bioenergy 22, 389–395.

85. Show, K.Y., Zhang, Z.P., Tay, J.H., Liang, T., Lee, D.J., Ren, N.Q., et al., 2010. Critical assessment of anaerobic processes for continuous biohydrogen production from organic wastewater. Int. J. Hydrogen Energy 35,13350–13355.

86. Lay, J.J., Fan, K.S., Chang, J., Ku, C.H., 2003. Influence of chemical nature of organic wastes on their conversion to hydrogen by heat-shock digested sludge. Int. J. Hydrogen Energy 28, 1361–1367.

87. Valdez-Vazquez, I., Rios-Leal, E., Esparza-Garcia, F., Cecchi, F., Poggi-Varaldo, H.A., 2005. Semi-continuous solid substrate anaerobic reactors for H-2 production from organic waste: mesophilic versus thermophilic regime. Int. J. Hydrogen Energy 30, 1383–1391.

88. Kim, S.H., Han, S.K., Shin, H.S.H.S., 2004. Feasibility of biohydrogen production by anaerobic co-digestion of food waste and sewage sludge. Int. J. Hydrogen Energy 29, 1607–1616.

89. Cheng, S.S., Chang, S.M., Chen, S.T., 2002. Effects of volatile fatty acids on a thermophilic anaerobic hydrogen fermentation process degrading peptone. Water Sci. Technol. 46, 209-214.

90. Cai, M.L., Liu, J.X., Wei, Y.S., 2004. Enhanced biohydrogen production from sewage sludge with alkaline pretreatment. Environ. Sci. Technol. 38, 3195–3202.

91. Wang, C.C., Chang, C.W., Chu, C.P., Lee, D.J., Chang, B.V., Liao, C.S., 2003. Producing hydrogen from wastewater sludge by Clostridium bifermentans. J. Biotechnol. 102, 83–92.

92. Venkata Mohan, S., 2009. Harnessing of biohydrogen from wastewater treatment using mixed fermentative consortia: process evaluation towards optimization. Int. J. Hydrogen Energy 34, 7460–7474.

93. Venkata Mohan, S., 2010. Waste to renewable energy: a sustainable and green approach towards production of

biohydrogen by acidogenic fermentation. In: Om, S., Steve, H (Eds.), Sustainable Biotechnology: Renewable Resources and New Perspectives. Springer, pp. 129–164.

94. Lalit Babu, V., Venkata Mohan, S., Sarma, P.N., 2009. Influence of reactor configuration on fermentative hydrogen production during wastewater. Int. J. Hydrogen Energy 34, 3305–3312.

95. Venkata Mohan, S., Bhaskar, Y.V., Sarma, P.N., 2007b. Biohydrogen production from chemical wastewater treatmen by selectively enriched anaerobic mixed consortia in biofilm configured reactor operated in periodic discontinuous batch mode. Water. Res. 41, 2652–2664.

96. Venkata Mohan, S., Mohanakrishna, G., Reddy, S.S., Raju B.D., Rao, K.S., Sarma, P.N., 2008d. Self-immobilization o acidogenic mixed consortia on mesoporous material (SBA 15) and activated carbon to enhance fermentative hydrogen production. Int. J. Hydrogen Energy 33, 6133–6142.

97. Venkata Mohan, S., Lalit Babu, V., Sarma, P.N., 2007a. Anaerobic biohydrogen production from dairy wastewater treatment in sequencing batch reactor (AnSBR): effect of organic loading rate. Enz. Microbial. Technol. 41 (4),506–515.

98. Venkata Mohan, S., Lalit Babu, V., Srikanth, S., Sarma, P.N. 2008a. Bio-electrochemical evaluation of fermentative hydrogen production process with the function of feeding pH. Int. J Hydrogen Energy 33 (17), 4533–4546.

99. Venkata Mohan, S., Mohanakrishna, G., Ramanaiah, S.V. Sarma, P.N., 2008f. Simultaneous biohydrogen production and wastewater treatment in biofilm configured anaerobic periodic discontinuous batch reactor using distillery wastewater. Int. J Hydrogen Energy 33 (2), 550–558.

100. Yokoi, H., Tokushige, T., Hirose, J., Hayashi, S., Takasaki, Y. 1997. Hydrogen production by immobilized cells of aciduric Enterobacter aerogenes strain HO-39. J. Ferment. Bioeng. 83, 481–484.

101. Venkata Mohan, S., Mohanakrishna, G., Veer Raghuvulu, S., Sarma, P.N., 2007c. Enhancing biohydrogen production from chemical wastewater treatment in anaerobic sequencing batch biofilm reactor (AnSBBR) by bioaugmenting with selectively enriched kanamycin resistant anaerobic mixed consortia. Int. J. Hydrogen Energy 32, 3284–3292.

102. Venkata Mohan, S., Bhaskar, Y.V., Krishna, T.M., Chandrasekhara Rao, N., Lalit Babu, V., Sarma, P.N., 2007d. Biohydrogen production from chemical wastewater as substrate by selectively enriched anaerobic mixed consortia: influence of fermentation pH and substrate composition. Int. J. Hydrogen Energy 32, 2286–2295.

103. Venkata Mohan, S., Mohankrishna, G., Sarma, P.N., 2008b. Integration of acidogenic and methanogenic processes for simultaneous production of biohydrogen and methane from wastewater treatment. Int. J. Hydrogen Energy 33, 2156–2166.

104. Luo, G., Xie, L., Zou, Z., Wang, W., Zhou, Q., Shim, H., 2010a. Anaerobic treatment of cassava stillage for hydrogen and methane production in continuously stirred tank reactor (CSTR) under high organic loading rate (OLR). Int. J. Hydrogen Energy 35, 11733–11737.

105. Ren, N.Q., Li, J.Z., Li, B.Q., Wang, Y., Liu, S.R., 2006. Biohydrogen production from molasses by anaerobic fermentation with a pilot-scale bioreactor system. Int. J. Hydrogen Energy 31, 2147–2157.

106. Barros, A.R., Cavalcante de Amorim, E.L., Reis, C.M., Shida, G.M., Silva, E.L., 2010. Biohydrogen production in anaerobic fluidized bed reactors: effect of support material and hydraulic retention time. Int. J. Hydrogen Energy 35, 3379–3388.

107. Chang, F.Y., Lin, C.Y., 2004. Biohydrogen production using an up-flow anaerobic sludge blanket reactor. Int. J. Hydrogen Energy 29, 33–39.

108. Hallenbeck, P.C., Ghosh, D., 2009. Advances in fermentative biohydrogen production: the way forward. Trends Biotechnol. 27, 287–297

109. Hafez, H., Baghchehsaraee, B., Nakhla, G., Karamanev, D., Margaritis, A., El Naggar, H., 2009. Comparative assessment of decoupling of biomass and hydraulic retention times in hydrogen production bioreactors. Int. J. Hydrogen Energy 34, 7603–7611.

110. Oh, S.E., Iyer, P., Bruns, M.A., Logan, B.E., 2004a. Biological Hydrogen Production Using a Membrane Bioreactor. In: A Wiley Company, vol. 87. Wiley Subscription Services, Inc., pp. 119–127.

111. Wu, S.Y., Hung, C.H., Lin, C.N., Chen, H.W., Lee, A.S., Chang, J.S., 2006. Fermentative hydrogen production and bacterial community structure in high-rate anaerobic bioreactors containing silicone-immobilized and self flocculated sludge. Biotechnol. Bioeng. 93, 934–946.

112. Chang, J.S., Lee, K.S., Lin, P.J., 2002. Biohydrogen production with fixed-bed bioreactors. Int. J. Hydrogen Energy 27,1167–1174.

113. Peintner, C., Zeidan, A.A., Schnitzhofer, W., 2010. Bioreactor systems for thermophilic fermentative hydrogen production: evaluation and comparison of appropriate systems. J. Clean. Prod. 18, S15–S22.

114. Zhang, Z.P., Show, K.Y., Tay, J.H., Liang, D.T., Lee, D.J., 2008a. Biohydrogen production with anaerobic fluidized bed reactors—a comparison of biofilm-based and granule-based systems. Int. J. Hydrogen Energy 33, 1559–1564.

115. Lee, K.S., Lo, Y.S., Lo, Y.C., Lin, P.J., Chang, J.S., 2004a. Operation strategies for biohydrogen production with a highrate anaerobic granular sludge bed bioreactor. Enzyme Microb. Technol. 35, 605–612.

116. Oh, Y.K., Kim, S.H., Kim, M.S., Park, S., 2004b. Thermophilic biohydrogen production from glucose with trickling biofilter. Biotechnol. Bioeng. 88, 690–698.

117. Basak, N., Das, D., 2010. Photofermentative hydrogen production using purple non-sulfur bacteria Rhodobacter sphaeroides

O.U.001 in an annular photobioreactor: a case study. Biomass Bioenergy 33, 911–919.

118. Chen, C.Y., Chang, J.S., 2006. Enhancing phototropic hydrogen production by solid-carrier assisted fermentation and internal optical-fiber illumination. Proc. Biochem. 41, 2041–2049.

119. Kondo, T., Wakayama, T., Miyake, J., 2006. Efficient hydrogen production using a multi-layered photobioreactor and a photosynthetic bacterium mutant with reduced pigment. Int. J. Hydrogen Energy 31, 1522–1526.

120. Liu, H., Grot, S., Logan, B.E., 2005. Electrochemically assisted microbial production of hydrogen from acetate. Environ. Sci. Technol. 39, 4317–4320.

121. Escapa, A., Manuel, M.F., Moran, A., Gomez, X., Guiot, S.R., Tartakovsky, B., 2009. Hydrogen production from glycerol in a membraneless microbial electrolysis Cell. Energy Fuels 23, 4612–4618.

122. Sun, M., Sheng, G.P., Zhang, L., Xia, C.R., Mu, Z.X., Liu, X.W., et al., 2008. An MEC-MFC-coupled system for biohydrogen production from acetate. Environ. Sci. Technol. 42, 8095–8100.

123. Brentner, L.B., Peccia, J., Zimmerman, J.B., 2010. Challenges in developing biohydrogen as a sustainable energy source: implications for a research agenda. Environ. Sci. Technol. 44, 2243–2254.

124. Fan, Y., Zhang, G., Guo, X., Xing, Y., Fan, M., 2006. Biohydrogen-production from beer lees biomass by cow dung compost. Biomass. Bioenergy. 30, 493–496.

125. Venkata Mohan, S., 2009. Harnessing of biohydrogen from wastewater treatment using mixed fermentative consortia: process evaluation towards optimization. Int. J. Hydrogen Energy 34, 7460–7474.

126. Venkata Mohan, S., 2010. Waste to renewable energy: a sustainable and green approach towards production of biohydrogen by acidogenic fermentation. In: Om, S., Steve, H.

(Eds.), Sustainable Biotechnology: Renewable Resources an New Perspectives. Springer, pp. 129–164

127. Venkata Mohan, S., Lalit Babu, V., Sarma, P.N., 2008e. Effect o various pretreatment methods on anaerobic mixed microflor to enhance biohydrogen production utilizing dairy wastewate as substrate. Bioresour. Technol. 99,59–67.

128. Zhu, H., Beland, M., 2006. Evaluation of alternative methods o preparing hydrogen producing seeds from digested wastewate sludge. Int. J. Hydrogen Energy 31, 1980–1988.

129. Venkata Mohan, S., Srikanth, S., Lenin Babu, M., Sarma, P.N. 2010a. Insight into the dehydrogenase catalyzed redox reaction and electron discharge pattern during fermentative hydroge production. Bioresour. Technol. 101,1826–1833.

130. van Ginkel, S., Sung, S.W., Lay, J.J., 2001. Biohydroge production as a function of pH and substrate concentration Environ. Sci. Technol. 35, 4726–4730.

131. Yokoyama, H., Ohmori, H., Waki, M., Ogino, A., Tanaka, Y. 2009. Continuous hydrogen production from glucose by usin extreme thermophilic anaerobic microflora. J. Biosci. Bioeng 107, 64–66.

132. Tang, G.L., Huang, J., Sun, Z.J., Tang, Q.Q., Yan, C.H., Liu, G.Q. et al., 2008. Biohydrogen production from cattle wastewater by enriched anaerobic mixed consortia: influence of fermentatior temperature and pH. J. Biosci. Bioeng. 106, 80–87.

133. Wang, J., Wan, W., 2009. Factors influencing fermentative hydrogen production: a review. Int. J. Hydrogen Energy 34,799-811.

134. Vijaya Bhaskar, Y., Venkata Mohan, S., Sarma, P.N., 2008. Effec of substrate loading rate of chemical wastewater on fermentative biohydrogen production in biofilm configured sequencing batch reactor. Bioresour. Technol. 99,6941–6948.

135. Logan, B.E., 2004. Extracting hydrogen and electricity from renewable resources. Environ. Sci. Technol. 38,160A–167A.

136. Logan, B.E., Regan, J.M., 2006. Microbial fuel cells-challenges and applications. Environ. Sci. Technol. 40, 5172–5180.

137. Ueno, Y., Haruta, S., Ishii, M., Igarashi, Y., 2001. Microbial community in anaerobic hydrogen-producing microflora enriched from sludge compost. Appl. Microbiol. Biotechnol. 57, 555–562.U.S. Department of Energy, 2007. Hydrogen, Fuel Cells and Infrastructure.

138. de Vrije, T., Claasen, P.A.M., 2003. Dark hydrogen fermentation. In: Reith, J.H., Wijffels, R.H., Barten, H. (Eds.), Bio-methane and Bio-hydrogen. Dutch Biological hydrogen foundation, The Hague, the Netherlands, pp. 103–123

139. Datsenko, K.A., Wanner, B.L., 2000. One-step inactivation of chromosomal genes in Escherichia coli K-12 using PCR products. Proc. Natl. Acad. Sci. USA 97, 6640–6645.

140. Akkerman, I., Janssen, M., Rocha, J.M.S., Reith, J.H., Wijffels, R.H., 2003. Photobiological hydrogen production: photochemical efficiency and bioreactor design. In: Reith, J.H., Wijffels, R.H., Barten, H. (Eds.), Biomethane and Biohydrogen: Status and Perspectives of Biological Methane and Hydrogen Production. Dutch Biological Hydrogen Foundation, Hague, the Netherlands

141. Benemann, J.R., 2000. Hydrogen production by microalgae. J. Appl. Phycol. 12 (3-5), 291–300. Boyles, D., 1984. Bioenergy Technology—Thermodynamics and Costs. Wiley, NewYork.

142. Benemann, J.R., 1997. Feasibility analysis of photobiological hydrogen production. Int. J. Hydrogen Energy 22 (10-11), 979–987.

143. Tredici, M.R., Zittelli, G.C., 1998. Efficiency of sunlight utilization: tubular versus flat photobioreactors. Biotechnol. Bioeng. 57, 187–197.

5

Biogas - Feedstocks, Enhancement Process and Economics

5.1 BIOGAS

Biogas has shown a great headway in the evolution of biofuels, its versatility contributes to the success in worldwide acceptance of biogas as an environmental friendly fuel and its sustainability as the clean fuel paved way for its advancement in production. Biogas quenches one of the biggest challenges of this century which is to obtain a better replacement for fossil fuels concerning the environmental issues. Besides the substitution of fossil fuels, it also helps in waste management and increasing agricultural productivity [1]. The generation of biogas is most widely done since it can be easily used as such with the existing technology [2]. Biogas is a mixture of combustible gases formed from a biological process called anaerobic digestion (AD) in which the decomposition of organic substances takes place in anaerobic and moist ambience. The existing biomass such as kitchen waste, cow dung, crop waste and organic residues or effluents from industries and municipal wastes contributes to some potent sources for biogas generation. The main products of AD are biogas and slurry. Biogas has been declared as an alternative to petrol and diesel to eliminate the import complexities. De-sulphurised and cleaned biogas can be utilized as natural gas. 1m^3 of biogas can replace about 0.6L of heating oil [3, 4]. Biogas is synthesized mainly for the production of heat, steam and electricity because they do not require any pre treatment of gas [4]. Methane is one of the useful components of biogas which constitute 40–75% and carbon di oxide 15–60% volume [4]. Nowadays the upgrading treatment aims at reducing the concentration of CO_2 in biogas which would yield an increased level of methane termed as biomethane, which is used as vehicle fuel [5,6]. CO_2 removal also increases the energy content of th gas. The lower heating value of biogas ranges between 15 and 25 MJ Nm³ for 50–60% vol. of gas. CH_4 content can be raised to values typica of natural gas (35 MJ/Nm³) when methane content reaches more tha 96–97% [5]. It has mean calorific value of 20MJ/m³ [7].

5.1.1 Biogas Production

The production of biomass follows a series of metabolism in which the organic matter present in various biological sources gets converted to methane and carbon di oxide in the presence of association of microorganisms which is depicted in (Figure 1). The effect o extracellular enzymes produced by microorganisms such as bacteri on the organic wastes makes it undergo hydrolysis, acidogenesis acetogenesis and methanogenesis [8, 9].

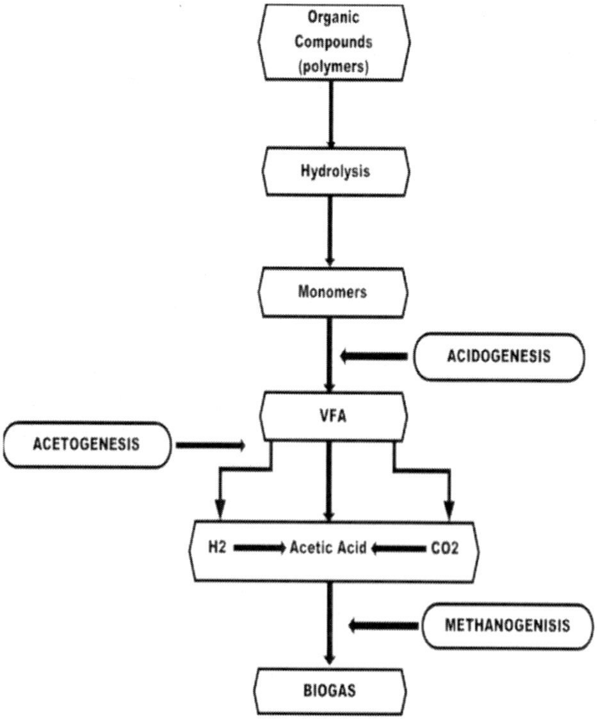

Figure 1 Series of steps involved in biosynthesis of methane

Hydrolysis or liquefaction is the first step in which the organic substances like carbohydrates, proteins, fats are depolymerised to the respective monomers or oligomers (e.g. sugars, glycerol; aminoacids; long chain fattyacids) by hydrolytic bacteria which release their enzyme extracellularly. These monomers undergo degradation reaction which produces acetic acid, long chain fatty acid and CO_2 with the help of fermentative acidogenic bacteria. The bacterial population claimed as *acetogens* later anaerobically oxidizes the long chain fatty acid, alcohols and carbohydrates to produce hydrogen, acetate and carbon di oxide, this phase is hence referred as *acetogenesis*. H_2 and CO_2 can further be converted to acetate in the presence of hydrogen-oxidizing acetogenic bacteria referred as *homoacetogens*, a process which is yet to be characterized [1, 8]. Methane along with CO_2 is produced from the action of methanogens on acetic acid, a process which is sensitive to oxygen concentration depending on the increasing level of the acids present in the environment making its pH lower than the threshold level of 7 and 8 [8]. Among the total CH_4 produced, about 70% originates from the decarboxylation of acetate, while the remaining CH_4 is produced from CO_2 reduction [10-12]. A minor contribution is made by formic, propionic and butyric acids from other organic substrates by methanogens [13]. Apart from the energy that goes for microbial growth (10% and 4% for fermentative and methanogenic bacteria) huge amount of energy is packed in the end product, methane [14,15].

5.1.2 Biogas Composition and Application

Biogas is a mixture of methane, carbon di oxide and trace amounts of water vapour, hydrogen sulfide and ammonia whose distinctive compositions are depicted in (Table 1). It is almost 20% lighter than air. Methane in biogas burns with a clean blue flame that is much hotter than fire that is used as traditional resources for cooking [16]. Like LPG, it cannot convert to liquid state under normal temperature [17].

Table 1 Composition of biogas [9]

Composition of biogas	Content (v/v)
Methane	55–65%
Carbon di oxide	25–55%

Composition of biogas	Content (v/v)
Hydrogen	traces
Nitrogen	0–5%
Moisture	1–5%
Hydrogen sulphide	0–5000 ppm
Ammonia	0–500 ppm

The sole component that contributes to the heating value is CH_4. When 1m^3 of raw biogas at standard temperature and pressure containing 60% CH_4 is burnt, it gives a heating value of 21.5MJ (5.97 kWh) but, in the same condition pure CH_4 gives 35.8MJ (9.94 kWh) [18]. Biogas has varied end uses such as cooking, lighting and heating in household and from the industrial aspect as Combined Heat and Power (CHP) generation, transportation fuel (after being upgraded to biomethane) or upgraded to natural gas quality for other purposes [19]. Upgradation of biogas to natural gas through biomethanation can extend the prospect of transporting the biogas to utilization areas [20]. The large scale plants are constructed in some developing countries mainly for electricity generation through fuel cells [20, 9] or through CHP engines. In India community biogas digester was successfully used to power a modified diesel engine and to run an electrical generator [21]. After removing CO_2, biogas can be compressed in cylinders to make it easily used as fuel for automobiles and biomethane which has equivalent amount of methane present in CNG can be used for all CNG applications [22, 23]. According to the world consumption of biogas the most common use of biogas goes for cooking followed by lighting. It is implemented with the help of exclusive gas mantle lamps that consume 0.07-0.14 m^3 of biogas per hour with an optimum functional gas pressure of 70–84 mm of water [24].

5.1.2.1 ***Application of Biogas Slurry and its Composition*** The Bio Digested Slurry [BDS] is free from weeds, foul smell and pathogens yet contain high quantity of microorganisms which is essential for soil and meets the requirements of plant nutrients. (Table 2) illustrates the comparison of essential plant nutrient concentration present in the digested slurry and conventional manures used.

Table 2 Comparison of nutrient contents present in digested slurry and conventional manure

Plant nutrients	Digested slurry	Farm yard manure	References
Nitrogen	1.8%	0.75%	[152]
Phosphorous	1.00%	0.8%	[152]
Potash	1.00%	0.75%	[152]
Manganese	188 ppm	137 ppm	[168]
Iron	3550 ppm	1788ppm	[168]
Zinc	144 ppm	34ppm	[168]
Copper	28 ppm	2.5 ppm	[168]

The benefits of BDS over other conventional manures used in farm are more retention of organic carbon and more availability of nitrogen due to less loss of nitrogen during decomposition. High status of humus helps in more penetration of essential soil microbes and the water holding capacity is fair enough for rain fed crops [25]. Thus in contrast to composting and direct burning, AD provides fuel and fertilizer, rather than simply one or the other. The spent digested slurry (digestate) exiting the biogas plant remains rich in both macro- and micro-nutrients, and improved flow properties, thus soil penetration is more which reduces the chance of nitrogen loss.

5.1.3 Anaerobic Digestion and Biogas Production: Historical Preview

Biogas production through anaerobic digestion was witnessed as a resolution for the ambiguity of rural development in India. Anaerobic digesters were installed in India in 1859 in order to provide illumination. During 1910 anaerobic digestion of cellulosic materials were tried to augment the fall of nitrogen in Indian agriculture and to effectively dispose the animal wastes [26]. In the year of 1935, biogas plants were developed to conserve nitrogen using animal dung for the above mentioned reason [27, 28]. This was considered the ideal route for procuring nitrogen rich manure before the dawn of nitrogenous fertilizers from fossil fuels. 1982 saw the implementation of a national level program on biogas development to hasten the broadcasting process. Simultaneously biogas plant as a vehicle for development was seen in China and other developing countries. Chinese biogas program that built about 5.5 million biogas plants by 1990 was a prominent one [29]. Seventies welcomed the community biogas plant due to the "fuel wood crisis". Later a shift from fuel wood to biogas was seen in rural kitchen [30]. It is inevitable to note that both in terms of environmental as well as economic virtues, AD is superior to composting, incineration or the combination of digestion and composting because of an improved energy balance and reduced emission of volatile compounds such as ketones, aldehydes, ammonia, and methane [31, 32]. Though the initial purpose of designing anaerobic digesters in developing countries were solely to meet rural energy needs due to the arising form of end users and infrastructural availability which gave shape to the demands on the type of design for anaerobic systems lead to the development of small anaerobic digesters (biogas plant) and large anaerobic waste treatment (solid and liquid wastes). The small ones for rural utilities used animal and agro waste as feedstock and were expected to provide 2-4 m^3/d biogas suffice to meet the cooking energy needs and to a prominent extent for illumination. India produces up to 44 million tons of urban wastes and 500 Mt of non-woody rural residues annually. The former consists of greater proportion of methanogenic fermentable materials whereas the latter ones are potent soil fertilizers [29].

5.1.4 Global View on Substrates Ideal for Biogas

The key lead of biomass synthesis is its extensive range of substrate availability. The most fitting raw materials would be agricultural biomass residues, such as animal manure, human excreta and the organic fraction of MSW, livestock residues, and wastewater from food industry. The main contemplation while opting for the substrate is its sustainability, energy, environmental and economic values. On the basis of net energy generation, biogas produced from these resources has been reported to be much more competitive than alcohol-based liquid biofuels [33, 34]. Developed countries are by now making use of millions of tons of organic wastes from municipal, industrial and agricultural processes for biogas production [33, 35], but mammoth volumes of organic waste stay under utilized in developing countries [9]. Various amounts of organic wastes and their availability are described in (Table 3). Animal manures as a feedstock for AD have 75-

92% of moisture content and 72-93% of volatile solids (VS) of the total solids (TS) present, good buffering capacity also adds to the point in claiming it as an ideal substrate [36, 37].

Table 3 Organic wastes and their availability in India

Organic wastes	Estimated quantity (million tons/year)
Municipal solid waste	30
Municipal liquid waste	3758.56
Food & fruit processing waste	4.5
Dairy industry waste	18.8

The diversity of microbial flora in animal manure helps in eliminating the step of inoculating the digester, yet due to slow digestion (low readily degradable organic content) and poisoning effect of high concentration NH_4 generated on methanogens, animal manures are pushed back in the line [38]. Co-digestion with a carbohydrate rich source and nitrogen free/fewer raw materials can surmount the disadvantage of animal manure by appreciably enhancing biogas production [9].

Animal excreta especially poultry droppings contain easily degradable organic materials than other agricultural waste products, but is also known to have a high content of about 40–50% of the TS of lignin and lignocelluloses which are biofibers [39]. Another alternative for improving biogas yield is co-digestion of poultry manure with other organic wastes which possess increased loading of readily biodegradable organics, improved balance of nutrients and C/N ratio [40, 41]. Digestion of poultry droppings have been carried out in combination with other organic wastes such as whey, rice and wheat straws, municipal solid wastes, hog manure, buffalo manure, dairy manure and sewage sludge [42-44]. Cattle dung is a widely used substrate which does not require any treatment before it is charged into the digester. This is mixed with hot water in the ratio of 1:1 and fed into the tank to give biogas and manure can be supplied at the rate of 2-3 per kg [3].

MSWs are chiefly dominated by organic matter which accounts for about 55% of organic solid waste. However MSW contains less readily fermentable substrates and deficient in N and P yet consists o relatively high CH_4 potential when digested properly [45]. Pretreatmen involving particle size reduction and co-digestion with other biomas like municipal sludge or animal manures substantially increases CH production from MSW [45,46]. Usually it is recommended to remove a certain amount of inorganic fraction (e.g., metals and glass) in raw MSW prior to anaerobic digestion for more efficient operation of anaerobi digesters and prevention of equipment failures involved in materia handling. Some pre-treatment methods may also be applied to increase the digestibility of the organic solids and increase the efficiencies o anaerobic digesters [47]. In the study reported by [48], a pre-compostin stage was used for the pulp mill sludge to obtain a slight degradation o organics to prevent fast acidification during anaerobic digestion. Afte 49 days experiment, the pretreated sludge had higher VS reduction (50% than the untreated sludge which had 34% VS reduction. Studies or mechanical pretreatment have also shown that a Rotary Drum Reacto (RDR) process provided an effective means for separating the organic from MSW using a combination of mechanical forces and biologica reactions [49, 50]. Thermophilic conditions have certain advantages over mesophilic conditions for digesting the organic wastes separatec from MSW, such as faster degradation rate, higher biogas productior rate, lower effluent viscosity, and higher pathogen destruction [51] Rice husk constitutes one of the major environmental bother as they form the major municipal solid waste heaps in the areas where they are disposed. Although, rice husk has various applications such as ir construction materials [52], for the adsorbent of heavy metals [53], anc in the design of gas furnace as an alternative to coal [54] or at the side of pulverized coal which is conducive to gas phase combustion [55], i has not yet become a practical option to use for energy synthesis. Mos of the rice husks generated during rice milling are burnt in the field This kind of traditional disposal method has resulted in a widespreac environmental concern since it is an air pollutant. Since field burning activities are not encouraged in some countries subsequent methods to dispose and use of agricultural residues such as rice straw and rice husk have shifted towards the global "waste to resource" agenda [56].

Agriculture comprises a major fraction of the national economy in developing countries. They produce about 1,680 million dry metric tons of crop residues annually [57], a significant portion of which can be used for biogas production. Although the composition of crop residues varies significantly with crop type, in general, crop residues have relatively low moisture content, high VS content and variable amounts of readily fermentable constituents. Because most non-leguminous crops have a low concentration of available nitrogen, co-digestion of crop residues with animal manure or municipal sludge substantially improves CH_4 production [58]. Perhaps, the best resources for biomethane production are food and food-processing wastes because of their high moisture (>80%) and VS (95% of TS) contents. With the exception of meat waste, most food processing wastes are poor in nitrogen content; but are rich in readily fermentable organic matter. The co-digestion of food-processing residue with nitrogen-rich feed stocks (e.g., municipal sludge and animal manures) thus enhances system stability and overall CH_4 production [59].

Inedible residues mostly lignocellulosic materials from food and energy crops, produced in large amount worldwide, presents great prospective. Among these materials, wheat straw is highly attractive, being the second richest agricultural residue in the world [60]. Nevertheless, the intricate composition of lignocelluloses, consisting of three main types of polymers—cellulose, hemicellulose and lignin—frequently restricts the biodegradability and consequently results in low biodigestion efficiency and methane production [61]. The biodegradability of lignocellulosic material, including straw, can be increased by a biological, chemical, thermal or mechanical pretreatment, or through their combination [62]. In case of straw, there is example in the literature of studies using these techniques for enhanced biogas production, mainly in batch systems. For example, biological treatment with softwood-degrading fungi has been reported to increase the methane potential of straw by 30% [63], while chemical pretreatment (i.e., acid and alkali-based) to solubilise lignocellulosic material has been found to increase methane production by 43% [64]. Thermal pretreatment with steam explosion has also been shown to increase the biodegradability of lignocellulosic material [61, 62, 65]. During steam explosion, organic material is heated at relatively high pressure followed by a rapid pressure drop, resulting in mechanical disruption of the fibers. Steam explosion has been shown to increase the methane potential of lignocellulosic materials, including straw by up to 20% [66,67]. There is a drawback with some pretreatment methods, and they result in solubilisation of lignin, e.g., steam explosion at temperatures above 160 °C following an acid pretreatment [61]. This results in the release of phenolic compounds and furans which can have an inhibitory effect on methanogens/archaea and thus can potentially affect biogas production [61]. Besides, biogas residues containing high concentrations of phenols can have an inhibitory effect on the soil microbial population if applied as fertilizer [68] thus pretreatment is an inevitable step which will have a great influence on end use of biogas [69]. Grassland biomass is suitable in many ways for producing energy. Currently it is used in practice as a feedstock for biogas production and as solid biofuel for combustion. Future pathways can include the production of lignocellulosic bioethanol, synthetic biofuels or synthetic natural gas. Feedstock from grassland will also be used as raw material for the bio-industry within Green Biorefineries [70]. Within Green Biorefinery concepts, biogas production is a key technology for the sustainable use of grassland biomass [71]. Each form of energetic utilization requires specific characteristics of the grassland biomass, being highly variable and depending mainly on grassland management. In practice, grass usually is co-digested with slurry and other crop feedstock using common anaerobic digestion technology without specific adaptations [72].

Rice straw, corn stalks and wheat straw are the top three crop straw wastes in China and account for 32.3%, 25.0% and 18.3% of the total crop straw output, respectively [73]. Thus, making use of these wastes for biogas generation can be quite significant. However, crop wastes cannot be effectively degraded due to an imbalance in nutrients for microorganism and a lack of buffering capacity for the chemical reaction [74]. This can often be overcome by co-digestion with livestock manure. Studies have been done to improve biogas production from co-digestion of livestock manure and other wastes by AD [75-77]. Compared with the digestion of single feedstock, co-digestion increases the biogas production rate because of the better nutrient balance and improvement of AD efficiency. Generally, livestock manure contains high total nitrogen, which decreases the carbon-to-nitrogen (C/N) ratios of single straw residues substrates and is beneficial to co-digestion with straw residues. Livestock manure is also helpful to achieve a suitable pH during anaerobic fermentation with the production of ammonia [78]. Hence, livestock manure is an excellent raw material for anaerobic co-digestion [79].

5.1.5 Biogas Generation, Purification and Bottling

The substrate is collected and fed into the primary digester after pretreatment if required, where the hydrolysis cum acidification takes place. The next digester is the main tank involved in biosynthesis of methane and BOD reduction. The treated biomass flows to the manure pits which can be supplied to the farmers at the rate of Rs. 2-3 per kg. The biogas is simultaneously collected in the dome of the main tank (gas holder). It is a drum-like structure made of either steel sheets or Fiber Glass Reinforced Plastic (FRP). It fits like a cap on the delta of the digester where it is sunken under water and rests in the ridge constructed inside the digester for its purpose. The drum thus collects the gas from the slurry produced inside the digester as it gets decomposed and rises up being lighter than air. Simple plants use one tank where bacteria breakdown matter until it can be decomposed by a second set of bacteria, which creates acetic acid and carbon. Methanogens then eat these compounds, primarily acetic acid and create methane. A two-tank system performs the basic breakdown and decomposition in one tank and then move the waste to another tank for the compound breakdown and methane production, which speeds up the process. Newer system involves three tanks that switch between the stages of methane production. Raw biogas is compressed to operational pressure of 8-10 bar and directed to purification tank or column, where water is flushed from the cooling system from the top of the column in the counter current direction to the flow of biogas. Scrubbing with water involves the physical absorption of CO_2 and H_2S in water at high pressures and regeneration takes place by release of pressure with negligible change in the temperature. Purification in general means enrichment of biogas' methane content. Primarily pressurized water scrubbing is used which balances both technical and economic sides. Enrichment of biogas to methane is discussed briefly in the following section. Enriched or purified biogas is made moisture free by passing through filters later, compressed to 200 bar pressure using 3 or 4 stage gas compressor. The compressed gas is stored at high pressure stee cylinder similar to CNG [3].

5.1.6 Various Types of Biogas Plants

There are essentially two types of bioreactor for biogas production– they are fixed dome biogas digester and floating dome biogas digeste made of different materials and in different shapes. However both ar anaerobic and are mostly used for energy crops such as maize silage An advanced anaerobic digester must have an external dry feede with a percent TS of 30% or greater which has to be injected into th digester. This is because biogas yield is totally dependent on percen TS and percent total organic solids. A mixer inside the digester enable it to operate with % TS more than 30%. Then an efficient Interna Combustion engine generator set in 40-43% range allows the systen to convert as much of methane possible into electricity [80]. Th simplest and most common reactor configuration for biogas plant are the continuous stirred tank reactor (CSTR). Nearly 90% of moder biogas plants are CSTRs [81]. The substrate is continuously fed to th plant and biogas is produced constantly. The advantages of CSTR are a simple reactor design and corresponding to that low investmen and operating costs. Further benefits are easy operation and seldon technical failure, since no advanced controller is necessary for a stabl operation. Drawbacks are high residence times, which are necessar for an efficient operation and the tendency to foam and scum formatio [82]. Most of the researches in the past focus on two-phase systems In the first reactor the first degradation step resulting in volatile fatt acids, carbon di oxide and hydrogen take place and in these methan is produced. It is stated that the principle benefit of two-phase system results from the possibility to implement different conditions (pH dilution rate) in each reactor, which is beneficial for the diverse microbia communities. Furthermore, the increased process stability is beneficia which results from controlling the first reactor which prevents fron over-loading, enrichment of toxic material and pH shock [83]. But thi is also a drawback, since controlling makes the operation more difficul and expensive [84]. In an ideal PFR no back mixing occurs, but in rea plants this is the case only in thin and long reactors. Often additiona installations such as baffles [85] or an inclination of the whole reacto

[86] are applied to increase the solid retention time. Several tubular reactors are operated semi-continuously to improve the process performance [87]. In general, the combined CSTR with PFR shows better performance than the comparable two CSTRs systems. Especially for two reactors of the same size, the improvement is significant [88]. Whereas in case of batch or semi-continuous reactor, the first reactor is filled with substrate and in the second phase, the reaction takes place and biogas is produced without feeding or removing liquid from or to the reactor. After that, the reactor is emptied and refilled with substrate. In the emptying phase, the digestate is not removed completely from the reactor, but a residue is kept to ensure the presence of microorganisms after the filling with substrate. Batch reactors are used in different fields with various kinds of substrates [82,87]. The main benefits of batch reactors according to these reviews are simple design and easy process control, furthermore the robustness towards coarse and heavy containments. Resulting from that, the investment costs are lower than incomparable CSTRs. High process flexibility in terms of cycle time and sequence brings another positive aspect. Drawbacks of the batch process are irregular gas production. It is stated that in batch reactors larger volumes are needed and a lower biogas yield is obtained [82]. In a semi-continuous operation, two different operating phases can be distinguished. In the feeding phase, a continuous operation is assumed by pumping substrate to the reactor and removing the digestate. In the second phase, the reaction takes place; only the biogas is removed without feeding any substrate to the reactor [88].

5.1.7 Enrichment of Biogas Production, Upgradation and Management

Biogas generated by anaerobic digesters has handsome energy equivalent for traditional uses like kitchen, furnace, boiler and power generation. However the impurities like CO_2 and H_2S do not allow the use of biogas to fulfill its complete potential. The other side effects of these gases have detrimental effect on the structure of downstream equipment like corrosion of engine and boiler tubes and steel chimneys. Also the high value of CO_2 results in lowering the heating value of biogas which makes the combustion difficult. If these impurities could be removed the purified gas would have a quality of methane thus it becomes ideal for vehicle fuel like CNG [89]. The amount of methan content after enrichment of biogas is given in (Table 4). Bioga enrichment involves removal of CO_2, H_2S and water vapour. One c the specialized techniques is use of pressurized water as an absorben liquid after which the biogas is fed to bottom of scrubber colum where water is sprayed from top. In another method the gases ar absorbed in down going water and methane goes up and is collecte in vessel. However water requirement is high. The biogas enrichmen unit consists of a scrubber usually packed bed absorption column wit supporting frame, water supply system and gas supply system whic in turn consists of biogas plant, single stage compressor, and pressur vessel and pipe fittings.

Table 4 Amount of biogas and biomethane production from potentia sources [90]

Raw Material	Biogas Production (Lts/Kg)	Methane Production (%)
Cattle dung	40	60
Food waste	160	62
Fruit waste	91	49.2
Non-edible oil seed cakes	242	67.5
Bagasse	330	56.9
Green leaves & twigs	100	65
Dry leaves	118	59.2

The compression unit consists of a three stage compressor, ultr filters, storage cylinders, rotameter and gas chromatograph. The cos accountable for 1000 NM^3/ day capacity is Rs.60 lakhs approximatel which includes purification and bottling charges [90]. Another metho which uses activated carbon adsorbents and molecular sieves is calle pressure swing absorption (PSA). Here the selectivity is based on th molecular size and weight. Absorption takes place at high pressur and desorption takes place at lower pressure. The simple design an operation, consistent purity of gas, clean process requiring very les floor area and dry gas production are the advantages of PSA process In this method water removal is carried out by condensation, dryin over silica gel or Al_2O_3 if low dew points need to be achieved o through absorption in glycol or hygroscopic salts. The most commo processes for removal of H_2S are carried out through catalytic removal, absorption on activated carbon or microorganisms. PSA biogas purification system is the gas separation system with specifically designed absorbent. These Molecular Sieves (MS) have micropores on their surface and absorb CO_2, N_2, H_2S and H_2O molecules selectively at certain pressure. Absorbent can be regenerated by depressurizing. This system can produce purified methane enriched gas continuously by repeating above steps [89]. Slurry can be dehydrated to enrich because, the excessive quantity of water (about 90%) makes the handling and transport of BDS difficult. Moreover it takes large space and more time for drying under sun. The most efficient method is drying, sedimentation and absorption of slurry and dry organic refuse and its filtration using sand bed. The loss of ammonia during sun drying could be reduced using low grade rock phosphate. After 10-12 days of sedimentation the settled sludge is removed with higher solid content from the bottom, having only 10-15% moisture. Total nitrogen content in organic manure does not exceed 1.5%. The bio-digested slurry can be mixed with oil cakes such as neem cake, castor oil cake which will enhance nutrient value of manure [168]. Co-digestion is also an appealing alternative for improving the yield of biogas through anaerobic digestion of solid wastes. The use of co-substrate is employed to improve the yield due to positive synergism established in the digestion medium and to provide lacking nutrients by co-substrates. In order to get an optimum result the co-substrate has to be chosen carefully [91].

5.2 CONCLUSION

In conclusion gaseous fuels are the basis of heat energy which is used directly as fuels for various engines. Both biohydrogen and biogas can be produced from easily available sources such as MSWs, industrial effluents, agricultural waste and domestic wastes respectively. Also these fuel gases are versatile and cleaner energy source with non-polluting characteristics purging the tribulations caused by conventional fossil fuels. In case of biogas removal of CO_2 provides an enriched biogas equal to the qualities of biomethane and thus it can also be utilized for the applications of biomethane. Compared to incineration or composting, AD is considered as a beneficial route for biogas production since it provides both fuel and fertilizer. The slurry can be enriched further through dehydration which would remov about 90% of water contained in it initially or by mixing with oil cake that enhance the manurial value of spent slurry.

5.3 REFERENCES

1. Khanal SK. Anaerobic biotechnology for bioenergy productior principles and applications. Ames, Iowa, United States: Joh Wiley & Sons, Inc.; (2008).

2. Schnurer A, Jarvis A. Microbiological Handbook for Bioga Plants. AVFALL SVERIGE utveckling; (2009).

3. K.S Popli; Biogas production, upgradation and slurr management V.K Vijay; Narosa publication (2012) p109-113.

4. Renato Baciocchi, Ennio Carnevale, Giulia Costa, Renat Gavasci, Lidia Lombardi, Tommaso Olivieri, Laura Zanchi Daniela Zingaretti; Performance of a biogas upgrading proces based on alkali absorption with regeneration using air pollutio control residues; Waste Management 2694–270533 (2013).

5. Ryckebosch, E., Drouillon, M., Vervaeren, H., Techniques fo transformation of biogas to biomethane. Biomass Bioenergy 35 1633–1645(2011).

6. Patterson, T., Esteves, S., Dinsdale, R., Guwy, A., An evaluatio of the policy and techno-economic factors affecting the potentia for biogas upgrading for transport fuel use in the UK. Energ Policy 39, 1806–1816(2011).

7. M.Kaltschmitt, D.Thran and K.R Smith, "renewable energ from biomass"; encyclopedia of physical sci and tech,ed. By R A. Meyers (academic press 0;203-228(2001).

8. Lastella .G, Testa.C,Cornacchia.G, Notornicola.M,Voltasio.F Vinod kumar Sharma; Anaerobic digestion of semi solid organi waste: biogas production and its purification; Energy con an management 43,63-75(2002).

9. K.C. Surendra, Devin Takara, Andrew G. Hashimoto, Sami Kumar Khanal Biogas as a sustainable energy source fo

developing countries: Opportunities and challenges Renewable and Sustainable Energy Reviews 31846–859(2014).

10. McCarty PL.The methane fermentation.In:Henkelkian H, Dondero NC, editors. Principles and application in aquatic microbiology. New York, United States: John Wiley and Sons, Inc.;(1964).

11. Gerardi MH. The microbiology of anaerobic digesters. New Jersey, United States:John Wiley and Sons Inc.;(2003).

12. Zieminski K,Frac M. Methane fermentation process as anaerobic digestion of biomass: transformations, stages and microorganisms.Afr J Biotechnol;11:4127–39(2012).

13. Woodard F. Industrial waste treatment handbook. 2nd ed. Oxford: Butterworth-Heinemann; (2006).

14. Schink B.Energetics of syntrophic cooperation in methanogenic degradation. Microbiol Mol Biol Rev;61:262–80(1997).

15. WestermannP.Temperatureregulationofanaerobicdegradationof organic matter. World J Microbiol Biotechnol;12:497–503(1996).

16. Singh KJ, Sooch SS.Comparative study of economics of different models of family size biogas plants for state of Punjab, India. Energy Convers Manag; 45:1329–41(2004).

17. Nina Johansson, production of liquid biogas, LPG, with cryogenic and conventional upgrading technology, institutionen for Teknik och Samhalle, Lunds Tekniska Hogskola.(2008)

18. Muylaert MS, Sala J, de Freitas MAV. Consumo de energia e aquecimento do planeta: Análise do mecanismo de desenvolvimento limpo (MDL) do Proto- colo de Quioto. Case studies. Rio de Janeiro: Post-graduate Engineering Programs Coordination (COPPE); (2000).

19. Kristensson, I. et al., Biogas på gasnätet utan propantillsats. Rapport SGC 176, 1102-7371, ISRN SGC-R-176-SE, pp. 6–18(2007).

20. J.B. Holm-Nielsen, T. Al Seadi, P. Oleskowicz-Popiel; The futur of anaerobic digestion and biogas utilization; Bioresourc Technology 100 5478–5484(2009).

21. Reddy AKN.Lessons from the Pura community biogas projec Energy Sustain Develop;8:68–73(2004).

22. Kapdi S.S., Vjay V.K., Rajesh S.K., Prasad R.; Biogas scrubbing compression and storage: perspective and prospectus in India context; Renewable Energy 30, 1195-1202(2005).

23. Persson, M., Jönsson, O., Wellinger, A., 2006. Biogas upgradin to vehicle fuel standards and grid injection. IEA Bioenergy, Tas 37 – Energy from Biogas and Landfill Gas,(2006).

24. Biogas Support Programme (BSP) and Netherlands Developmen Organiza- tion (SNV/N). Advanced courses in bio-gas technology Lalitpur, Nepal;(2001).

25. D.S. Bhandari; Biogas production, upgradation and slurr management V.K Vijay; Narosa publication(2012). p 73-74.

26. Fowler, J and Joshie, M. Fermentation of cellulose; Bull. Ind Inst. Sci., 2:21; (1921).

27. Acharya, C.N.; Studies on the anaerobic decomposition of plan materials- 1. The anaerobic decomposition of rice straw *(Oryza sativa)*. Biochem.J, 29,528-535 (1935).

28. Desai S.V., Biswas S.C.; Manure and gas production by anaerobic fermentation of organc wastes. Ind. Farming. 6, 67-68 (1945).

29. Chanakya HN, Biogas production, upgradation and slurry management V.K Vijay; Narosa publication(2012) p1-2.

30. Reddy, Rajaraman I., Subramanian DK. and Rajabapaiah P A Community biogas plant system of Pura village. Technica Report. Karnataka State Council For Science And Technology Banglore(1979).

31. Edelmann W, Joss A, Engeli H.Two step anaerobic digestior of organic solid wastes. In:Mata-Alvarez J, Tilche A Cecchi F. editors. Proceedings of the second internationa

symposium on anaerobic digestion of solid wastes. Barcelona, Spain;vol.1.p.153–60(1999).

32. De Baere L.Anaerobic digestion of solid waste:state-of-theart. In:Mata- AlvarezJ,Tilche A,Cecchi,F.editors.Proceedings of the second international symposium on anaerobic digestion of solid wastes. Barcelona, Spain; vol. 1.p.290–99(1999).

33. Chynoweth DP, Owens JM, Legrand R. Renewable methane from anaerobic digestion of biomass. Renew Energy;22:1–8(2001).

34. Edelmann W, Schleiss K, Joss A. Ecological, energetic and economic comparison of anaerobic digestion with different competing technologies to treat biogenic wastes.Water Sci Technol; 41:263–73 (2000).

35. Gunaseelan VN.Anaerobic digestion of biomass for methane production: a review. Biomass Bioenergy;13:83–114(1997).

36. Fujino J,MoritaA,MatsuokaY,SawayamaS.Vision for utilization of livestock residue as bioenergy resource in Japan.Biomass Bioenergy;29:367–74(2005).

37. Muller JA,WinterA,Strunkmann G. Investigation and assessment of sludge pre-treatment processes.Water Sci Technol;49:97–104(2004).

38. Kayhanian M. Ammonia inhibition in high-solids biogasification: an overview and practical solutions. Environ Technol;20:355–65(1999).

39. Triolo, M.J., Sommer, G.S., Moller, B.H., Weisbjerg, R.M., Jiang Y.X.,." A new algorithm to characterize biodegradability of biomass during anaerobic reaction: Influence of lignin concentration on methane production potential." Bioresour. Technol. 102(20), 9395–9402 (2011).

40. Wang, X., Yang, G., Li, F., Feng, Y., Ren, G., Han, X.,. Evaluation of two statistical methods for optimizing the feeding composition in anaerobic codigestion: mixture design and central composite design. Bioresour. Technol. 131, 172–178(2013).

41. Wang, X., Yang, Y., Feng, Y., Ren, G., Han, X.,. Optimizing feedin composition and carbon–nitrogen ratios for improved methan yield during anaerobic codigestion of dairy, chicken manur and wheat straw. Bioresour. Technol. 120, 78–83(2012).

42. Callaghan, F.J., Wase, D.A.J., Thayanithy, K., Forster, C.F., Continuous codigestion of cattle slurry with fruit and vegetabl wastes and chicken manure. Biomass Bioenergy 27, 71–77(2002)

43. Gelegenis, J., Georgakakis, D., Angelidaki, I., Mavris, V., Optimization of biogas production by co-digesting whey wit diluted poultry manure. Renewable Energy 32, 2147–2160(2007)

44. Borowski, S., Weatherley, L. Co-digestion of solid poultr manure with municipal sewage sludge. Bioresour. Technol. 142 345–352(2013).

45. Del BorghiA,ConvertiA,PalazziE,DelBorghiM.Hydrolysis an thermophilic anaerobic digestion of sewage sludge and organi fraction of municipal solid waste.Bioprocess Eng; 20:553–6 (1999).

46. Gomez-Lahoz C, Fernandez-Gimenez B, Garcia-Herruzo F Rodriguez-Maroto JM, Vereda-Alonso C. Biomethanizatio of mixtures of fruits and vegetables solid wastes and sludg from a municipal waste water treatment plant. J Environ Sc Health;42:481–7(2007).

47. Bernal, O., Llabres, P., Cecchi, F., Mata-Alvarez, J. A comparativ study of the thermophilic biomethanization of putrescibl organic wastes. Wastewaters 1, 197–206(1992).

48. Capela, I.F., Azeiteiro, C., Arroja, L., Duarte, A.C. Effects of pre treatment (composting) on the anaerobic digestion of primar sludges from a bleached kraft pulp mill. In: Mata-Alvarez J., Tilche, A., Cecchi, F. (Eds.), Proceedings of the Secon International Symposium on Anaerobic Digestion of Soli Wastes, vol. 1. Barcelona, pp. 113–120(1999).

49. Hayes, P. Expanding the potential of In-Vessel Co-composting Biocycle 45, 38–42(2004).

50. Spencer, R.L. Rotary digester operational realities. Biocycle 47, 53–55(2006).

51. Cooney, C.L., Wise, D.L. Thermophilic anaerobic digestion of solid waste for fuel gas production. Biotechnol. Bioeng. 17, 1119–1135(1975).

52. Yang HS, Kim DJ, Kim HJ. Rice straw-wood particle composite for sound absorbing wooden construction materials. Bioresource Technology ;86:117-21(2003).

53. Lakshmi UR, Srivastava VC, Mall ID, Lataye DH. Rice husk ash as an effective adsorbent: evaluation of adsorptive characteristics for indigo carmine dye. Journal of Environmental Management;90:710-20(2009).

54. Jain AK, Goss JR. Determination of reactor scaling factors for throatless rice husk gasifier. Biomass and Bioenergy;18:249-56(2000).

55. Chen WH, Wu JS. An evaluation on rice husks and pulverized coal blendsusingadroptubefurnaceandathermogravimetricanalyzer for application to a blast furnace. Energy;34:1458-66(2009).

56. Lim JS, Manan ZA, Wan Alwi SR, Hashim H. A review on utilization of biomass from rice industry as a source of renewable energy. Renewable and Sustainable Energy Reviews; 16:3084-94(2012).

57. Food and AgricultureOrganization(FAO).Better utilization of crop residues and by-products in animal feeding: research guidelines – 1.State of knowledge. FAO animal production and health Paper 50.1;1985. http://www.fao. org/DOCREP/003/X6553E/X6553E00.HTM [retrieved 25.10.11].

58. Lehtomaki A, Huttunen S, Rintala JA. Laboratory investigations on codigestion of energy crops and crop residues with cow manure for methane production: effectofcroptomanureratio. Resour,ConservRecycl ;51:591–609(2007).

59. Kim HW, Han SK, Shin HS. Anaerobic co-digestion of sewage sludge and food waste using temperature-phased anaerobic digestion process. Water Sci Technol;50:107–14(2004).

60. Wang, G., Gavala, H.N., Skiadas, I.V., Ahring, B.K., 2009. We explosion of wheat straw and codigestion with swine manure effect on the methane productivity. Waste Manag. 29, 2830 2835.

61. Hendriks, A.T.W.M., Zeeman, G., 2009. Pretreatments to enhanc the digestibility of lignocellulosic biomas. Bioresour. Technol 100, 10–18.

62. Chandra, R.P., Bura, R., Mabee, W.E., Berlin, A., Pan, X. Saddler, J.N., 2007. Substrate pretreatment: the key to effectiv enzymatic hydrolysis of lignocellulosics? Adv. Biochem. Eng Biothechnol. 108, 67–93.

63. Müller, H.W., Trösch, W., 1986. Screening of white-rot fungi fo biological pretreatment of wheat straw for biogas production Appl. Microbiol. Biotechnol. 24, 180–185.

64. Sambusiti, C., Monlau, F., Ficara, E., Carrère, H., Malpei, F. 2013. A comparison of different pre-treatments to increase methane production from two agricultural substrates. App Energy. 104, 62–70.

65. Horn, S.J., Nguyen, Q.D., Westereng, B., Nilsen, P.J., Eijsink V.G.H., 2011a. Screening of steam explosion conditions fo glucose production from non-impregnated wheat straw. Biomass Bioenergy. 35, 4879–7886.

66. Bauer, A., Bösch, P., Friedl, A., Amon, T., 2009. Analysis of methane potentials of steam-exploded wheat straw and estimation of energy yield of combined ethanol and methane production. J. Biotechnol. 142, 50–55.

67. Vivekanand, V., Ryden, P., Horn, S.J., Tapp, H.S., Wellner, N. Eijsink, V.G.H., Waldron, K.W., 2012. Impact of steam explosion on biogas production from rape straw in relation to changes in chemical composition. Bioresour. Technol. 123, 608–615.

68. Levén, L., Nyberg, K., Schnürer, A., 2012. Conversion of phenols during anaerobic digestion of organic solid waste – a review of important microorganisms and impact of temperature. J Environ. Manage. 95, 99–103.

69. Kajsa Risberg a, Li Sun a, Lotta Levén c, Svein Jarle Horn b, Anna Schnürer; Biogas production from wheat straw and manure – Impact of pretreatment and process operating parameters; Bioresource Technology 149 (2013) 232–237.

70. Kromus, S., Kamm, B., Kamm, M., Fowler, P., Narodoslawsky, M., 2006. The green biorefinery concept – fundamentals and potential. Biorefineries – Industrial Processes and Products 12, 235–285.

71. Bauer, A., Hrbek, R., Amon B., Kryvoruchko, V., Bodiroza, V., Wagentristl, H., Zollitsch, W., Liebmann, B., Pfeffer, M., Friedl, A., Amon, T., 2007. Potential of biogas production in sustainable biorefinery concepts. In: Proceedings of the Fifth Research and Development Conference of Central- and Eastern European Institutes of Agricultural Engineering. Part 2, pp. 20–31.

72. A. Prochnow, M. Heiermann, M. Plöchl, B. Linke, C. Idler, T. Amon, P.J. Hobbs; Bioenergy from permanent grassland – A review: 1. Biogas; Bioresource Technology 100 (2009) 4931–4944.

73. MOA, 2011. National crop straw resources survey and evaluation reports. Agric. Eng. Technol. (Renew. Energy Ind.) 02, 2–5.

74. Babaee, A., Shayegan, J., Roshani, A., 2013. Anaerobic slurry co-digestion of poultry manure and straw: effect of organic loading and temperature. J. Environ. Health Sci. Eng. 11, 15.

75. Astals, S., Nolla-Ardevol, V., Mata-Alvarez, J., 2013. Thermophilic co-digestion of pig manure and crude glycerol: process performance and digestate stability. J. Biotechnol. 166, 97–104.

76. Chen, J.H., Lin, C.C., Wang, K.S., 2013. Potential of methane production by thermophilic anaerobic co-digestion of pulp and paper sludge with pig manure. J. Biobased Mater. Bioenergy 7, 300–304.

77. Saidu, M., Yuzir, A., Salim, M.R., Salmiati, Azman, S., Abdullah, N., 2013. Influence of palm oil mill effluent as inoculum on anaerobic digestion of cattle manure for biogas production. Bioresour. Technol. 141, 174–176.

78. Ashekuzzaman, S.M., Poulsen, T.G., 2011. Optimizing feed composition for improved methane yield during anaerobic digestion of cow manure based waste mixtures. Bioresour Technol. 102, 2213–2218.

79. Jiang Li, Luoyu Wei, Qiwu Duan, Guoquan Hu, Guozhi Zhang Semi-continuous anaerobic co-digestion of dairy manure with three crop residues for biogas production; Bioresource Technology 156 (2014) 307–313.

80. Sanjay Patel; Biogas production, upgradation and slurry management V.K Vijay; Narosa publication (2012); p116-121.

81. Weiland, P., 2006.Biomass digestion in agriculture: a successful pathway for the energy production and waste treatment in Germany. Engineering in Life Sciences 6(3), 302–309.

82. Nizami, A.S.,Murphy, J.D.,2010.What type of digester configurations should be employed to produce biomethane from grass silage? Renewable and Sustainable Energy Reviews 14(6),1558–1568.

83. Demirel, B.,Yenigun,O.,2002.Two-phase anaerobic digestion processes: a review. Journal of Chemical Technology and Biotechnology77(7),743–755.

84. Lettinga, G.,1995. Anaerobic digestion and waste water treatment systems. Antonie van Leeuwenhoek, International Journal of General and Molecular Microbiology 67 (1), 3–28.

85. Barber, W.P., Stuckey,D.C.,1999.The use of the anaerobic baffled reactor (ABR) for waste water treatment:a review.Water Research 33(7),1559–1578.

86. O' Neill, C., Hawkes, F.R., Esteves, S.R.R.,Hawkes, D.L., Wilcox S.J., 1999. Anaerobic and aerobic treatment of a simulated textile effluent. Journal of Chemical Technology and Biotechnology 74(10),993–999.

87. B o u a l l a g u i H.,Touhami,Y.,Cheikh,R.B.,Hamdi,M.,2005.Bioreactor performance in anaerobic digestion of fruit and vegetable wastes.Process Biochemistry 40 (3–4), 989–995.

88. Astrid Bensmann, Richard Hanke-Rauschenbach, Kai Sundmacher; Reactor configurations for biogas plants – a model based analysis; Chemical Engineering Science104(2013)413–426

89. Sanjay Sharma; Biogas production, upgradation and slurry management V.K Vijay; Narosa publication (2012); p61-63.

90. Virendra Kumar Vijay; Biogas production, upgradation and slurry management V.K Vijay; Narosa publication (2012); p55-60.

91. Khandelwal, Vibha gupta, Dhussa AK; Biogas production, upgradation and slurry management V.K Vijay; Narosa publication (2012); p25-46.

6

Microalgal Biorefineries - The Choice for Clean Fuel Production

6.1 INTRODUCTION

The uncertainty in the price of petroleum crude and the fact of depleting resources forces the search of new and renewable energy. Biodiesels are assumed to be promising alternative to traditional petroleum fuels. Biodiesel or fatty acid methyl esters (FAME) can be produced from vegetable oils. Large scale production of biodiesel increased in the past decade, but the usage of conventional vegetable oils for the production of biodiesel ends in food vs fuel crisis. The approach of biodiesel production from microalgal oil acts as a suitable alternative to these situations, which again leads to the waste disposal problems. These production facilities leave enormous amounts of byproducts which are a threat to the environment resulting in problem over the disposal of these effluents and shooting up the cost of biodiesel. On the other hand usage of conventional vegetable oils for the production of biodiesel ends in food vs fuel crisis. Integrated biorefinery systems can be a suitable alternative to overcome the pitfalls, in simple a biorefinery can be termed as zero waste technology. Several researches are under progress to study the efficiency of various types of biorefinery approach. This chapter discusses about the various options available in the area of microalgal biorefineries for the effective utilization of biomass feedstock thereby reducing the wastes generated.

The price of petroleum crude in world market is fluctuating and trading around \$110 per barrel, which is imposing a strong impact on developing nations and slow down of the industrial output leading to a huge economic decline. On the other hand the demand for th petroleum oil is rising significantly. More than 95% of the crude oi requirement is consumed by transportation sector. Alternative fuel derived from biomass feedstocks promises a suitable alternative to th conventional energy resources. Fermentative production of alcohol and conversion of certain vegetable oils and oils from other source to biodiesel yields special interest because of their potential to use a a substitute for conventional fuel without any major modification i the system, and also it was proved that these oils release low emissio when compared to the conventional fuels.

6.2 MICROALGAL BIODIESEL

Microalgae are unicellular microorganisms found in water source either in fresh water or marine. Algae can act as a suitable alternate t the use of food crops such as corn and soybean for fuel [1, 2] (Chisti 2007, Hu et al., 2008). Microalgae have high potentials in biodiese production compared to other oil crops. First, the cultivation o microalgae does not need much land as compared to that of terraneou plants [3] (Guan Hua Huang et al., 2010, Chisti Y. 2007). Even thoug algae can substitute such crops it requires plenty of water for thei growth, their water demand is as high as 11–13 million L ha^{-1} year for cultivation in open ponds [4] (Senthil Chinnasamy et al., 2010) They can be cultivated in waste water sources from which they utiliz nitrogen, phosphorus and other inorganic compounds for their growth Microalgae are photosynthetic organisms, they can utilize the CO generated from industries and other sources for their optimal growth thereby paving way for carbon mitigation (Shakeel A. Khan et al, 2010)

They tend to produce various metabolic products [5] (Wijffels et al. 2010) which can be utilized for the production of value added products It was reported that most of the microalgal species were rich in lipid and proteins [6] (Carioca et al., 2009), more particularly they have th ability to synthesize triacylglycerols (TAGs). TAGs are considered to b the potential feedstock for the production of biodiesel [7] (Shakeel A Khan et al, 2010). They can produce and accumulate large quantity o lipids which vary from 20-50% on dry weight basis. Various species o microalgae have various capacities of lipid production.

Table 1 Oil content of microalgae (Chisti, 2007)

Microalga oil content	(% dry wt)
Botryococcus braunii	25–75
Chlorella sp.	28–32
Crypthecodinium cohnii	20
Cylindrotheca sp.	16–37
Dunaliella primolecta	23
Isochrysis sp	25–33
Monallanthus salina	>20
Nannochloris sp.	20–35
Nannochloropsis sp.	31–68
Neochloris oleoabundans	35–54
Nitzschia sp.	45–47
Phaeodactylum tricornutum	20–30
Schizochytrium sp	50–77
Tetraselmis suecica	15–23
B. braunii	25–75

It was reported that the marine species Nannochloropsis sp. can produce more oil when compared to other species studied. Radolfi et al reported the result of his study with 30 different microalgal strains in which Nannochloropsis sp. F&M-M26 produced 61 mg/L/day of lipid.

Table 2 Efficiency of various microalgal species (Table adapted from Rodolfi et al [13])

Microalgae	Biomass productivity (g/L/day)	Lipid content (% biomass)	Lipid productivity (mg/L/day)
	Marine strains		
Porphyridium cruentum	0.37	9.5	34.8
Tetraselmis suecica F&M-M33	0.32	8.5	27.0
Tetraselmis sp. F&M-M34	0.30	14.7	43.4
Tetraselmis suecica F&M-M35	0.28	12.9	36.4

Microalgae	Biomass productivity (g/L/day)	Lipid content (% biomass)	Lipid productivity (mg/L/day)
Phaeodactylum tricornutum F&M-M40	0.24	18.7	44.8
Nannochloropsis sp. F&M-M26	0.21	29.6	61.0
Nannochloropsis sp. F&M-M27	0.20	24.4	48.2
Nannochloropsis sp. F&M-M24	0.18	30.9	54.8
Nannochloropsis sp. F&M-M29	0.17	21.6	37.6
*Ellipsoidion sp.*F&M-M31	0.17	27.4	47.3
*Nannochloropsis sp.*F&M-M28	0.17	35.7	60.9
Nannochloropsis CS246	0.17	29.2	49.7
Isochrysis sp. (T-ISO)CS177	0.17	22.4	37.7
Pavlova salina CS49	0.16	30.9	49.4
Pavlova lutheri CS182	0.14	35.5	50.2
Isochrysis sp. F&M-M37	0.14	27.4	37.8
*Skeletonema sp.*CS252	0.09	31.8	27.3
Thalassiosira pseudonana CS173	0.08	20.6	17.4
Skeletonema costatum CS181	0.08	21.1	17.4
Chaetoceros muelleri F&M-M43	0.07	33.6	21.8
Chaetoceros calcitrans CS178	0.04	39.8	17.6
Fresh Water			
*Chlorococcum sp.*UMACC112	0.28	19.3	53.7
*Scenedesmus sp.*DM	0.26	21.1	53.9
Chlorella sorokiniana IAM-212	0.23	19.3	44.7
*Chlorella sp.*F&M-M48	0.23	18.7	42.1

Microalgae	Biomass productivity (g/L/day)	Lipid content (% biomass)	Lipid productivity (mg/L/day)
Scenedesmus sp. F&M-M19	0.21	19.6	40.8
Chlorella vulgaris F&M-M49	0.20	18.4	36.9
Scenedesmus quadricauda	0.19	18.4	35.1
Monodus subterraneus UTEX151	0.19	16.1	30.4
Chlorella vulgaris CCAP211/11b	0.17	19.2	32.6

).3 MICROALGAL BIOREFINERIES

Even though there is huge advancement in the technology for the naximum production, problems such as waste generation and heir disposal problems shoots up the whole production cost. More particularly the waste or byproducts are generated in large quantity luring the refining process of the produced biofuel. Much more research activities are under progress for the effective utilization of these waste or byproduct. The concept of microalgal biorefinery systems can be a suitable alternative to mitigate this crisis.

The biorefinery is similar in concept to the petroleum refinery, except that it is based on conversion of biomass feedstocks rather than crude oil. Biorefineries in theory would use multiple forms of biomass o produce a flexible mix of products, including fuels, power, heat, chemicals and materials. In a biorefinery, biomass would be converted nto high-value chemical products and fuels (both gas and liquid). By-products and residues, as well as some portion of the fuels produced, would be used to fuel on-site power generation or cogeneration facilities. A biorefinery is a facility that integrates biomass conversion processes and equipment to produce fuels, chemicals, feed, materials and energy from biomass. The objective of a biorefinery is to optimize the use of resources and minimize wastes, thereby maximizing benefits and profitability. Inspite of using vegetable oils, microalgal oil was studied to be a better alternative for the production of biodiesel. It is mainly obtained by the transesterification of microalgal oils in th presence of a catalyst by a primary alcohol (usually methanol) leadin to a fatty acid methyl ester (FAME) and byproducts with major amoun of glycerol.

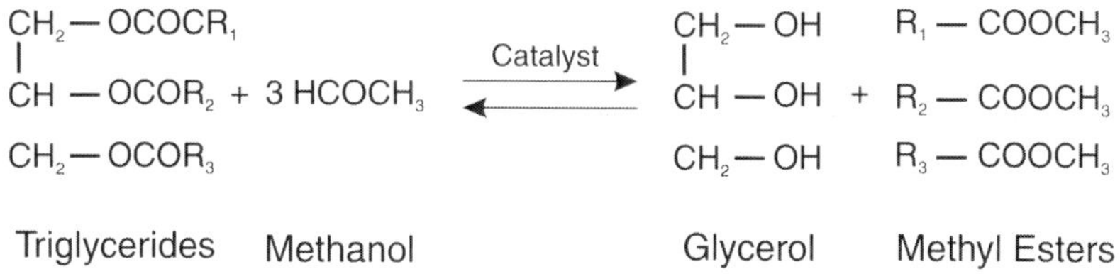

It was estimated that at approximately 10 wt% of glycerol i generated as waste in biodiesel and oleochemical industry [8, 9] (D Silva et al, Dasari et al). These glycerol can be used as substrate fo the biological production of various products like 1,3-propanedio 1,2-propanediol, dihydroxyacetones, hydrogen, polyglycerols, succini acid, and polyesters which can aid the reduction of total production cos of biodiesel [10] (Naresh Pachauri et al., 2006). Biomass is converte to liquid fuel using glycerol that can be blended with gasoline as a alternative fuel [11] (Demirbas, 2000). Mixed culture fermentation o glycerol synthesizes short and medium chain polyhydroxyalkanoat blends [12] (Koller et al., 2005). Tokuma Fukuoka et al., [14 synthesized a novel branched type PLA containing glyceric acid whic is a breakthrough in utilization of waste glycerol for the production o value added bioproducts.

6.4 CONCLUSION

Microalgal biorefineries are in general found to be promising technolog for the production of biodiesel, which can be an integrated facilit to produce various bioproducts as listed above. They have severa advantages which includes: these biorefineries can provide highe surface area for microalgal oil production, like other technologies i does not need more cultivable area thereby can manage food vs fue crisis, and it can be a source to mitigate CO_2 and other toxic gases Even though microalgal biorefineries have several advantages there ar some pitfalls as in the current situation the technology is in initial stag and needs improvements in the process development.

.5 REFERENCES

. Chisti Y, 2007. Biodiesel from microalgae. Biotechnol Adv.25(3):294-306

2. Qiang Hu, Milton Sommerfeld, Eric Jarvis, Maria Ghirardi, Matthew Posewitz, Michael Seibert and Al Darzins. 2008. Microalgal triacylglycerols as feedstocks for biofuel production: perspectives and advances. The Plant Journal. 54 (4): 621–639.

3. Guan Hua Huanga, Feng Chen, Dong Wei, Xue Wu Zhang, Gu Chen. 2010. Biodiesel production by microalgal biotechnology. Applied Energy 87, 38–46

4. Senthil Chinnasamya, Ashish Bhatnagara, Ryan W.Hunt, K.C.Das. 2010. Microalgae cultivation in waste water dominated by carpet mill effluents for biofuel applications. Bioresource Technology. 101, 3097–3105

5. Wijffels, R.H., Barbosa, M.J., 2010. An outlook on microalgal biofuels. Science 329 (5993), 796–799

6. Carioca, J.O.B., HiluyFilho, J.J., Leal, M.R.L.V., Macambira, F.S., 2009. The hard choice for alternative biofuels to diesel in Brazil.Biotechnol. Adv. 27 (6), 1043–1050.

7. Shakeel A. Khanl, Rashmi, 2010. Algal Biorefinery: a Road towards Energy Independence and Sustainable Future. International Review of Chemical Engineering (I.RE.CH.E.), Vol.2, N. l, 63-68

8. Da Silva, G.P., Mack, M., Contiero, J. 2009. Glycerol: A promising abundant carbon source for industrial microbiology. Biotechnol, Adv. 27, 30-39

9. Dasari, M.A., P.P. Kiatsimkul, W.R. Sutterlin, and G.J. Suppes 2005. Low-pressure hydrogenolysis of glycerol to propylene glycol. Applied Catalysis A: General 281(1-2): 225-231.

10. Naresh Pachauri, Brian He. 2006. Value-added Utilization of Crude Glycerol from Biodiesel Production: A Survey of Current Research Activities. 2006 ASABE Annual International Meeting

11. Demirbas, A. 2000.Conversion of biomass using glycerin to liquid fuel for blending gasoline as alternative engine fuel. Energy Conversion and Management 41(16): 1741-1748.

12. Koller, M., R. Bona, G. Braunegg, C. Hermann, P. Horvat, M. Kroutil, J. Martinz, J. Neto, L. Pereira, and P. Varila. 2005. Production of polyhydroxyalkanoates from agricultural waste and surplus materials. Biomacromolecules 6(2): 561-565.

13. Liliana Rodolfi, Graziella Chini Zittelli, Niccolo Bassi, Giulia Padovani, Natascia Biondi, Gimena Bonini, Mario R. Tredici1 2008. Microalgae for Oil: Strain Selection, Induction of Lipid Synthesis and Outdoor Mass Cultivation in a Low-Cost Photobioreactor. Biotechnology and Bioengineering, 102(1): 100-112.

14. Tokuma Fukuoka, Hiroshi Habe, Dai Kitamoto, KejiSakaki 2011. Bioprocessing of glycerol into glyceric acid for use in bioplastics monomer. Journal of Oleo Science. 60 (7) 369-373.

7

Aquatic Biomass for Biofuels

7.1 INTRODUCTION

Energy security is a major issue faced by countries all over the world. The increasing energy consumption has dropped the fuel resource by maximum. In 1985, total global petroleum consumption was 2807 million tons, but in 2008, the figure hiked 3928 million tons [1] and is still increasing due to intensified energy consumption. In 2007, there were 806 million cars and light trucks [2] which is expected to increase up to 1.3 billion by 2030 and 2 billion by 2050 [3]. Currently, one-fifth of the global CO_2 emission is due to transportation and trucking [4]. It is critical to realize the negative impacts imparted on the global environment by fossil fuels that has drifted the exploitation of alternate fuels. The green replacement of fossil-based petro fuel is the trending strategy that has gained much attention from scientists all over the world. Biofuels have the potential to replace existing conventional fuels, reinforce energy security and reduce the emission of both greenhouse gases (GHGs) and other air pollutants. Biofuels are considered sustainable, renewable and environment friendly fuels. Biofuels such as bio-ethanol, bio-butanol and bio-diesel are produced from sugar beet, sugar molasses, soybean or rapeseed [5]. Biodiesel is alkyl esters obtained from transesterification of fatty acids obtained from renewable biomass. Biodiesel is a proven fuel and its production and processing techniques are known for more than 5 decades [6-13]. Biomass fuel potential include wood, short-rotation woody crops, agricultural wastes, short-rotation herbaceous crops, animal wastes, and a host of other materials [14, 15]. Various sources of hydrocarbons used in commercial biodiesel production include canola oil, animal fat, palm oil, corn oil, jatropha oil and waste cooking oil [11, 12, 16]. However using agricultural crops as biofuel feedstock denotes competing with food production as large quantity of fresh water an land supply are needed which raises serious issues on raw materia sustainability [5]. Currently 95% of world's biodiesel is produced fror edible oils [17]. This concern is discussed as food versus fuel disput in long term vision where food scarcity will be the consequence i considering agro crop based biofuel production [18, 19]. Though ther are several other non-edible oil sources, sustainable production of lan based non-edible biomass is difficult to feed large scale biodiesel plants Due to rolling food cost from the competition, global economies hav become unstable since 2006 and the global food stocks of major grain such as rice, wheat, and corn were at their lowest for the past 20 year [20]. For countering these reasons, algae based biofuel productio progressed recklessly in the midst of biofuel research.

Autotrophic micro-organisms such as algae seem to be a promisin way out for unceasing energy appetite [21]. Algae can be directl converted to energy. The hydrocarbon content of algae distinctivel the fatty acids (FAs) and acyl glycerides [22] have the potential t counter the diminishing fossil assets [23, 24]. The oil extracted fron algae can be used for biodiesel production. The residual biomas rich in sugar fraction can be used for production of bio-butanol an bio-ethanol by fermentation. The algal cells suspended in nutrien rich water acts as a reliable bio-mechanism that efficiently convert nutrients and CO_2 to hydrocarbons. Microalgae are photosyntheti micro-organisms that convert sunlight, water and carbon di-oxide t algal biomass [25]. Microalgae are being explored at a faster rate a potential oil source for biodiesel production. About 25,000 specie are reported out of which only 15 are employed as commercial oi producers [26]. Active researches are promoted in the field of micr algal biodiesel in the scope of evaluating it as a potential source fo sustainable biodiesel production. On the other hand, macroalga commonly called as seaweeds remain as untapped marine mysterie for fuel production. Only few investigations have been reported fo macro-algae based biodiesel. Marine macroalgae is a potential biomas for biofuel production [27 -29] because, macroalgae have highe productivity rates than terrestrial biomass such as corn and switcl grass. [20]. Annual biomass production of macroalgae is too high [28 and the ease of degree of maintenance in large scale cultivation o marine seaweed serves as a key factor for using marine macroalga as biofuel feedstock. Although many researchers have been trying t ıtilize lignocellulosic biomass that is not used for food, this biomass :an still incur the same environmental consequences associated with and use and water consumption [30, 31]. Thus, terrestrial biomass-ɔased bio-refinery seems not to be a sustainable choice for scale up at ɔresent due to environmental as well as economic impacts. Microalgae ınd marine macroalgae have the high potentials to fully and partly lisplace terrestrial biomass and produce sustainable bioenergy and ɔiomaterials. This chapter focuses on the technical aspects of algae ɔased biofuel production and its potential to replace fossil fuels.

'.2 ALGAE

Algae are a diverse group of photosynthetic organisms ranging from ınicellular (microalga) to multicellular (macroalgae) forms. They have :hlorophyll as primary photosynthetic pigment and do not have a :ommon ancestor. Commonly, algal population falls under two broad :ategories (1) Microalgae: Microscopic algae that grows in fresh water ınd marine environment and (2) Macroalgae: Comparatively large, nulticellular organisms that grows in marine environment. There are wo main populations of algae: filamentous and phytoplankton algae. These two species, in particular phytoplankton, intensifies rapidly to form ılgal blooms [32]. Though the main storage compound of these algae is ;tarch, oil can also be produced or induced to accumulate within the ɔiomass. The faster growth rate and greater lipid content of microalgae :ompared to oilseed crops urge researchers to develop technologies 'or algae utilization in biodiesel production instead of plant oils [33, 21]. Algae based biofuel production has very less degree of intrusion in the Food vs Fuel dispute of tomorrow which is an added advantage [34].

'.2.1 Microalgae

Microalgae are microscopic, unicellular and phototrophic organisms :hat falls under the categories (1) Diatoms (*Bacillariophyceae*), (2) Green algae (*Chlorophyceae*) and (3) Golden algae (*Chrysophyceae*). Cyanobacteria (*Cyanophyceae*) such as *Arthrospira platensis* and *Arthrospira maxima* are also referred to as microalgae. Diatoms are the dominant life forms in phytoplankton and probably represent the largest group of biomass producers on earth. Green algae are abundantly found in fresh water than in marine waters. The golden algae are similar to diatoms and produce oils and carbohydrates Microalgae are efficient producers of lipids and other great metabolite that work by utilizing nutrients in the presence of solar energy. Th algal species found suitable for biofuel research includes *Chlorella sp.* *Dunaliella sp.*, *Botryococcus braunii*, *Nannochloropsis sp.*, and man others [5]. Autotrophic micro- algae are capable of using carbon dioxid and solar energy to synthesize organics such as protein and lipid fo their growth. Most of the production of autotrophic microalgae fo biodiesel production occurs in indoor photobioreactors that consume heavy illumination for photosynthesis. In comparison, heterotrophi microalgae are more flexible for the cultivation condition (can grov under light free condition), and was found capable of accumulatin higher lipid in the cells. The lipid content of heterotrophic *Chlorell protothecoides* was 3 times higher than that of the autotrophic one [35, 36, 37]. Microalgae commonly double their biomass within 24 and biomass doubling time during exponential growth can be as shor as 3.5 h. And, under specific cultivation conditions, their oil conten can exceed 50% by weight of dry biomass [21]. Microalgae require les land for cultivation than terrestrial crops, can grow in non-potabl water, and do not displace food crops [22, 38]. Strategies such a selection of carbon source and nitrogen source for enhancing lipi accumulation in microalgae have been reported [12, 13, 36, 39, 40] Nanomaterials are found capable of enhancing the microbial activitie and hence, it could be considered that the addition of nanomaterial to microalgae cultivation medium could impact on lipid accumulatio [41, 42]. Nano-materials act as carriers of chemicals to cell thereb performing the role of organic solvents. They also have a significan role in immobilization of lipase in enzymatic transesterification [43] Productivities of microalga and plants are discussed in Table 1.

7.2.1.1 ***Microalgae Diversity and Collection Centers*** It is estimate that 50,000 species of micro algae exists, but only 15 were use commonly [45]. About 20,000+ species are yet to be isolated an analyzed. National Institute for Environmental Studies Collectio (NIES), in Japan holds a collection of about 2150 strains and 70 species of different algae. Fresh water microalgae collection of Coimbr (Portugal) accommodates 4000 strains and 1000 species. The CSIR Collection of Living Microalgae (CCLM) in Australia holds about 80 strains from Australian waters. Collection of Goettingen University ha bout 2213 strains and 1273 species. University of Texas Algal Culture Collection lodges about 2300 different species of microalgae [46]. Several other microalgae collection centers are being established to upport algal process and products.

.2.1.2 Cultivation Methods There are many types of cultivation nethods available for microalgal farming that includes open ponds or aceways, closed ponds, photobioreactors (PBRs), plastic bag systems, vell systems, etc. As of now, raceways and PBRs are the well-established nethods promoting photobiological reactions at affordable cost. Bioreactors favoring desired biochemical conversions are selected or efficient biomass production. Selection of culture systems greatly lepends on the metabolic nature of algae. Autotrophic members need llumination facilities where illuminated raceways, tubular PBRs are great choice for cultivation. Heterotrophic cultures can be grown in conventional bioreactors and where photoheterotrophs need PBR and open ponds for normal growth. Mixotrophic cultures can be cultured in aceways or PBR. Time period of exposure to light also has great effect on productivity of biomass and other co-products in autotrophic and nixotrophic cultures. Depending on the species, growing conditions and growth stage, microalgae have been shown to produce various types of lipids including triglycerides, phospholipids, glycolipids and betaine ipids [47]. Productivity is a measure of how much algal biomass is produced per area per unit of time. Production of up to 127,000kgha^{-1} yr^{1} can be achieved in high-rate raceway ponds [21]. Productivity rates between 20 and 30gm^{-1} day^{-1} (73–109,000kgha^{-1} yr^{1}) are in the range of usual open raceway performance [48]. Productivity is estimated to be a function of photosynthetically active radiation coupled with data for nsolation and radiation [49].

Table 1 Estimated productivity of microalgae compared with other plants [44].

Land plants/microalgae production systems	Annual productivities (Mg DM/ha)
C3 land plants	8–10
C4 land plants	10–30
Open raceway ponds (paddle wheels)	20

Land plants/microalgae production systems	Annual productivities (Mg DM/ha)
Tubular reactor (mixing via air and CO_2)	60
Tubular reactor (dilution of light)	80
Flat panels (intense mixing, short light–dark periods)	100

Algal cell concentrations can be determined using a particle size and enumerator (Multi-Sizer III, Beckman-Coulter). Cell-free sample of filtered culture were taken daily for the determination of residua nutrient concentrations. Few milliliter samples can be filtered throug a 0.22 µm filter and can be stored frozen for analysis and data evaluatio in later stages of experimentation. Biomass productivity can b measured by weighing dry biomass and optical density measurements The mass balance based on system input, output relating growth oxygen evolution rate, nutrients, and carbon dioxide consumption rat are given [50, 51, 52].

Biomass productivity can be measured using the below formula,

$$\frac{dc_x}{dt} = (r_x) - \frac{c_x}{\tau} = (r_x) - DC_x$$

Where C_x (g L^{-1}) is biomass conc., r_x is mean biomass volumetri growth rate, τ is mean residence time and D (d^{-1}) is dilution rate.

Dilution rate (D, d^{-1}) or specific growth rate (μ, d^{-1}) is given by

$$D = \mu = \frac{\text{Volumetric liters harvested } (\textit{Sample})}{\textit{Total Volume} \times T_n}$$

T_n denotes the no. of days taken to harvest the sample (If 1 liter o the sample is harvested in 1 day, T_n = Volume harvested = 1).

Volumetric productivity (P, g $L^{-1}d^{-1}$) is obtained by product o specific growth and biomass concentration (C_x, g L^{-1}),

$$P = \mu \times C_x$$

Notable fall in the biomass concentration during night due to bsence of light can be observed. This can be calculated from below ormula,

$$\textit{Average Biomass loss during night time} = \frac{\text{In}\,(C_{\ll(0)}) - \text{In}\,(C_{(\,)})}{\Delta t_{night}}$$

Where $C_{x(0)}$ denotes concentration of biomass at time 0 (initial). $C_{x(t)}$ denotes concentration of biomass at time t (final), Δt_{night} denotes ime interval between sunset and sunrise in hours (h).

Dissolved ammonium and phosphate in the cultures can be letermined using commercially available kits such as colorimetric assay its with portable spectrophotometer (e.g., TNT830 and TNT843; Hach Co., Loveland, CO) [53]. Post cultivation harvesting process involves biomass recovery from the culture medium that may contribute to 20-30% of the total biomass production cost [54]. Medium cost plays a significant role in cultivation of algal population for biofuels. In case of biodiesel production, the glycerol byproduct can be recycled as feed for nixotrophic and heterotrophic cultures [55]. However there are some effective methods and cost-effective mediums established and being esearched (Table 2 & 3).

Table 2 Cost-efficient medium for microalgae cultivation

Substrate	Biomass (g/L)	Oil content (%)	Oil yield (g/L)	Conversion ratio of biomass/ sugar	Conversion ratio of oil/sugar	Ref.
Glucose	51.2	50.3	25.7	ND	ND	[56]
MHL	97.1	57.1	55.4	0.542	0.318	[56]
MHD	70.9	57.6	40.8	0.481	0.277	[56]
Sugarcane	121.3	45.0	54.8	0.409	0.185	[56]
Cassava	53.6	53.0	28.4	ND	ND	[56]
Sorghum	3.6	52.5	ND	ND	ND	[57]

MHL - Molasses hydrolysate N-limited medium = molasses hydrolysate + base medium
MHD - molasses hydrolysate is the sole component in the medium

Table 3 Requirements and risk factor in farm scale cultivation of microalgae

Type of Cultivation	Biomass production	Risk factor	Maintenance	Investment and operational cost	Agitation and flow	Particular requirements	Common requirements
Open ponds	++	++++	Easy	Low	Paddle wheel, Water jet, Air pump	Water proof sheets (Preferably white), Frame setups for raised wells or wells in excavated pits, Green house,	Nutrients, Mother culture, Cell counter, Circulation pump, heating/ cooling setup, filter harvesters and pH meter, exit gas analyzer.
Closed ponds	+++	+++	Easy	Low (but higher than open ponds)			
Photobioreac-tors	++++	+	Difficult	High	Compressible circulators, Air pumps and Spargers	Acrylic tubes, Glass construc-tions, Feed mixing tank, Automated control-lers for medium, Temp and pH.	
Plastic bags	+++	++	Simple	Low	Air pumps	Plastic bags and Construction frames, Hangers.	
Well systems	+++	+++	Trouble-free	Average	Agitators and air pumps with spargers	Excavated pits with metal tanks, Water proof sheets (Preferably white).	

7.2.1.2.1 *Open Ponds and Raceways* Currently, suspend-based open ponds and enclosed PBRs are commonly used for algal-biofuel production [58]. Open ponds are a very efficient and cost-effective method of cultivating algae, but they become contaminated with unwanted species very quickly [59]. About 98% of commercial algae biomass production is currently with open ponds, even for high value nutritional products, that are sold for prices over hundreds or thousands fold higher than allowable biofuels [60]. Open ponds constructed in excavated pits or raised above ground level is equipped with isolation material such as thick silpauline sheets to carry water and nutrients in tones. Paddle wheel is the commonly used mechanism for mixing the culture in raceways. Other mechanisms include water jet and air pump for mixing. Raceways can be closed by building greenhouse over it to ensure preliminary safety against species contamination and other contaminants. Raceways are advised to be launched in highly selective environments since there is high possibility of other algal species and protozoa to thrive in. Though the biomass productivity of raceways is fair, light utilization is poor. These ponds can be operated as batch or semi-continuous routine. Ponds use more energy to homogenize nutrients and the water level cannot be kept much lower than 15 cm (or 150 L/m^2) for the microalgae to receive enough light [61]. Gas utilization by autotrophic cultures is good at the surface but poor at depth. Atmospheric air has 0.03-0.06% of CO_2 which is very less for good mass transfer [46]. Evolved gas cannot be collected and analyzed. Hence this method proves to be incorrect for bio-hydrogen collection by algae accomplishing co-photolysis. Biomass productivity is poor compared with photobioreactor systems. Area/ Volume ratio is too low as the system requires large surface area and the risk factor is directly proportional to the size of pond. Evaporation results in significant changes of ionic composition in the medium [62] and temperature maintenance due to seasonal variations [21] are the main problematic outlooks. Low capital, operational costs, simple and easy maintenance of ponds adds up for scale up feasibility.

7.2.1.2.2 *Photobioreactors* Photobioreactors can be tubular or flat, serpentine or manifold, helical, inclined, vertical or spiral, pyramids made of acrylic glass. PBRs are flexible for biomass production where operational conditions can be easily matched with optimal growth condition of cultures (Figure 1). PBRs are exclusive for cultivation c phototrophs and mixotrophs. Tubular PBR provides good biomas productivity in outdoor environment. Tubular and flat plate PBR provide large surface area at relatively cheaper cost than column PBR: Flat plate and column reactors provide low hydrodynamic stress o algal cell's nutrient penetration [63]. Volumetric productivity is foun to be 8 times and cell concentration is 16 times higher than open pon production [61]. Sophisticated construction, difficulty in maintenanc of reactor, high capital investment, bio-fouling, illumination ar drawbacks counterbalanced by automation availabilities, hig productivity, less contamination, low oxygen buildup, good mas transfer, etc. (Figure 1). Cultivation of microalga in ponds and raceway for products other than biofuels are given in Table 4.

Table 4 Microalgae cultivation, mode of production and useful bio-proc ucts obtained

Species	Production	Mode	Annual production (in dry wt.)	Product	Area of Application	Price (€/kg)
Spirulina	OP, PBRs	P	3000 tonnes	SCP, Phycobiliproteins, γ – Linolenic acid	Health care, cosmetics	11–35
Chlorella	OP, PBRs	P, H	2000 tonnes	Protein isolates, feed for aquatic organisms	Aquaculture, health care	36–50
Dunaliella salina	OP, PBRs	P, H	1200 tonnes	B-carotene	Health care and cosmetics	215–2150
Aphanizomenon flos-aquae	OP, PBRs	P	500 tonnes	Nutritional additive	Human nutrition	-
Haematococcus pluvialis	OP, PBRs	P, H	300 tonnes	Carotene, Astaxanthin	Health care	501–7150
Crypthecodinidium cohnii	PBRs	H	240 tonnes	Dihydroxy acetone oil	Health care and nutrition	43
Shizochytrium	PBRs	H	10 tonnes	Dihydroxy acetone oil	Health care and nutrition	43
Porphyridium	PBRs	P, H	-	Arachidonic acid	Human nutrition	-

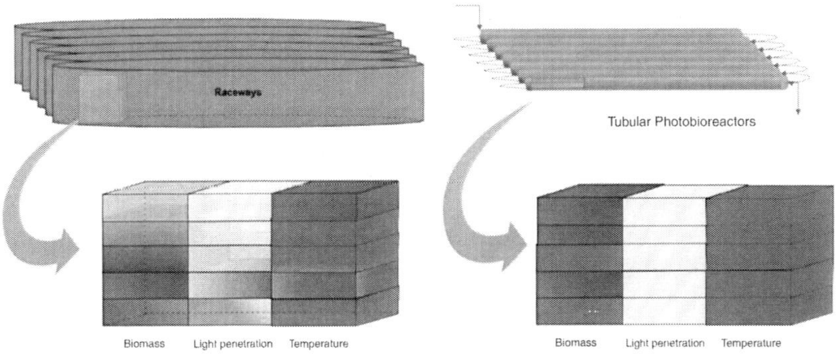

Figure 1 Graphical illustration showing gradient concentration of biomass, light penetration and temperature across a unit volume of open algal pond in comparison with photobioreactor. Photobioreactor superiors open ponds by providing equal biomass concentration, light and temperature throughout the reactor by intense and equal mixing.

7.2.1.3 *Proposed Location for Construction of Microalgae Farms* Though microalgal biofuel production is possible in laboratory scale, meaningful scale up is viable only if cost reducing technologies are adopted and integrated with the existing technologies. Co-location of algal plants with industries producing recommended resources will be the only intelligent way to reduce process economics. Power plants and other chemical industries can be chosen for co-location of algal plants since they produce large amount of heat and CO_2 which is very much required for algal production and processing. The proposed locations are already being investigated for efficiency and may bring fortunate improvements towards cost-effective algal cultivation.

7.2.1.3.1 *Co-location with sugar mills* The ability of cane mills to provide CO_2, water and heat are already being discussed [80]. Sugar is the main product of cane industry with bagasse and molasses as important co-products. Molasses, a thick dark colored liquid, strongly acidic in pH with high BOD and COD [81], is a rich source of glucose which can be fed to algae for higher biomass yield. Bagasse is a fiber residue that can be burnt to heat boiler for electricity generation. Heat produced by boiler is used to convert water to steam and rotate turbine for electricity generation. Since drying the algae up to 90% expenses 60% of process economy [82], a proportion of heat lost in boiler can be driven to heat the wet algal biomass and the electricity produce can be optionally used for illumination of algal reactors. Drying th wet algal biomass is the serious issue faced in post-harvest processin as the energy content of algal biomass has lower energy content tha required for its drying. C. Lohrey and V. Kochergin 2012 investigate sugar mill integrated microalgal biodiesel production and reporte the requirements as 2.5 kg CO_2, 3.4 kWh energy and 1.9 L/kg wate to produce per kg algae dry wt. For sustainable biodiesel productior algal meal can be utilized for co-energy generation in boilers [83 along with glycerol byproduct obtained in biodiesel production tha has heating value of 5.8 kWh/kg [84]. Flue gas from boiler heatin setup is a rich source of CO_2 recommendable as carbon source fo biomass cultivation. D. Yan et al. 2011[85] cultured heterotrophi *C.protothecoides* in molasses hydrolysate medium and the oil fror algae is converted to biodiesel which had palmitic acid methyl esters linolenic acid methyl esters and oleic acid methyl esters contributing t total of 80% comparable with that of glucose-fed microalgal biodiese [86]. Further advancements and researches targeting integratio of these two technologies can bring out significant counts of viabl microalgal biofuel industries (Figure 2).

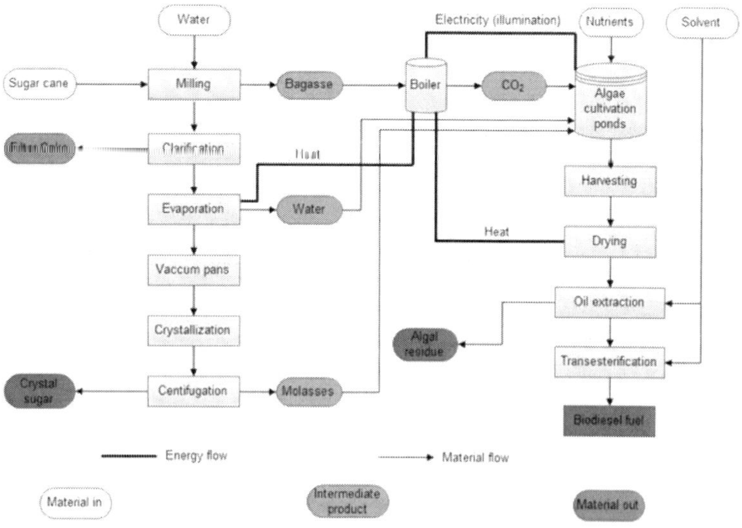

Figure 2 Possible material and energy flow in an algal biofuel plant inte grated with sugar mill. Data adopted [83] and modified in th present form.

.2.1.3.2 ***Cultivation in Waste Water*** Municipal waste water is found to be rich in N, P, K and hence can be used as cultivation medium for microalgae [87,88]. Aquaculture of microalgal strains is a promising method for economical algal cultivation coupled with bioremediation advantage. Waste water with high organic content and inorganic salts left in environment can cause eutrophication thus leading to decrease in dissolved oxygen content of water bodies. Serious environmental issues are being faced with waste water disposal. Biological treatment of waste water can be incorporated in the concept of microalgae cultivation. Some of the members such as *Chlorella sp.* [89,90], *Arthospira sp.* [91], and *Botryococcus sp.* [92] are investigated. At an average *Chlorella vulgaris* can remove about 72% of nitrogen and 28% of phosphorous from waste water [93]. 14 potential strains have been investigated by Li et al 2011 [94] where *C. kessleri* and *C. protothecoides* were found to be capable of mixotrophic growth when cultivated on centrate with high biomass yield of 2.01 g/l and 1.30 g/l respectively. Higher light intensity and exogenous CO_2 concentration with longer lighting period promotes good biomass growth and nutrient accumulation in waste water. Comparative efficiency of microalgae production in piggery waste water and recycled harvest water was evaluated for *Chlorella zofingiensis* [95]. *Neochloris oleoabundans* was investigated for its growth ability on anaerobically digested dairy waste for biodiesel production [96]. Coastal microalgae culturing plants enjoy unlimited water supply that suits mass cultivation of marine microalgae. Use of sea/sewage water can reduce nitrogen usage by 94% and eliminates addition of potassium, magnesium and sulfur [97]. Arid regions will be of good choice for culture systems requiring large surface area such as open ponds. Cultivation of algae in alternate mediums other than conventional nutrient medium is discussed in Table 5.

Table 5 Cultivation of microalgae in industrial, synthetic and farm waste water.

Waste water	Microalgal species	Biomass productivity (mg/L/d)	Lipid productivity (% Dry cell wt.)	Lipid productivities (mg/L/d)	Ref.
Carpet mill	*Botryococcus braunii*	-	13.2	4.5	[98]
	Chlorella saccharaphila	-	18.1	4.2	[98]

Waste water	Microalgal species	Biomass productivity (mg/L/d)	Lipid productivity (% Dry cell wt.)	Lipid productivities (mg/L/d)	Ref.
	Duniella tertiolecta	-	15.2	4.3	[98]
	Pleurochrysis carterae	-	12.0	4.0	[98]
Nitrate and ammonia	*Scenedesmus sp.*	-	12.8	16.2	[99]
Science industrial park	*Chlamydomonas sp. TAI-2*	-	20.9	31.0	[100]
Brewery industry	*Chlorella vulgaris*	-	18	26.8	[101]
	Chlorella sp.	-	20	40.0	[101]
Thiocyanate	*Algae-bacteria mixed*	-	15.2	40.1	[102]
Road waste water reclamation facility	*Nannochloropsis salina*		-	-	[103]
Digested dairy waste	*Neochloris oleoabundans*	88.3 ± 7.9	-	2.57 ± 0.9	[104]
		55 ± 5.5	-	16.5 ± 2.5	[104]
Piggery waste water	*Chlorella zofingiensis*	86.30 ± 6.61	-	-	[105]
		266.66 ± 18.82	-	-	[105]

7.2.2 Macroalgae

Macro-algae or "seaweeds" are fast growing multicellular plant: growing in salt or fresh water that can reach sizes of up to 60 m ir length [106]. Seaweeds are lower level plants with undifferentiatec roots, leaves and stems. Seaweeds are classified into three broad group: based on their pigmentation: (1) brown seaweed (Phaeophyceae) (2 red seaweed (Rhodophyceae) and (3) green seaweed (Chlorophyceae [107]. Macroalgae are photoauxotrophic and thus produce, and store organic carbons (i.e., carbon sources for biorefinery) by utilizing eithe atmospheric CO_2 or HCO_3. [108]. Most macroalgae directly uptake HCO_3 rather than CO_2 for their growth because the diffusion rate of CO_2 is found to be extremely slow in seawater [109]. Due to the igh photosynthetic ability of macroalgae, they have the potential to generate and store sufficient carbon resources needed for biorefinery. Evaluating macroalgal species as feedstock for biofuel is the emerging rend in biorefinery research that has grabbed attention world-wide ince one can grow the same in enormous amount where the globe vill share approximately 3/4th area for it. Advantages of macroalgae as iofuel feedstock includes atmospheric CO_2 mitigation, entrapment of HCO_3 in the water bodies there by reducing the acidic nature of water odies and acid rain hazards, promoting green fuel for green earth, etc. Analysis of macroalgal biomass as fuel feedstock is given in Table 6.

Table 6 Analysis of biomass composition of marine macroalgae [28, 110]

S. No.	Analysis and results					
1.	**Proximate analysis (Different basis)**					
	Volatile matter	Fixed carbon	Moisture	Ash	Sum	
	45.1	23.1	10.7	21.1	100.0	
2.	**Proximate analysis (Dry basis)**					
	Volatile matter	Fixed carbon	Moisture	Ash	Sum	
	50.5	25.9	00.0	23.6	100.0	
3.	**Ultimate analysis (Dry ash-free basis)**					
	Carbon	Oxygen	Hydrogen	Nitrogen	Sulfur	Sum
	43.2	45.8	6.2	2.2	2.60	100.00

7.2.2.1 ***Macroalgae Diversity and Collection Centers*** 90% of marine plants are algae contributing 40% of global photosynthesis [111]. Global diversity and count of seaweed species still remains mysterious (Figure 3). Macroalgal Culture Collection of Kobe University, Japan had about 95 genera, 161 species, 338 strains of macroalgae of which most of them are native to Japan and some foreign countries. This includes 45 genus and 71 species of Phaeophyceae, 42 genus and 63 species of Rhodophyceae, 7genus and 26 species of Chlorophyceae and 1 genus and 1 species of Schizocladiophyceae according to 2011 update. Currently, the center holds 600 species. Research Culture Collection of Institute of Biotechnology (VAST), Vietnam holds about

177 strains comprising 32 genera and 81 species of fresh and marin microalgae, cyanobacteria and macroalgae isolated from soil, fres water and marine water of Vietnam. Culture Collection of Baltic Alga belonging to Institute of Oceanography, University of Gdańsk store 300 species of which 200 species are abroad collections and 100 ar novel collections. Only few established collection centers are availabl for macroalgae that preserves seeds by cryopreservation technique.

Figure 3 Macroalgae diversity 1. *Caulerpa racemosa* [112], 2. *Gracilari fergusonii* [112], 3. *Gracilaria pygmaea* [112], 4. *Hypnea musci formis* [112], 5. *Padina pavonica* [112], 6. *Callophyllis variegat* [113], 7. *Delisea pulchra* [113], 8. *Himantothallus grandifoliu* [113], 9. *Palmaria decipens* [113], 10. *Sargassum duplicatum* [112], 11. *Kappaphycus alvarezii* [114], 12. *Durvillaea antarctic* [113].

7.2.2.2 *Culture Methods* Few cultivation methods have bee established and commercialized for mass cultivation of macroalga biomass. Of all the macroalgae species reported so far, only few hav been cultivated for commercial uses [115]. Approximately 200 specie are known worldwide of which around 10 are cultivated massively

Some of the intensively cultured species include *Laminaria japonica, Undaria pinnatifida, Monostroma sp., Enteromorpha sp., Porphyra sp., Eucheuma sp., Kappaphycus, Gracilaria,* etc. [116]. Small seaweeds such as *Enteromorpha, Ulva, Sargassum, Gracilaria,* etc., can be cultured in artificial systems by pumping seawater in the pond-like aquatic farms. Potential cultivation of seaweeds coupled with offshore wind farm has already been studied to economize the process [117]. Large kelps are difficult to cultivate in artificial environment because of their monstrous growth. Cultivation of seaweeds, growth and biomass intensification depends on various factors such as nutrient composition, pH, temperature, climate, salinity, etc. Seaweeds are mainly produced for these end uses in Asian countries such as China, Philippines, North and South Korea, Japan and Indonesia. The USA, Canada and European countries such as France, Germany and the Netherlands are attempting to establish large-scale seaweed cultivation [118, 119, 120]. China cultivates *Laminaria japonica* in large amounts [116]. *Kappaphycus alvarezii* are being cultivated by fisherwomen in Rameswaram and other southern parts of India. Galician sea accommodates about 17500 species of macroalgae some of which are used in agriculture traditionally [121]. For offshore culture, sandy clay bottom is required for less organic load. Water should be 0.3–0.5m deep with specific gravity around 1.010 (unit). Optimal growth occurs at alkaline pH (7.0–8.0) and temperature between 15-30°C. Pools, crab ponds, bays and straights are chosen as algae cultivation sites. Spores are naturally released from the existing cultures and on favorable conditions, they germinate. Shells of oysters and other molluscs can be used to act as cheap substrates for attachment. Culture techniques should be designed as per nature of the seaweed growth. For suspension culture, rope hangs can be chosen whereas (for) seaweeds requiring bottom supports off-bottom setups, long lines, etc., can be chosen. Kelp cultivation in China is done on artificial floating rafts. These species have low wet to dry biomass weight ratio. Temperature control has greater effect of growth. Spore release does not occur if temperature goes below 5°C. Control of culture conditions and barriers against natural calamities are uncontrollable in macroalgal farming. Artificial pond culture of seaweeds in off-marine areas requires artificial seawater. Composition of artificial seawater has direct effect on algal biomass growth. P. Kaladharan (2000) achieved significant increase in the yields of *Gracilaria cortica and Ulva lactuca* of about 15.5% an 18% respectively [122]. Composition of seawater is given in Tabl 7. Cultivation techniques focusing enhanced biomass productivit towards biofuel production has not been investigated in enoug numbers. Economics of macroalgae cultivation for bioenergy remain untested but evidenced as efficient by majority of production in UK an Ireland. Macroalgae biomass harvest may be possible by wild harves or artificial cultivation in ponds. Possibly, the environmental impact of mass cultivation of macroalgae may be reduced by coupling it witl compatible aquaculture systems [123]. Marine macroalgae farmin suffers few drawbacks such as introduction of invasive species, grazin of weeds by fishes, fishing difficulties, changes in nutrient compositio at the cost of prolonged cultivation in a same location, etc.

Table 7 Seawater composition for alga cultivation (natural and artificial ly prepared)

Ion	Natural seawater [a]	Artifical seawater [b]
Cl^-	19.353	19.353
Na^+	10.76	10.764
SO_4^{2-}	2.712	2.701
Mg^{2+}	1.294	1.297
Ca^{2+}	0.413	0.406
K^+	0.387	0.387
HCO^{3-}	0.142	0.142
Br^-	0.067	0.066
Sr^{2+}	0.008	0.014
H_3BO_3	0.026	0.026
F^-	0.001	0.001

[a] Salinity 35% [124]
[b] Salinity 35% [125]

.3 HARVEST TECHNOLOGIES

Harvesting refers to segregation of solid content. Ratio of mass of liquid to solid should be low at the time of harvest to ensure peak biomass produced. Macroalgae can be hand harvested manually. Microalgae harvesting and dewatering involves use of technical procedures such as filtration, vacuum flow filtration, tangential flow filtration, etc. [126]. Centrifugation is the mostly preferred technique of harvesting [54, 127,128]. Centrifugation uses centripetal acceleration against gravity to settle down the algal cells depending on their densities. Continuous feed fixed bowl centrifuge can be used to obtain 15-20% dry matter followed by vacuum dry around 70°C, 500 torr for dry powder [129]. Higher energy consumption [130] is a limiting factor for centrifugal separation though it gives >95% separation efficiency [131]. Algal paste can be obtained by flocculation or coagulation. The algal paste thickening can be done by ammonium sulphate assisted coagulation for which algae concentration should be around 0.206 g/l [132]. Coagulated sludge can be allowed to settle in settling tanks and later separated. Oil rich microalgae naturally float on water surface and can be harvested easily using flat filter harvesters. Magnetophoretic harvesting technique involves attachment of magnetic nano particles coated with cationic substances by electrostatic interaction. Biomass harvesting is completed in few hours by application of external electric field in harvesting electrodes [133]. Kyubock Lee et al. (2013) adopted this technique using biocompatible chitosan-Fe_3O_4 for recovering 90% of oleaginous *Chlorella sp.* [134]. Starch and chitosan are also investigated as bioflocculants [135, 136]. Chitosan is considered as a widely used bioflocculant and effective in biomass recovery when used in culture medium at pH 10 [137]. Ammonium sulphate, ferric sulphate, ferric chloride, etc., are chemical flocculants advised for large scale harvesting but known to have toxic effects on microalgae [138]. Ultra sonic aggregation is a non-hazardous, non-toxic efficient method for continuous biomass harvesting [130]. Flocculation is followed by filtration, flocculation or gravimetric separation [54]. Microalgae can also be immobilized in alginate beads and cultivated to reduce difficulties in harvesting but growth hinders to a notable extent. Wet biomass is generally sun-dried, low pressure shelf-dried or drum-dried [139]. For valuable product extraction, spray drying [140], fluidized bed drying [141] and freeze drying [142] is recommended. Drying dehydrates the biomass and avoids spoilage. Microalgal drying seems to consume more of the total energy utilized in the process [54, 127]. Renewable energy driving devices such as solar panels, heat reflecting windows can be used for drying. High capital investment is the set up hindering constraint.

7.4 POST-HARVEST PROCESSING TECHNOLOGIES

The biomass collected from cultivation site is processed prior to fuel conversion steps. This includes (1) drying for dewatering the biomass, (2) extraction of products & co-products and (3) grinding for homogenization of biomass components with biological/chemical reactants during conversion.

7.4.1 Cell Disruption Techniques

Number of techniques has been practised for cell lysis which broadly falls under (i) mechanical disruption (ii) chemical disruption and (iii) enzymatic disruption. Mechanical press is the conventional method similar to grinding that involves disruption of cell by mechanical shear stress. Average of 65% oil can be obtained by this method when operated manually or engine powered. Filtration and degumming are necessary post extraction treatments [143]. Cell debris often intrudes with the targeted product and mechanical press suits only for macroalgae. Dried cells can be disrupted and powdered using ball milling method. Electroporation done by electric charge variation can vary the size of pores on cell wall and possibly break if the charge variation exceeds than advised. High pressure homogenization seems energy expensive method. Osmotic shock can also be employed to outflow the cellular contents to aqueous medium. Polymyxin, lysine strong mineral acids such as HCl and H_2SO_4, organic acids such as acetone and methanol, Dimethyl Sulfoxide (DMSO) and all cationic agents can be used for chemical degradation of cell wall [144]. Addition of chemicals makes downstream costly and can impart toxic nature to the product. Possibility of quality degradation is high in this case and enzymatic treatment is considered comparatively a safer method reatment using papain, pectinase, neutrase, cellulose, etc., [145] can ffectively break cell wall in a product friendly manner but remains ostly. Product extraction can also be made possible without cell death nown by a process called milking. This method suits microalgae roduct extraction using suitable solvents [146]. However upcoming echnologies like ultrasound and microwave assisted cell lysis are ppropriate for algal disruption.

.4.2 Oil and Biomass Recovery

Bligh–Dyer method is one among the simplest and earliest discussed nethods for lipid extraction [147]. For soxhlet extraction, wet/dry iomass can be loaded directly after proper chopping and pulverization. Dry biomass can be ball milled to crack cell wall and improve extraction fficiency. Chloroform: methanol in the ratio 2:1 is the mostly preferred olvent composition for lipid extraction [148]. Hexane is used to extract eutral lipids [129]. Selection of solvent is based on polarity difference xisting between the solvent and compound to be extracted. General pan of lipid extraction is 6-12 hours where some unusual biomass oading may require 12-24 hrs or even more. In traditional solvent based ipid extraction, chlorophyll and magnesium associated compounds re usual contaminants spoiling the oil and biodiesel quality [149]. The xtracted solvent along with lipid and coloring pigments are treated vith charcoal and evaporation to recycle solvent and complete oil xtraction. Wet lipid extraction combines the use of centrifugation nd solvent extraction after acid/alkaline treatment of biomass [150]. Quality oil can be obtained but tedious steps in extraction supports only nalytical purposes. Solvents are costly, volatile and hazardous. Use of n-hexane in extraction is environmentally hazardous as it releases oxic water as waste. Eco-friendly extraction is achieved by employing nzymes as molecular machines. But prolonged mining of products rom biomass and the enzyme cost has limited the establishment of his method [151].

Super critical carbon dioxide method is an alternate green echnology that eliminates solvent assistance for oil extraction [160]. Super critical CO_2 facilitates a safe extraction due to its low toxicity, ow flammability, and lack of reactivity [161]. Super critical methanol for direct conversion of oil from wet algae to alkyl esters was als investigated [162]. Optimal operation conditions would be as high a 50°C and 200-250 bar pressure for which energy allocation is difficu in algae processing. Ultrasound Assisted Extraction (UAE) is th emerging and promising method for extraction of cell lysate [151 Microwave Assisted Extraction (MAE) of lipids was identified as th simplest and efficient methods of lipid extraction [163]. UAE and MA are considered for efficient and quality extraction in short processin time [164, 165]. UAE and MAE can be performed by taking biomas solvent in w/v ratio of 1:100 at temperature ranging between 30°C–50° for about 5-30 min [148]. These methods are non-toxic, energy efficien environmental friendly and hence the choice of many. To eliminat the oil extraction step in the case of biodiesel production, direc conversion also called as *in situ* transesterification of oil in biomas has been proposed [166,167]. Direct conversion of biomass showe much reduced phospholipid content as result of efficient extractio followed by transesterification [168]. Various oil extraction technique and corresponding yields are listed in Table 8.

Table 8 Various oil extraction methods and lipid yield [152-160]

Extraction method	Technique adopted	Conditions	Lipid yield (%)
Chemical method	n-hexane in soxhlet	-	95-99
	Chloroform, ethanol, deionized water	8hr	49±2.4
	Aqueous oil extraction	2hr	38
	Ultrasound assisted aqueousvoil extraction	50°C, pH 9, 6hr	67
	1st acetone, n-hexane	48hr	-
	Subcritical ethanol extraction	Ethanol:alga 20:1 (v/w), 105°C, 100 min	72.82
Enzymatic method	Aqueous enzymatic oil extraction with cellulase/hemicellulose	60°C, pH 4.5, 2hr	73

Extraction method	Technique adopted	Conditions	Lipid yield (%)
	Aqueous enzymatic oil extraction with alkaline protease	60°C, pH 7, 2hr	86
		50°C, pH 9, 6hr	64
	Ultrasound assisted alkaline protease treatment	50°C, pH 9, 6hr	74
Mechanical methods	Engine driven	-	68
	Screw press	-	80
		-	79
	Ram press	-	62.5
Microwave method	B20 as co-solvent	80°C,1.2KW, 2.45GHz, 15 min hold, 30 min, cool-down	13.1±0.8
		100°C ,1.2KW, 2.45GHz, 15 min hold, 30 min, cool-down	16.5±1.6
		120°C ,1.2KW, 2.45GHz, 15 min hold, 30 min, cool-down	11.8±2
	B40 as co-solvent	80°C, 1.2KW, 2.45GHz, 15 min hold, 30 min, cool-down	32.2±5.9
		100°C ,1.2KW, 2.45GHz, 15 min hold, 30 min, cool-down	38.2±7.7
		120°C ,1.2KW, 2.45GHz, 15 min hold, 30 min, cool-down	56.6±7.9
	Chloroform + ethanol	80°C, 1.2KW, 2.45GHz, 15 min hold, 30 min, cool-down	15.5±0.7
		100°C ,1.2KW, 2.45GHz, 15 min hold, 30 min, cool-down	45.5±2.2

Extraction method	Technique adopted	Conditions	Lipid yield (%
		120°C ,1.2KW, 2.45GHz, 15 min hold, 30 min, cool-down	53.1±2.6
Super critical method	Super critical CO_2	80°C, 250 bar	14.1

7.4.3 Fuel Conversion Processes

Fuel conversion strategies are generally thermochemical or biochemica in nature. Thermochemical conversion includes gasificatior liquefaction, pyrolysis and direct combustion. Biochemical metho covers biological production of biogas, fermentable production o alcoholic solvents and biophotolysis for hydrogen gas.

7.4.3.1 ***Oil Transesterification*** Oil having triacylglycerides (TAG) an free fatty acids (FFA) is transesterified with alcoholic solvents in th presence of suitable catalyst (Figure 4). Prior to transesterification oi is washed with water to remove impurities. 0.88% KOH can be use to wash oil to reduce lipid loss during washing [169]. Washing of oi is followed by pretreatment to reduce FFA content. Methanol is th best acyl acceptor known to give peak conversion and quality biodiese [16]. Other acyl acceptors include ethanol, isopropanol, n-butanol n-propanol, isobutanol, isoamyl alcohol, etc. Poor solubility of shor chain alcohols may have inhibitory effect on the active site of enzyme fo which solvents like tert-butanol, petroleum ether, n-hexane can be use [170]. Solvents enhance mutual solubility of hydrophobic TAG wit hydrophilic alcohols. Batch wise addition of acyl acceptor is advise for solvent free system [171]. Use of novel solvents such as dimethy carbonate eliminates complex glycerol separation [172, 173]; ethy acetate [174] and methyl acetate [175] are other novel acyl acceptor eliminating glycerol production and micro-emulsion formation durin biodiesel synthesis.

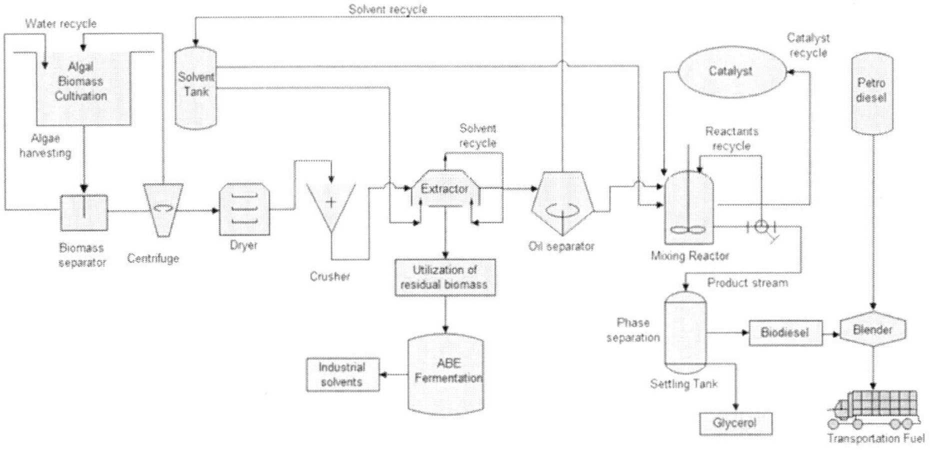

Figure 4 Algal Biodiesel production and processing

Unfortunately, solvent cost remained an inhibitory factor for the process commercialization. Effect of both homogeneous and heterogeneous acid/base catalysts has been exclusively reviewed [176]. Homogeneous (H_2SO_4, trifluoro acetic acid, HCl, etc.) and heterogeneous (SO_4/SnO_2, ZrO_2/SO_4^{2-}, S-ZrO_2, $Al_2O_3/ZrO_2/WO_3$, etc.) acid catalysts are used for conversion of oil with high Free Fatty Acid (FFA) content [34], 177, 178]. Alkali catalysts (NaOH, KOH, KF/Al_2O_3, CaO/Al_2O_3, Lithium-doped ZnO, $Ca(OCH_2CH_3)_2$, etc.) are conventionally used for biodiesel production from low FFA oil. Enzymes are good possible biological catalysts to produce biodiesel. They can easily treat fatty acids as well as triglycerides to produce biodiesel from non-edible or waste oil, reaching high conversion [179, 180]. Use of immobilized enzyme and whole cell catalyst gives clear glycerol and biodiesel products [181]. Application of nanomaterials in biodiesel transesterification is the current trend under experimentation. Rhizopus is the most widely used whole cell catalyst in biodiesel transesterification to achieve good conversion of oil [16]. For aquatic microalgal biomass, when the solid content reaches 90%, transesterification similar to soybean oil can be performed [48]. It is also estimated that 2–10 l of water/l of biodiesel is needed [31]. Esterification of oil from various sources and resulting ester production is extensively reviewed [181]. Main parameters affecting biodiesel production includes FFA content, moisture, acyl acceptor, nature & concentration of catalyst, organic co-solvents, reaction temperature, pH, reaction time, agitation, mode of reactio system, etc. [16, 181-192]. Biodiesel transesterification methods ar listed in depth by Marchetti (2009) provided in Table 9.

Table 9 Comparison of different technologies for biodiesel productio [193]

Variable	Base	Enzyme	Supercritical	Monolithic	Resin	Acid
Temp. [°C]	60–70	30–50	200–350	50–180	60–180	50–80
Products from FFA	Soaps & Esters	Esters	Esters	Esters	Esters	Water & Esters
Effect of water [a]	↓	↓			↓	
Yield to ester	Normal	High	High	Normal	Good	Normal
Purification of glycerol	Difficult	Simple	Simple	Simple	Simple	Difficult
Reaction time [b]	1–2 h	8–70 h	4–10 min	6 h	Variable	4–70 h
Ester purification	Difficult	Simple	Simple	Simple	Simple	Difficult
Cost	Cheapest	Expensive	Expensive	Medium	Medium	Cheaper
Amount of equipment	High	Low	Low	Low	Low	High

a In this case the down arrow means that water is a drawback while the line means that is not effected and the system will be able to treat a raw material with some amount of water.

Base - down arrow - saponifies the fatty acid.

Enzyme - down arrow - water in small quantity activates enzyme, but large quantity water deactivates the catalyst.

Resin - down arrow as well as a line - water has different effects over different solid catalysts.

Monolithics - line - leaching is not caused by water but for a non-stability of the catalyst.

b The reaction time set in this table is what it is most likely, however, it is important to point out that other times for the same technology could be found in the open literature.

Biodiesel quantification can be done by given formula,

For, direct single step conversion (direct SCM, UAE, MAE assiste esterification)

$$\text{FAAE (\% by wt of algae loaded)} = \frac{\text{wt of FAAE obtained}}{\text{wt of algae taken}} \times 100$$

For, two step biodiesel production (oil extraction followed b separate transesterification)

$$\text{FAAE (\% by wt of oil taken)} = \frac{\text{wt of FAAE obtained}}{\text{wt of oil taken}} \times 100$$

Integrated microalgae photobioreactor equipped with biogas and biodiesel production facility was developed and economy analysis was also done [126]. Unreacted free fatty acids in biodiesel can be removed by EFAR process [194]. Glycerol byproduct from biodiesel production can be valorized for economic benefit [126]. Direct supercritical methanol and *in situ* transesterification reduces the need for lipid extraction step but bounces drop in yield %. Blending of biodiesel with petro-diesel is recommended for use in engines. B6-B20 blends are covered by American Standards for Testing of Materials (ASTM D7467) [195].

7.4.3.2 ***Alcoholic Fermentation*** ABE fermentation is a high valued biochemical process known to produce acetone, butanol and ethanol in the ratio 3:6:1 [196-199]. Two significant phases in ABE fermentation are (1) Acidogenesis: Production of organic acids in exponential phase, and (2) Solventogenesis: Production of solvents from acids. Aldehyde dehydrogenase, butanol dehydrogenase and alcohol dehydrogenase are the enzymes involved in production of butanol. Marine macroalgae and microalgae have been utilized as a good source of saccharides for ethanol fermentation [200, 201]. For this purpose, conventional batch reactors remain simple and consistent [202, 203] but suffers severe disadvantage like prolonged lag phase, product inhibition, etc. [204]. To overcome substrate inhibition and increase biomass concentration, fed batch fermentation is advised [205]. Further increase in biomass productivity and product yield can be achieved by using cell recycle or immobilized bioreactors [206] where the reactant and product stream flows continuously [207]. Immobilized reactor allows the flow of broth alone arresting the movement of microbial strain [205] preventing cell damage from shear forces of the flowing broth [208] enhancing cell reuse [209]. Recommended strains for ABE fermentation are *Clostridium acetobutylicum*, *Clostridium beijerinckii*, *Clostridium saccharoperbutylacetonicum* and *Clostridium saccharobutylicum* [210]. Hydrolysate of green *Ulva lactuca* was used for the production of acetone, butanol and ethanol (ABE) by *C. acetobutylicum* and *C. beijerinckii*. Hydrolysate-based media were fermentable without nutrient supplementation. *C. beijerinckii* utilized all sugars in the hydrolysate and produced ABE at high yields (0.35 g ABE/g sugar consumed), while *C. acetobutylicum* produced mostly organic acid (acetic and butyric acids) [211]. Prior to fermentation, pretreatmer of biomass is done to hydrolyse the complex cellulosic biomas Ammonia fiber expansion (AFEX) process decrystallizes cellulos depolymerises hemicelluloses, and depolymerises and removes ligni from cellulose/hemicelluloses. This method is found to be an effectiv pre-treatment for promoting enzymatic hydrolysis of a wide variet of biomass sources [212-215]. *Aspergillus niger, Aspergillus oryza Fusarium solani, Penicillium occitanis, Trichoderma reesi, Clostridiu acetobutylicum, Bacillus subtilis,* and *Streptomyces lividans* are highl recommended for biological pretreatment. NaOH is the stronge alkaline used in pretreatment [216]. Hot water treatment also calle autolysis is economic and degradation of xylose is less [217]. Aci treatment involves the use of sulphuric acid for desaccharificatio from cellulose. Release of glucose from cellulosic substrate increase with removal of lignocellulosic fraction [218]. Assortment of algae a ethanol feedstock is done by evaluation of sugar fractions, lignin an relative sugar content in the biomass. Lignin content can be evaluated b method discussed by Hill et al. (1998) [219]. Carbohydrate content ca be determined by sulphuric acid–phenol assay [220]. Yield of ethan from *Ulva reticulata* is estimated to be about 60 l/tonne after hydroge peroxide pretreatment after acid hydrolysis of cellulosic biomass [221 More of the macroalgae species has to be investigated for promisin ethanol outcome from algal biorefinery.

7.4.3.3 *Anaerobic Digestion* Anaerobic digestion promotes energ recovery from sunlight via photosynthetic system in the form of bioga (CO_2 and CH_4). Methanogenic digestion of oil extracted algal biomas was proposed to reduce external energy demands [48]. Concept o combining anaerobic digestion to economize the microalgal cultivatio was first proposed in 1959 [222]. Anaerobic digestion is appropriate fo biomass with high moisture content (80–90%) and thus can be use for microalgae [223]. Anaerobic digestion can convert the carbon an nitrogen content of delipidized algal biomass to methane worth th energy obtained from extracted lipids [224] in a few steps [225].

(1) Biopolymers to saccharides (hydrolyzation),

(2) Saccharides to carboxylic acids and alcohol (fermentation),

(3) Acids and alcohol to acetate, hydrogen and carbon dioxide (acetogenesis), and

(4) Acetogenic products to methane and carbon dioxide (methanogenesis).

Biomass ($C_\alpha H_\beta O_\gamma N_\omega$) → Methane ($CH_4$) + Ammonia ($NH_3/NH^+_4$) + Carbon dioxide ($CO_2$) + Residue

Biomethane production can be calculated from the formula,

$$M = M_{BG} \times (\%CH_4) \times (M_{BM} + M_S + M_C + M_G)$$

Where, M is the mass of biomethane produced, M_{BG} is mass of biogas produced by anaerobic digestion of biomass, and M_{BM}, M_S, M_C and M_G are the mass of biomass, sludge, waste paper and glycerine respectively.

Good carbon to nitrogen (C/N) ratio is expected in the biomass to operate the fermentation process with positive impressions [223]. Rate of degradation of organic matter (volatile solids) is of major importance in determining the process performance which is 45% for algae and 60% for waste water sludge [226]. Delipidization leaves the algal cells with good C/N ratio thus making it more suitable for anaerobic fermentation. More optimal C/N ratio can be obtained by mixing sewage sludge, waste paper pulp, and glycerine with biomass [227, 228].

C/N ratio can be calculated from formula,

$$CN_{min} \leq (a \times CN_{BM} + b \times CN_S + d \times CN_C + k \times CN_G) \leq CN_{max}$$

$$(a; b; d; k;) = 1 \text{ and } 0 \leq a; b; d; k \leq 1.$$

Subscripts BM, S, C and G stand for biomass, sludge, waste paper and glycerin, respectively. a, b, d, and k are the different percentages of the respective feedstocks.

Efficient and economic biomethane production depends on algal concentration after cultivation. Unfermented residues can be used as fertilizer or animal feed. The fermented remains of biomas are proposed to undergo recycling in algal cultivation to conserve th outflow of nutrients from the system [222, 225]. The pH of fermente residue lies between 6.6 and 7.6 implying nearer to normal pH thu ready for nutrient recycle after solid-liquid extraction [229-231 Resultant biogas is mainly composed of methane and carbon dioxid but also traces of hydrogen sulphide, dinitrogen, dihydrogen and othe volatile compounds [236]. Methane has low solubility in water tha H_2S and CO_2. Hence water is passed through a packed bed pressurize with biogas to dissolve out gases other than methane [237]. Methan yield from various biomass are given in Table 10.

Table 10 Yield of methane produced from different feedstocks

Biomass	Methane yield ($m^3\ kg^{-1}$)	Reference
Laminaria sp.	0.26–0.28	[232]
Gracilaria sp.	0.28–0.4	[233]
Sargassum sp.	0.12–0.19	[233]
Macrocystis	0.39–0.41	[232]
L. digitata	0.5	[234]
Ulva sp.	0.20	[235]
Water hyacinth	0.13–0.21	[232]
Sorghum	0.26–0.39	[232]
Poplar	0.23–0.32	[232]
Food waste	0.54	[232]
Tetraselmis suecica	0.20–0.41	[126]
Microalgae – ACAD model	0.54	[228]

* Biomethane yields depend highly on the fermentation conditions, therefore the data derived in the same studies are comparable, but not necessarily the data between individual studies.

7.4.3.4 ***Gasification*** Biomass oxidation at high temperature (betwee 800–1000°C) results in syngas formation [238]. Syngas contains CO CO_2, H_2, N and CH_4 in varying compositions [239]. Biomass react with oxygen and steam (H_2O) to produce syngas. Nitrogen conten volves out after mineralization as ammonia. Gasification can be done o any biomass for syngas generation. Gasification of spirulina at 000°C yielded 0.64g of methane per gram of biomass [240]. *Chlorella ulgaris* is gasified and by adopting novel nitrogen recycle mechanism, mmonia rich fertilizer grade biomass is produced [241]. Energy spent or gasification and drying seems not to be significantly beneficial. Iowever, if biomass to syngas conversion is efficient, gasification may urn out to be advantageous and much favored for biogas outlet.

'.4.3.5 ***Pyrolysis*** Pyrolysis also known as thermal cracking is hermal decomposition of biomass in the absence of oxygen [242]. It s a promising method of biomass conversion technique but complex ind influenced by various factors [243]. Pyrolysis is a conventional echnique for biomass-liquid fuel conversion potential for large scale iofuel production [244]. This process is simple, effective and complete 245], yielding products with good physical and chemical properties 246]. Any lignocellulosic biomass will have three components: Iemicellulose, cellulose and lignin in 20-40, 40-60 and 10-25% wt. espectively. Thermal cracking involves exposure of biomass to igh temperature (up to 700°C) in the presence of suitable catalyst. Conversion is a four-step process [247]: (i) moisture evolution, (ii) emicellulose decomposition, (iii) cellulose decomposition and (iv) ignin decomposition. Variety of organic compounds such as alkenes, lkanes, carboxylic acids and other aromatics are derived in this rocess [143]. Biomass conversion efficiency peaks 95.5% [244] and uality products are obtained by adopting hydrogenation and catalytic racking treatments [239]. Pyrolytic treatment of cellulose, D-glucose, hlorogenic acid and xylan biomass are analyzed using FTIR for gas ompounds evolved from the process [248, 249]. Pyrolytic yields of sugar raction-rich biomass are given in Table 11. Pyrolytic experiments with *Chlorella protothecoides* are much extensive and microalgal pyrolysis lata has reached enough entries [251] ready for commercialization. Bio-oil yield of *Chlorella protothecoides* is found as 51.8% (39.7 MJ kg^{-1}) 244] and 57.9% (41 MJ kg^{-1}) [35]. Ash and carbon residue disposal are major issues involved with the process.

Table 11 Gas components evolved by pyrolysis of hemicellulose, cellu lose and lignin [250]

Sample	Gas product yield (milli mol/g-biomass)					
	H_2	CO	CH_4	CO_2	C_2H_4	C_2H_6
Hemicellulose	8.75	5.37	1.57	9.72	0.05	0.37
Cellulose	5.48	9.91	1.84	6.58	0.08	0.17
Lignin	20.84	8.46	3.98	7.81	0.03	0.42

7.4.3.6 ***Biophotolysis*** Algae are capable of producing biohydroge by photolytic reaction of water against photonic energy from sunligh [252]. Hydrogen is a naturally occurring molecule carrying good energ and so categorized as efficient fuel [163]. There are various method to produce biohydrogen from algal biomass (Table 12 & 13). Method include pyrolysis, gasification, microbial conversion, thermolysis electrolysis, etc. [178] but, biophotolysis and photofermentatio occur with live condition of algae. Algae produce hydrogen by direc photolysis of water while some species of bacteria photofermentat organic acids to give hydrogen [253–255]. Microalgal metabolism o organic compounds can also release H_2 along with CO_2 by indirec photolysis.

(a) Direct photolysis: $2H_2O \rightarrow \text{light} \rightarrow 2H_2 + O_2$

(b) Indirect photolysis:

(1) $12H_2O + 6CO_2 \rightarrow \text{light} \rightarrow C_6H_{12}O_6 + 6O_2$

(2) $C_6H_{12}O_6 + 12H_2O \rightarrow 12H_2 + 6CO_2$

Table 12 Biohydrogen production systems

Production method	Advantage	Disadvantage
Direct photobiolysis	Simple and average yield	Deactivated by excess oxygen, separation of oxygen and hydrogen from outstream is difficult.
Substrate catabolism	Sustainable production, suits indoor and does not need illumination or light.	Expensive, scale up difficulties, complexities in process control and maintenance.

Two-stage biophotolysis	Sustainable process for intensive H_2 production	Process parameters are difficult to evaluate and imply in the reaction environment.

Table 13 Various species of microalgae employed in photobiological hydrogen production

Species	Mode of cultivation	Type of culture	Reactor	Time	H_2 production	Ref.
Chlamydomonas reinhardtii (CC124)	Batch	Photomixotrophic	Tubular	48h	0.6ml L^{-1} h^{-1}	[265]
Chlamydomonas reinhardtii (CC124)	Batch	Photomixotrophic	Stirred tank	192h	1.5370.05 mL L^{-1} h^{-1}	[266]
Chlamydomonas reinhardtii (CC124)	Batch	Photomixotrophic	Stirred glass bottles	140h	102 ml	[267]
Chlamydomonas reinhardtii (CC124)	Continuous	Photomixotrophic	Stirred glass bottles	4000h	0.58ml h^{-1} L^{-1} PBR	[268]
Chlamydomonas reinhardtii (Dang 137Cmt+)	Batch	Photomixotrophic	Stirred glass bottles		175ml L^{-1}	[258]
Chlamydomonas reinhardtii (Dang 137 Cmt+)	Batch	Photoautotrophic	Flat glass bottles	100h	7173ml L^{-1}	[269]
Platymonas subcordiformis	Batch	Photoautotrophic	Bottle flask	50h	11,720nL h^{-1}	[270]
Platymonas subcordiformis	Batch	Photoautotrophic	Bottle flask	8h	0.339mL h^{-1} L^{-1}	[272]
Chlorella pyrenoidosa C-101	Batch	Photoautotrophic	Bubble column	60-65h	6.9×10^{-2} m^3 kg^{-1} cell	[271]
Chlorella vulgaris MSU 01	Batch	Photoautotrophic	Stirred tank	6-7 days	26 ml	[273]
Spirulina platensis (NIES-46)	Batch	Photoautotrophic	Erlenmeyer flask	20hr	2 mmol $mgDW^{-1}$	[274]
Tetraspora sp. CU2551	Batch	Photomixotrophic	Vials	>24 h	17.3–61.7 mmol mg^{-1} Chl a h^{-1}	[275]
Anabaena variabilis (ATCC29413)	Continuous	Photoautotrophic	Vials	18-100h	to5nmolh^{-1} mg $Chla^{-1}$	[276]

Species	Mode of cultivation	Type of culture	Reactor	Time	H_2 production	Ref.
Anabaena variabilis (CCAP 1403/4B)	Continuous	Photoautotrophic	Stirred container	35 days	12–14 mLg^{-1} DW h^{-1}	[277]
Anabaena variabilis (ATCC29413)	Batch	Photoautotrophic	Panel	50h	40-50ml	[278]

During photosynthesis, water is broken to give hydrogen ion (H^+) and oxygen (O_2). Hydrogenase enzyme converts hydrogen ions t hydrogen [256]. Photosynthetic oxygen production inhibits the activit of hydrogenase enzyme resulting in decreased hydrogen productio [257, 258]. In simultaneous production of photosynthetic oxygen an hydrogen, electrons released upon photosynthetic break down of wate is directly converted to hydrogen by hydrogenase [252]. Generall hydrogen production by photolysis is a two-step reaction where alga is cultivated photosynthetically [257] and subjected to anaerobi conditions for stable hydrogen production [259]. By this two stag technique, green alga is assumed to produce 198 kg H_2 h^{-1} per da theoretically [259]. Biohydrogen yield and conversion efficiency can b determined by using the formula below [260],

$$\text{Biohydrogen yield (\%)} = \frac{\text{Amount of hydrogen produced (mol)}}{\text{Amount of substrate supplied (mol)}}$$

$$\text{Conversion efficiency (\%)} = \frac{\text{Amount of substrate utilized}}{\text{Total substrate supplied}} \times 100$$

Inhibitors can also be employed to increase H_2 generatio but considered inferior to the above process [260]. Some specie of *Platymonas* and *Chlamydomonas* are insensitive against abov described sulphur deprivation where inhibitors and nitroge deficiency can bring out enhanced hydrogen production [255, 261] Photoproduction of seaweed *Platymonas helgolandica* has also bee investigated for hydrogen producing potential [255]. This method o biological production of hydrogen is environmentally safe, efficien enough and non-hazardous [262]. Co-cultivation of algae with bacteri can effect intensive hydrogen production [263] but the system will b difficult to understand as it involves complex biological reaction [264] Though cultivation under stressed condition gives good results, geneti ngineering application in biohydrogen production will be a novel .ttempt and expected to bring notable fortunes [255].

'.5 ALGAL BIOFUELS & CHARACTERISTICS

:urrently, algal biofuels are non-viable due to cost factors, but expected o be practically viable in the near future. Various fuel products are btained from algae of which methane, ethanol, butanol and biodiesel .re most significant. Other fuel products include bio-oil, hydrogen and ther hydrocarbon derivatives (Figure 5).

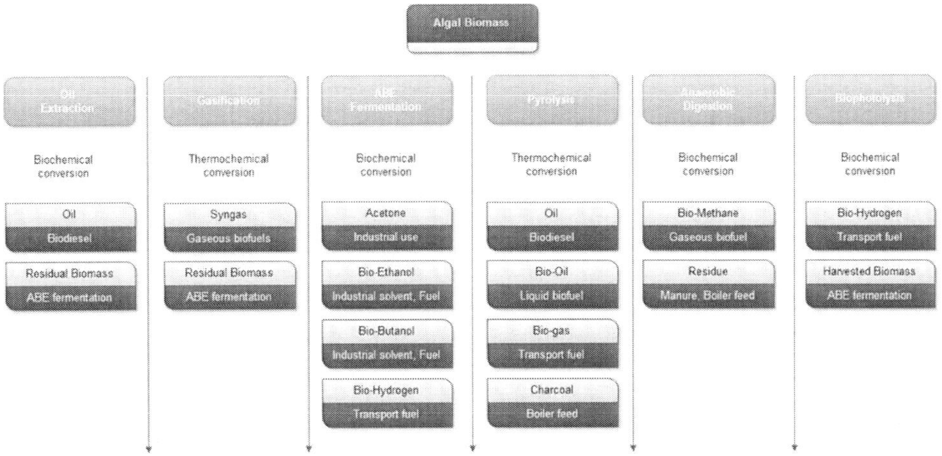

'igure 5 Available conversion processes and corresponding fuel products obtained from algal biomass [279–283].

'.5.1 Algal Oil and Biodiesel

3iodiesel is a high valued mono alkyl ester product considered as an merging alternate energy resource. Biodiesel obtained by various nethods such as pyrolysis, dilution, transesterification, microemulsion nd supercritical processes have different qualities and properties. 'roperties of biodiesel and petro-diesel are given in Table 14. Quality f biodiesel product greatly depends on the feedstock quality and xtent of reaction. Cetane number of biodiesel depends on fatty acid omposition of oil directly affecting the fuel combustion.

Table 14 Reported properties of algal biodiesel compared with 1st generation biofuels and ASTM biodiesel standards [129, 143].

Test	1st generation Biodiesel	Microalgae Biodiesel	Macroalgae Biodiesel	Biodiesel ASTM D6751standards	Petro-diesel ASTM D975 standards	ASTM Methods
Flash point (°C)	100-170	>160	149	Min. 100-170	60-80	D93
Distillation, 90% recovery	~339	-	371	360°C Max.	-	D1160
Carbon residue (wt %)	59.5 - 81	-	0.018	77	84-87	D4530
Total glycerine	0.2 – 0.3	0.102	0.169	Max. 0.24 % (m/m)	-	D6584
Free glycerine	Min. 0.001 - Max 0.009	0.014	0.006	0.02 % (m/m)	-	D6584
Water and sediment	<0.005	≤0.005	<0.005	Max. 0.05 (vol %)	0.05 (vol %)	D2709
Sulphur, total	0.0-0.0015	0.6	8.43	Max. 0.05% (m/m)	0.05% (m/m)	D5453
Cetane number	45-65	-	71.67	Min. 47	40-55	D613
Cloud point (°C)	-2 to -12	3.9	-16	-3 to -12	-20	D2500
Sulphated ash	<0.005	≤0.005	0.008	Max. 0.02% (m/m)	-	D874
Copper strip corrosion	1	1	1	Max. 3	1	D130
Acid number	<0.6	0.003	0.01	Max.0.50	0.062	D664
Kinematic viscosity at 40°C (cSt)	3.6 – 9.48	-	9.8	1.9 – 6.0	2.6	D445
Cold soak filtration	-	-	126	-	-	D6217
Phosphorus	<0.1	≤0.1	<1	Max.0.001	-	D4951

Moderate cetane number of 50 is preferred for proper operation of ignition engines with good cold start properties [284]. Low cost short chain methanol and ethanol yields quality biodiesel [143]. Long and branched chain alcohol can also be employed but costs higher [17, 285 - 288]. Biodiesel has good combustion characteristics, reduces smoke and carbon dioxide emissions by 78% compared to petro-diesel [10, 289]. Biodiesel production and utilization can promote us forward for a safer living, generate employment opportunities and light rural development [279].

7.5.2 Biobutanol and Bioethanol

Biobutanol and bioethanol are clean burning fuels and considered environmentally safe as greenhouse gas emission is comparatively lesser than fossil fuels. On combustion, they emit carbondioxide and water.

$$C_2H_60 + 30_2 \rightarrow 2CO_2 + 3H_2O \text{ (ethanol combustion)}$$

$$C_4H_{10}O + 60_2 \rightarrow 4CO_2 + 5H_2O \text{ (butanol combustion)}$$

Increasing food and energy security issues increased interest in alcohol fuel production from renewable sources and other conventional agricultural feedstocks [280]. Quantity of alcohols obtained from algae is much dependent on the reaction parameters [290]. Micro and macro algae are fast growing carbon accumulating biomass venturing alcoholic fuel production [149, 280, 291]. Bioethanol from microalgae can be produced by fermentation of biomass or direct cellular reactions [292, 293]. Extracted algal biomass can be pretreated to convert starch to glucose after which biological hydrolysis is done to produce ethanol [294]. Genetic engineering of fermentating organisms enabling utilization of fixed carbon to directly excrete biosolvents in fermentation medium is also considered [295]. *Zymomonas mobilis* was introduced into microalgal culture assuming that bacterium will efficiently utilize the CO_2 for ethanol conversion [295, 296]. This method of ethanol production reduces downstream processing of algal biomass and novelty lies with product excretion in the culture medium [292, 297]. Biobutanol production is done by biochemical reactions facilitated by clostridium species [25, 298 - 300].

Fuel properties of butanol and ethanol are given in Table 15.

Table 15 Properties of bioethanol and biobutanol as vehicular fuels

Properties	Butanol	Ethanol
Melting point (°C)	- 89.3	-114
Specific gravity	0.810– 0.812	0.79
Ignition temperature(°C)	35–37	276-456
Auto-ignition temperature (°C)	343–345	422
Flash point (°C)	25–29	12.77
Relative density (water: 1.0)	0.81	0.805-0.812
Critical temperature (°C)	287	239.85
Explosive limits (vol. % in air)	1.4–11.3	3.3-19
Vapor pressure (kPa at 20°C)	0.5	5.95
Boiling point (°C)	117–118	78
Density at 20°C (g/ml)	0.8098	0.7851
Energy density (MJ/l-1)	27–29.2	19.6
Energy content/value (BTU/gal)	110000	84000
Liquid Heat capacity (Cp) at STP (kJ/k-mol.°K)	178	112.3
Research octane number	96	129
Motor octane number	78	102
Viscosity (10^{-3} Pa.s)	2.593	1.078

7.5.3 Biomethane

Biomethane from biogas is a fuel source that can be converted to hea and electrical energy on combustion [280, 301, 302]. Combustion c methane in Internal Combustion (IC) engines happens by oxidation c methane. Combined heat and power gas (CHG) engine burns methan to CO_2 completely [237].

$$CH_4 + 2O_2 \rightarrow CO_2 + 2H_2O + \text{Heat}$$

Carbon dioxide from biogas and flue gas from combustion c biomethane can be recycled back to the culture system as nutrients [238

31, 232, 302]. Recycling flue gas and CO_2 from biogas as feed for carbon nitigating autotrophs can take the process in a feasible way (Figure). Recycling the culture water can reduce nutrient usage by 55% [97]. Syngas obtained by gasification method is a typical source of methane. Syngas has low calorific value of 4-6 MJ m^{-3} that still can be used to run ngines for heat, power and drive turbine for electricity [303].

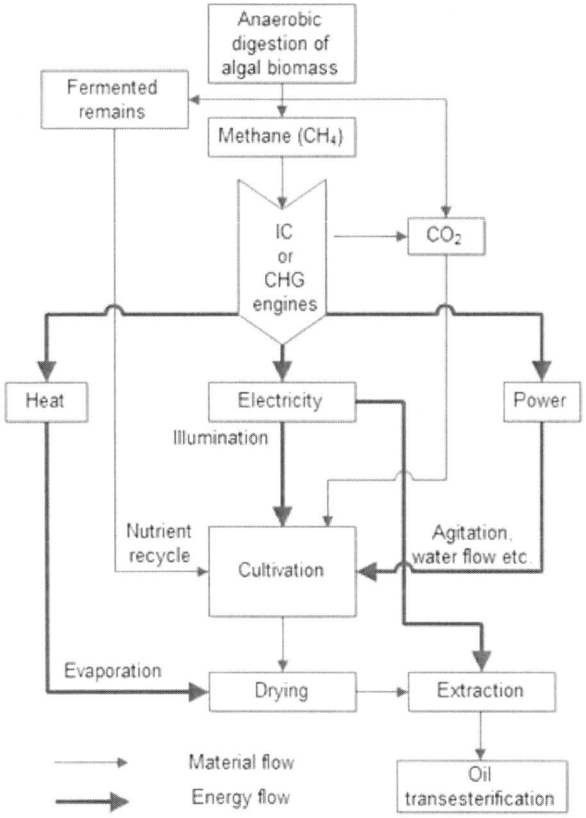

Figure 6 Reutilization of resources in integrated algal biodiesel–biogas plant describing material and energy flow possibilities

7.5.4 Bio-Oil

Bio-oil, the liquid content obtained by thermal treatment of biomass such as pyrolysis, thermal liquefaction, etc., shares much similar properties with petroleum oil. Bio-oil possesses various fuel compounds but as a whole has a good heating value ranging from 29-46.5 MJ kg^{-1}.

Density of oil is nearly equal to petroleum oil. Viscosity factor greatl varies according to composition and purity of bio-oil. Though man useful products are recovered from bio-oil, the reactor for therma liquefaction of biomass and fuel feed are complex and expensive [303 Initial capital cost limits the process scale up, but operational cos can be overcome by promising energy outcomes. Energy outcome t input ratio such as 6.67:1 [304] and 2.94:1 [305] denotes conversion c biomass to bio-oil is a viable option for fuel conversion. Characteristic and composition of bio-oil samples are given in Table 16.

Table 16 Composition and properties of bio-oils obtained by fast pyro ysis [35, 181, 286, 306]

Properties	**Typical values**			**ASTM testing methods**
	Bio-oils		**Petroleum oil**	
	Wood	**Microalgae**		
C (%)	56.4	62.07	83.0–87.0	-
H (%)	6.2	8.76	10.0–14.0	-
O (%)	37.3	11.24	0.05–1.5	-
N (%)	0.1	9.74	0.01–0.7	-
Density (kg l^{-1})	1.2	1.06	0.75–1.0	ASTM D-6751
Viscosity (Pas)	0.04–0.20 (at 40°C)	0.10 (at 40°C)	2–1000	ASTM D-445
Higher Heating Value (MJ kg^{-1})	21	29 – 45.9	42	ASTM D-6751

7.5.5 Biohydrogen

Biohydrogen is a zero emission fuel considered to be much safer tha all other fuels. Hydrogen carries energy that can be used to replac gasoline in the vehicles. On combustion, it reacts with oxygen to giv water.

$$H_2 + O_2 \rightarrow H_2O \text{ (water)}$$

Water produced is radiant and energy obtained is lesser than spen in production. Combustion in internal engines can drive out powe nd electricity useful for beneficial activities. Hydrogen as fuel is used n space craft propulsion as it has the highest heating value. Efficient tilization of hydrogen fuel for running vehicles is being considered s the hot topic in hydrogen combustion science. Electric power of HW is generated by utilizing hydrogen at volumetric flow rate of 19.7 mol/h using proton exchange membrane and fuel cell [307]. Biohydrogen production is coupled with fuel cells to harvest energy roficiently [308, 309]. Possibly, hydrogen can be used in Fuel Cell Vehicles (FCV) (Figure 7) and Hydrogen Internal Combustion Engines H-ICE). Hydrogen with wide range of flammability draws much ttention towards safer storage and handling. High flame speed, lower lensity, high diffusivity and low ignition energy provides increased uel-power ratio improving fuel economics. Biohydrogen production trategies are still on the verge of exploitation and offers improvement pportunities. Commercialization of biohydrogen fuel may take few ears if supported by novel research activities. Technical approaches ddressing scale up and operational issues can fortunately improve roduction economics for viable biohydrogen fuel in the future.

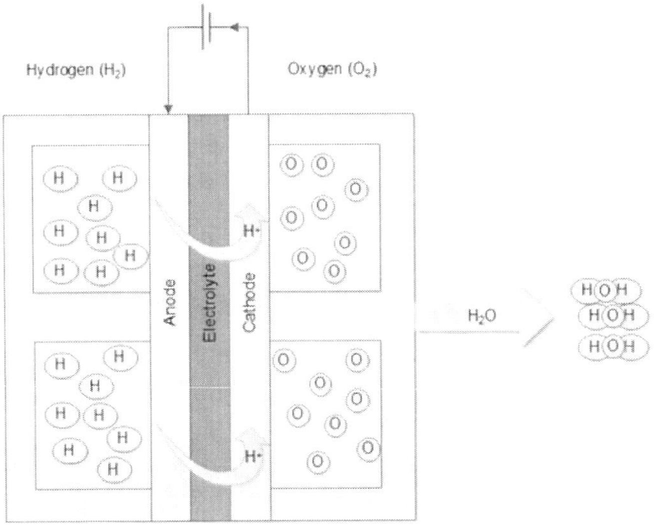

Figure 7 Fuel cell consuming hydrogen and oxygen to produce water and electrical energy

7.6 SIGNIFICANCE OF ALGAL BIOMASS FOR BIOFUELS

Aquatic biomass could also be used as raw material for co-firing t produce electricity, for liquid fuel (bio-oil) production via pyrolysi or for bio-methane generation through fermentation [107]. Cell wal of diatoms have been composed of polymerized silica and accumulat oil and chrysolaminarin. The fresh water green algae *Haematococcu pluvialis* is commercially important as a source for astaxanthin, *Chlorell vulgaris* as a supplementary food product, and the halophilic alga *Dunaliella* species as a source of -carotene. Extracted biomass propose to be used as fertilizer or animal feed is significant in adding economi value to the process. Co-products such as pigments, agar, carrageena and other bioactive compounds are value added products that can b removed before fuel conversion of biomass [69]. Methane, butanol an ethanol can be effectively produced by fementative digestion of residu algal biomass by selective microorganisms [310]. Microalga utilizes 1.8 kg of carbon/kg of dry weight of biomass [21]. Though algae are aquati they use less amount of water than terrestrial plants [311]. No herbicid or pesticide is recommended in algal cultivation [38]. Growth of alg population in waste water promotes them as dual purpose choice fo biofuel production and organic load degradation [256]. Biochemic composition of algal biomass can be changed by varying the growt parameters thus inducing richness of specific biomolecular fractions i resultant biomass steering to the purpose of cultivation [312].

7.7 COMPROMISES AND FUTURE OUTLOOK OF ALGAL BIOFUELS

Future of algal fuels is based on developing cost effective strategies b adopting suitable technologies such as sugar mill integration, biogas biodiesel production, co-cultivation of algae with other living aquati entities, promoting co-product recovery, marketing, etc. Working o technical aspects of cultivation systems thus making it more efficien for intensified biomass production, reduced material and energy los by technically advanced system fabrication are expected as next leve improvements for profitable fuel production. Exploiting algal specie with higher CO_2 mitigation rate can be effective enough in reducin greenhouse gas thus bringing mother earth back to naturalistic manne

'uel studies are made within the limited species of interest considering hemresearchfriendly. Heterotrophicalgaethoughsuperiorsautotrophic nes in accumulating mass lipids, it consumes high carbon, nitrogen nd phosphorous nutrients for which sewage water is proposed as an lternate cradle. Selecting outdoor culture systems for mixotrophic nembers can eliminate illumination requirements and promotes energy aving. Contamination risk of outdoor ponds and capital investment of losed photobioreactors limits the construction of commercial plants. strain selection should be done by considering climatic conditions revailing around the production facility. Appropriate reactor design nd processing strategies should be designated pondering the biology f algal biomass. Progress in biochemical and genetic engineering can timulate higher productivity of targeted product and co-products. lgal biomass-based biofuel production will emerge as a multi-purpose rocess if suitable strategies are developed and placed properly in a rocess sequence for sustainable outcome of value added co-products long with main fuel products.

'.8 CONCLUSION

'his review highlights the technical advancements in upstream/ ownstream processes of algal fuel production. Co-extraction of 'aluable products can take the process in a feasible way thus promoting ndustrialization and commercialization of the process. Versatility of lgae is a promising parameter for them being prompted to be used as a iofuel resource. Easy adaptability nature of algae for surviving in wide 'ange of climatic conditions make them most wanted in the scientific narket spread around the globe for the purpose of biofuel research nd implementation. Distinguished hybrid reactors are fortunate utcomes of ground breaking strategies and technology advancements 'or promoting aquatic biofuels. Sustainable supply of oil can be made ossible by combining microalgal oil with macroalgal oil in case of iodiesel transesterification. Algal oil can be possibly combined with ther plant oils even attempting absolute use of all oil feedstocks. Transesterification standards are to be evaluated and established for ommercialization of mixed oil conversion. Though technologies 'or efficient use of residual biomass are already developed, more of ts other side is to be made effective for efficient downstreaming of products. Cost effective algal biofuel is possible only when the energ content of the same remains higher than consumed in production an processing strategies. Sooner in the future, advancements in technolog boosted by convention of knowledge from chemical, biotechnology an marine science can turn out algal energy into a sustainable, renewabl economical and green fuel for safer world. Novel approaches utilizin alga and other fuel feedstock effectively have guaranteed market in th scientific community, leading to a possibly quicker progress to emerg first in the race for budgeting biofuels.

7.9 REFERENCES

1. British Petroleum (BP), BP Statistical review of world energ 2009. BP plc, London, June 2009.
2. Plunkett JW. Plunkett's automobile industry almanac 200ᵋ automobile, truck and speciality vehicle industry marke research, statistics, trends & leading companies, Housto (Texas): Plunkett research ltd, 2007.
3. World Business Council for Sustainable Development (WBCSD Mobility 2030: meeting the challenges to sustainability, Th Sustainable Mobility Project, Geneva (Switzerland), 2004.
4. Goldemberg J. Environmental and ecological dimension of biofuels, Int Conference on the ecological dimensions c biofuels, Washington (DC), 2008 (march 10).
5. Talebi AF, Mohtashami SK, Tabatabaei M, Tohidfar M, Bagheri A Zeinalabedini M. et al. Fatty acids profiling: A selective criterio for screening microalgae strains for biodiesel production. Alga Research 2013; 2:258–267.
6. Knothe G, Dunn RO, Bagby MO. Biodiesel: the use of vegetabl oils and their derivatives as alternative diesel fuels. ACS Sym Ser 1997; 666:172–208.
7. Fukuda H, Kondo A, Noda H. Biodiesel fuel production b transesterification of oils. J. Biosci. Bioeng. 2001; 92:405–16.

. Barnwal BK, Sharma MP. Prospects of biodiesel production from vegetables oils in India. Renew. Sustain. Energy Rev. 2005; 9:363–78.

. Demirbas A. Biodiesel production from vegetable oils via catalytic and non-catalytic supercritical methanol transesterification methods. Prog. Energy Combust. Sci. 2005;31(5–6):466–87.

0. Van Gerpen J. Biodiesel processing and production. Fuel Process Technol. 2005;86:1097–107.

1. Felizardo P, Correia MJN, Raposo I, Mendes JF, Berkemeier R, Bordado JM. Production of biodiesel from waste frying oil. Waste Manage. 2006;26(5):487–94.

2. Kulkarni MG, Dalai AK. Waste cooking oil — an economical source for biodiesel: A review. Ind. Eng. Chem. Res. 2006;45:2901–13.

3. Meher LC, Vidya Sagar D, Naik SN. Technical aspects of biodiesel production by transesterification — a review. Renew. Sustain. Energy Rev. 2006;10:248–68.

4. Demirbas A. Current technologies for thermo-conversion of biomass into fuels and chemicals. Energy Sour. 2004;26:715–30.

5. Xiu S, Shahbazi A. Bio-oil production and upgrading research: A review. Renew. Sustain. Energy Rev. 2012;16:4406–14.

6. Balasubramanium B, Perumal AS, Jayaraman J, Mani J, Ramanujam P. Comparative analysis for the production of fatty acid alkyl esterase using whole cell biocatalyst and purified enzyme from *Rhizopus oryzae* on waste cooking oil (sunflower oil). Waste Manage. 2012;32:1539-47.

17. Gui MM, Lee KT, Bhatia S. Feasibility of edible oil vs non edible oil vs waste edible oil as biodiesel feedstock. Energy 2008;33:1646-53.

18. Refaat AA. Correlation between the chemical structure of biodiesel and its physical properties. Int. J. Environ. Sci. Technol. 2009;6:677-94.

19. Chhetri AB, Tango MS, Budge SM, Watts KC, Islam MR Nonedible plant oil as a new sources for biodiesel production Int. J. Mol. Sci. 2008;9:169–80.

20. Heady D, Fan S. Anatomy of a Crisis: The Causes an Consequences of Surging Food Prices (IFPRI Discussion Pape 00831). The International Food Policy Research Institute 2008. Available from: <http://www.ifpri.org/sites/default/ files publications/ifpridp00831.pdf>.

21. Chisti Y. Biodiesel from Microalgae. Biotechno. Adv. 25 (2007 294–306.

22. Hu Q, Sommerfeld M, Jarvis E, Girardi M, Posewitz M, Seiber M et al. Microalgal Triacyl Glycerols as feedstocks for biofue pr.oduction: perspectives and advances. The Plant J. 2008;54:621 39.

23. Guschina I, Harwood J, Lipids and lipid metabolism i eukaryotic algae. Prog. Lipid Res. 2006;45:160-186.

24. Hasheminejad M, Tabatabaei M, Mansourpanah Y, Khatamifa M, Javani A. Upstream and Downstream strategies to economiz biodiesel production. Bioresour. Technol. 2010;102:461-68.

25. Chisti Y. Biodiesel from microalgae beats bioethanol. Trend Biotechnol. 2008;26:126-31.

26. Raja R, Hemaiswarya S, Kumar NA, Sridhar S, Rengasamy R A perspective on the biotechnological potential of microalgae Crit. Rev. Microbiol. 2008;34:77-88.

27. Tamilarasan S, Nagarajan NG, Sahadevan R. Production of alga biodiesel from marine macroalgae Enteromorpha compressa by two step process: Optimization and kinetic study. Bioresour Technol. 2013;128:392–400.

28. Ross AB, Jones JM, Kubacki ML, Bridgeman T. Classification o macroalgae as fuel and its thermochemical behavior. Bioresour Technol. 2008;99:6494–504.

9. Maceiras R, Rodrı´guez M, Cancela A, Urrejola S, Sanchez A. Macroalgae: Raw material for biodiesel production. Appl. Energy. 2011;88:3318–23.

0. Dominguez-Faus R, Powers SE, Burken JG, Alvarez PJ. The water footprint of biofuels: a drink or drive issue? Environ. Sci. Technol. 2009;43:3005–10.

1. Fargione J, Hill J, Tilman D, Polasky S, Hawthorne P. Land clearing and the biofuel carbon debt. Science 2008;319:1235–38.

2. Demirbas A. Production of biodiesel from algae oils. Energy Sources Part A 2009;31:163–8.

3. Meng X, Yang J, Xu X, Zhang L, Nie Q, Xian M. Biodiesel production from oleaginous microorganisms. Renew. Energy 2009;34:1–5.

34. Oncel SS. Microalgae for a macroenergy world. Renew. Sustain. Energy Rev. 2013;26:241–264.

35. Miao X, Wu Q. High yield bio oil production from fast pyrolysis by metabolic controlling of Chlorella protothecoides. J. Biotechnol. 2004;110: 85–93.

36. Miao X, Wu Q. Biodiesel production from heterotrophic microalgal oil. Bioresour. Technol. 2006;97:841–6.

37. Shen Y, Yuan W, Pei Z, Mao E. Heterotrophic culture of Chlorella protothecoides in various nitrogen sources for lipid production. Appl. Biochem. Biotechnol. 2010;160:1674–84.

38. Rodolfi L, Zittelli GC, Bassi N, Padovani G, Biondi N, Bonini GB et al. Microalgae for oil: strain selection, induction of lipid synthesis and outdoor mass cultivation in a low-cost photobioreactor. Biotechnol. Bioeng. 2009;100–112.

39. Cheng Y, Lu Y, Gao C, Wu Q. Algae-based biodiesel production and optimization using sugarcane as the feedstock. Energy Fuels 2009;23:4166–73.

40. Li X, Xu H, Wu Q. Large-scale biodiesel production fro microalga Chlorella protothecoides through heterotroph cultivation in bioreactors. Biotechnol. Bioeng. 2007;98:764–71

41. Shan G, Xing J, Zhang H, Liu H. Biodesulfurization dibenzothiophene by microbial cells coated with magneti nanoparticles. Appl. Environ. Microbiol. 2005;71:4497–502.

42. De Windt W, Aelterman P, Verstraete W. Bioreductive depositio of palladium (0) nanoparticles on Shewanella oneidens with catalytic activity towards reductive dechlorination polychlorinated biphenyls. Environ. Microbiol. 2005;7:314–25.

43. Zhang XL, Yan S, Tyagi RD, Surampalli RY. Biodiesel productio from heterotrophic microalgae through transesterification an nanotechnology application in the production. Renew. Sustai Energy Rev. 2013;26:216–223.

44. Rosch C, Skarka J, Patyk A. Microalgae – opportunities an challenges of an innovative energy source. In: 17th Europea biomass conference and exhibition, Hamburg, Germany; Jun 29–July 3, 2009.

45. Foidl N, Foidl G, Sanchez M, Mittelbach M, Hackel S. Jatroph curcas as a source for the production of biofuel in Nicaragu Bioresour. Technol. 1996;58:77–82.

46. Mata TM, Martins AA, Cetano NS. Microalgae for biodies production and other applications: A review. Renew. Sustai Energy Rev. 2010;14:217-32.

47. Greenwell H, Laurens L, Shields R, Lovitt R, Flynn K. Placin microalgae on the biofuels priority list: a review of th technological challenges. J. Royal Society Interface 2010;703 726.

48. Lardon L, Hlias A, Sialve B, Steyer JP, Bernard O. Life-cycl assessment of biodiesel production from microalgae. Enviro Sci. Technol. 2009;43:6475–81.

49. Clarens AF, Resurreccion EP, White MA, Colosi LM Environmental life cycle comparison of algae to other bioenerg feedstocks. Environ. Sci. Technol. 2010;44:1813-19.

0. Cornet JF, Favier L, Dussap CG. Modeling stability of photoheterotrophic continuous cultures in photobioreactors. Biotechnol. Prog. 2003;19:1216–1227.

1. Pruvost J, Van Vooren G, Le Gouic B, Couzinet-Mossion A, Legrand J. Systematic investigation of biomass and lipid productivity by microalgae in photobioreactors for biodiesel application. Bioresour. Technol. 2011;102:150–158.

2. Pruvost, J, Cornet JF, Legrand J. Hydrodynamics influence on light conversion in photobioreactors: an energetically consistent analysis. Chem. Eng. Sci. 2008;63:3679–3694.

3. Dickinson KE, Whitney CG, McGinn PJ. Nutrient remediation rates in municipal wastewater and their effect on biochemical composition of the microalga Scenedesmus sp. AMDD. Algal Research 2013;2:127–134.

4. Grima ME, Belarbi EH, Acien Fernandez FG, Medina RA, Chisti Y. Recovery of microalgal biomass and metabolites: process options and economics. Biotechnol. Adv. 2003;20:491-515.

5. Pyle D, Garcia R, Wen Z. Producing docosa hexaenoic acid (DHA) rich algae from biodiesel-derived crude glycerol: effects of impurities on DHA production and algal biomass composition. J. Agric. Food Chem. 2008;56:3933–3939.

6. Dong Yan, Yue Lu, Yi-Feng Chen, Qingyu Wu. Waste molasses alone displaces glucose-based medium for microalgal fermentation towards cost-saving biodiesel production. Bioresour. Technol. 2011;102:6487–6493.

7. Chunfang Gao, Yan Zhai, Yi Ding, Qingyu Wu, Application of sweet sorghum for biodiesel production by heterotrophic microalga *Chlorella protothecoides*. Appl. Energy 2010;87:756–761.

8. Wen Z, Johnson MB. Microalgae as a feedstock for biofuel production. Communications and Marketing, College of Agriculture and Life Sciences. Virginia Polytechnic Institute and State University. Blacksburg VA 2009; 442-886.

59. Schenk PM, Thomas-Hall SR, Stephens E, Marx UC, Mar JH, Posten C et al. Second generation biofuels: high-efficienc microalgae for biodiesel production. Bioenergy Res. 2008;1:20 43.

60. Benemann JR. Opportunities and challenges in algae biofuel production. In: Algae world 2008 conference, Singapore November 17–18, 2008.

61. Richmond A. Handbook of microalgal culture: biotechnolog and applied phycology. Blackwell Science Ltd; 2004.

62. Pulz O. Photobioreactors: production systems for phototrophi microorganisms. Appl. Microbiol. Biotechnol. 2001;57(3):287 93.

63. Liam Brennan, Philip Owende. Biofuels from microalgae—a review of technologies for production, processing, and extractions of biofuels and co-products, Renew. Sustain. Energ Rev. 14 (2010) 557–577.

64. Pulz O, Gross W. Valuable products from biotechnology o microalgae. Appl. Microbiol. Biotechnol. 2004;65:635–48.

65. Medina AR, Grima EM, Gimenez A, Gonzales MJI. Downstream processing of algal polyunsaturated fattyacids. Biotechnol. Adv 1998;16:517–80.

66. Vilchez C, Garbayo I, Lobato MV, Vega JM. Microalgae-mediated chemicals production and wastes removal. Enzyme Microb Technol. 1997;20:562–72.

67. Huerlimann R, Nys R, Heimann K. Growth, lipid content productivity and fatty acid composition of tropical microalga for scale-up production. Biotechnol. Bioeng. 2010;107:245–57.

68. Li Y, Horsman M, Wu N, Lan CQ, Calero ND. Biofuels from microalgae. Biotechnol. Prog. 2008;24:815–20.

69. Spolaore P, Joannis-Cassan C, Duran E, Isambert A. Commercial applications of microalgae. Journal of Biosci. Bioeng 2006;101(2):87– 96.

0. Amin Hejazi M, Wijffels RH. Milking of microalgae. Trends Biotechnol. 2004;22(4):189–94.

1. Lorenz RT, Cysewski GR. Commercial potential for Haematococcus microalgae as a natural source of astaxanthin. Trends Biotechnol. 2000;18(4):160–7.

2. Ratledge C. Fatty acid biosynthesis in microorganisms being used for single cell oil production. Biochimie 2004;86(11):807–15.

3. Camacho Rubio F, Acie´n Ferna´ndez FG, Sa´nchez Pe´ rez JA, Garcı´a Camacho F, Molina Grima E. Prediction of dissolved oxygen and carbon dioxide concentration profiles in tubular photobioreactors for microalgal culture. Biotechnol. Bioeng. 1999;62(1):71–86.

4. Carlozzi P. Dilution of solar radiation through "culture" lamination in photobioreactor rows facing south-north: a way to improve the efficiency of light utilization by cyanobacteria (Arthrospira platensis). Biotechnol. Bioeng. 2003;81(3):305–15.

5. Setlik I, Veladimir S, Malek I. Dual purpose open circulation units for large scale culture of algae in temperate zones. I. Basic design considerations and scheme of a pilot plant. Algologie Studies (Trebon) 1970;1(11).

6. Huntley M, Redalje D. CO_2 mitigation and renewable oil from photosynthetic microbes: a new appraisal. Mitigation and Adaptation Strategies for Global Change 2007;12(4):573–608.

7. Eriksen N. Production of phycocyanin—a pigment with applications in biology, biotechnology, foods and medicine. Appl. Microbiol. Biotechnol. 2008;80(1):1–14.

8. Morais MG, Radmann EM, Andrade MR, Teixeira GG, Brusch LRF, Costa JAV. Pilot scale semicontinuous production of Spirulina biomass in southern Brazil. Aquaculture 2009;294(1 2):60–4.

9. Richmond A, Lichtenberg E, Stahl B, Vonshak A. Quantitative assessment of the major limitations on productivity of Spirulina platensis in open raceways. J. Appl. Phycol. 1990;2(3):195–206.

80. Kochergin V, Day D, Suhr M. Algae production and sugar can mills: is there a synergy? In: 32nd symposium on Biotechnolog for Fuels and Chemicals, Clearwater Beach, FL 2010;19–22.

81. Satyawali Y, Balakrishnan M. Wastewater treatment i molasses-based alcohol distilleries for COD and color remova a review. J. Environ. Manage. 2008; 86:481–497.

82. Cooney MJ, Young G, Pate R. Bio-oil from photosyntheti microalgae: case study. Bioresour. Technol. 2011;102 (1):166 177.

83. Lohrey C, Kochergin V, Biodiesel production from microalga Co-location with sugar mills, Bioresour. Technol. 2012;108:76 82.

84. Johnson DT, Taconi KA. The glycerin glut: options for the valu added conversion of crude glycerol resulting from biodies production. Environ. Prog. 2007;26 (4):338–348.

85. Dong Yan, Yue Lu, Yi-Feng Chen, Qingyu Wu. Waste molasse alone displaces glucose-based medium for microalga fermentation towards cost-saving biodiesel productior Bioresour. Technol. 2011;102:6487–6493.

86. Xu H, Miao XL, Wu QY. High quality biodiesel production fror a microalga Chlorella protothecoides by heterotrophic growt in fermenters. J. Biotechnol. 2006;499–507.

87. Li Y, Chen YF, Chen P, Min M, Zhou W, Martinez B et a Characterization of a microalga Chlorella sp. well adapted t highly concentrated municipal wastewater for nutrient remova and biodiesel production. Bioresour. Technol. 2011.

88. Pittman JK, Dean AP, Osundeko O. The potential of sustainabl algal biofuel production using wastewater resources. Bioresou Technol. 2011;102(1):17–25.

89. Gonzales LE, Canizares RO, Baena S. Efficiency of ammoni and phosphorus removal from a Colombian agroindustria wastewater by the microalgae Chloreallа vulgaris an Scenedesmus dimorphus. Bioresour. Technol. 1997;60:259–62.

0. Lee K, Lee CG. Effect of light/dark cycles on wastewater treatments by microalgae. Biotechnol. Bioprocess Eng. 2001;6:194–9.

1. Olguı´n EJ, Galicia S, Mercado G, Perez T. Annual productivity of Spirulina (Arthrospira) and nutrient removal in a pig wastewater recycle process under tropical conditions. J. Appl. Phycol. 2003;15:249–57.

2. Jimenez-Perez MV, Sanches-Castillo P, Romera O, Fernandez-Moreno D, Perez-Martinez C. Growth and nutrient removal in free and immobilized planktonic green algae isolated from pig manure. Enzyme Microb. Technol. 2004;34:392–8.

3. Aslan S, Kapdan IK. Batch kinetics of nitrogen and phosphorus removal from synthetic wastewater by algae. Ecol. Eng. 2006;28(1): 64–70.

4. Yecong Li, Wenguang Zhou, Bing Hu, Min Min, Paul Chen, Roger R. Ruan, Integration of algae cultivation as biodiesel production feedstock with municipal wastewater treatment: Strains screening and significance evaluation of environmental factors. Bioresour Technol. 2011:102:10861–10867.

5. Zhu LD, Takala J, Hiltunen E, Wang ZM. Recycling harvest water to cultivate Chlorella zofingiensis under nutrient limitation for biodiesel production. Bioresour. Technol. 2013;144:14–20.

6. Levine RB, Costanza-Robinson MS, Spatafora GA. Neochloris oleoabundans grown on anaerobically digested dairy manure for concomitant nutrient removal and biodiesel feedstock production. Biomass Bioenergy 2011;35:40 – 49.

7. Yang J, Xu M, Zhang X, Hu Q, Sommerfeld M, Chen Y. Life-cycle analysis on biodiesel production from microalgae: Water footprint and nutrients balance. Bioresour. Technol. 2011;102:159–165.

8. Chinnasamy S, Bhatnagar A, Hunt RW, Das KC. Microalgae cultivation in a wastewater dominated by carpet mill effluents for biofuel applications. Bioresour. Technol. 2010;101(9):3097–3105.

99. Voltolina D, Cordero B, Nieves M, Soto LP. Growth c Scenedesmus sp. in artificial wastewater. Bioresour. Techno 1999;68(3):265–268.

100. Wu LF, Chen PC, Huang AP, Lee CM. The feasibility of biodies production by microalgae using industrial wastewater. Bioresou Technol. 2102;113:14–18.

101. Farooq W, Lee YC, Ryu BG, Kim BH, Kim HS, Choi YE et a Twostage cultivation of two Chlorella sp strains by simultaneou treatment of brewery wastewater and maximizing lipi productivity. Bioresour. Technol. 2013;132:230–238.

102. Ryu BG, Kim J, Yoo G, Lim JT, Kim W, Han JI et al. Microalga mediated simultaneous treatment of toxic thiocyanate an production of biodiesel. Bioresour. Technol. 2014;158:166–173

103. Dong B, Ho N, Ogden KL, Arnold RG. Cultivation c Nannochloropsis salina in municipal wastewater or digeste centrate. Ecotoxicol. Environ. Saf. 2014;103:45–53.

104. Levine RB, Costanza-Robinson MS, Spatafora GA. Neochlori oleoabundans grown on anaerobically digested dairy manur for concomitant nutrient removal and biodiesel feedstoc production, Biomass Bioenergy 2011;35:40-49.

105. Zhu LD, Takala J, Hiltunen E, Wang ZM. Recycling harvest wate to cultivate Chlorella zofingiensis under nutrient limitation fo biodiesel production. Bioresour. Technol. 2013;144:14–20.

106. McHugh DJ. A guide to the seaweed industry. Rome, FAO. FA Fisheries Technical paper 2003;441.

107. Demirbas A, Demirbas FM. Importance of algae oil as a sourc of biodiesel. Energy Convers. Manage. 2011;52:163–170.

108. Gao K, McKinley K. Use of macroalgae for marine biomas production and CO_2 remediation: a review. J. Appl. Phyco 1994;6:45–60.

109. Lobban CS, Harrison PJ, Duncan MJ. The Physiologica Ecological of Seaweed. Cambridge University Press. 1985.

10. Tite MS, Shortland A, Maniatis Y, Kavoussanaki D, Harris SA. The composition of the soda-rich and mixed alkali plant ashes used in the production of glass. J Archaeolog Sci 2006;33:1284–92.

11. Andersen RA. Diversity of eukaryotic algae. Biodiversity Conservation 1992;1: 267-292.

12. Domettila C, Brintha TSS, Sukumaran S, Jeeva S. Diversity and distribution of seaweeds in the Muttom coastal waters, south-west coast of India. Biodiversity J. 201;4 (1):105-110.

13. Wells E, Brewin P, Brickle P. Intertidal and Subtidal Benthic Seaweed Diversity of South Georgia, Report for the South Georgia Heritage Trust and Joint Nature Conservation Committee, Shallow marine surveys group, Survey September 2011.

14. http://www.seaweed.ie/aquaculture/eucheuma_introduction.php

15. Critchley AT, Ohno M, Largo DB, Gillespie RD. Seaweed Resources of the World. Japan International Cooperation Agency, 1998.

16. Luning K, Pang SJ. Mass cultivation of seaweeds: current aspects and approaches. J. Appl. Phycol. 2003;15:115-119.

17. Stanley M, Black K, Batty R, Fox C, Wilson B, Wilding T, et al. NERC proof of concept study for marine bioenergy. Nat Environ Res Counc 2008.

18. Pérez R. Ces algues qui nous entourent. Conception actuelle, role dans la biosphere, utilizations, culture. Editions Efrimer, 1997.

19. Buck BC, Buchholz CM. The offshore ring: A new system design for the open ocean aquaculture of macroalgae. J. Appl. Phycol. 2004;16:355-369.

20. Reith JH, BDeurwaarder BP, Hemmes K, Curvers APWM, Brandeburg W, Zeeman G. Bio-offshore. Grootschalige teelt

van zeewieren in combinatie met offshore windparken in d
Noordzee. ECN, 2005.

121. Maceiras R, Rodrı´guez M, Cancela A, Urrejola S, Sanche
A. Macroalgae: Raw material for biodiesel production. App
Energy2011;88:3318–3323.

122. Kaladharan P. Artificial seawater for seaweed culture. Indian .
Fish., 2000;47(3):257-259.

123. Hughes AD, Black KD, Campbell I, Heymans JJ, Korr K, Stanle
MS et al. Comments on 'Prospects for the use of macroalga
for fuel in Ireland and UK: An overview of marine managemer
issues, Marine Policy 2013;38:554–556.

124. Culkin F. The major constituents of seawater. In J.P.Riley an
G.Skirrow [eds.], Chemical oceanography, v.l. Academic, Londo
1965;121-161.

125. Lyman J, Flemming RH. Composition of seawater. J. Marin
Res. 1940;3:134-146.

126. Harun R, Davidson M, Doyle M, Gopiraj R, Danquah M, Ford
G. Technoeconomic analysis of an integrated microalga
photobioreactor, biodiesel and biogas production facilit
Biomass Bioenergy 2011;35:741 – 747.

127. Benemann JR, Koopman BL, Weissman JC, Eisenberg D, Goebe
R. Development of microalgae harvesting and high rate pon
technologies in California. In: Shelef G, Soeder CJ, editors. Alga
biomass: production and use. Amsterdam: Elsevier Biomedica
Press 1980;457-95.

128. Harun R, Singh M, Forde GM, Danquah MK. Bioproces
engineering of microalgae to produce a variety of consume
products. Renew. Sustain. Energy Rev. 2010;14:1037-47.

129. Krohn BJ, McNeff CV, Yan B, Nowlan D. Production of algae
based biodiesel using the continuous catalytic Mcgyan process
Bioresour. Technol. 2011;102:94–100.

30. Bosma R, Van Spronsen WA, Tramper J, Wijffels RH. Ultrasound, a new separation technique to harvest microalgae. J. Appl. Phycol.2003;15(2):143–53.

31. Heasman M, Diemar J, O'Connor W, Sushames T, Foulkes L. Development of extended shelf-life microalgae concentrate diets harvested by centrifugation for bivalve molluscs—a summary. Aquaculture Research 2000;31(8–9):637–59.

32. Huntley M, Redalje D. CO_2 mitigation and renewable oil from photosynthetic microbes: a new appraisal. Mitigation and Adaptation Strategies for Global Change 2007;12:573–608.

33. Kim J, Ryu BG, Kim K, Kim BK, Han JI, Yang JW. Continuous microalgae recovery using electrolysis: Effect of different electrode pairs and timing of polarity exchange. Bioresour. Technol. 2012;123:164–170.

34. Lee K, Lee SY, Na JG, Jeon SG, Praveenkumar R, Kim DM, Chang WS, Oh YK. Magnetophoretic harvesting of oleaginous Chlorella sp. by using biocompatible chitosan/magnetic nanoparticle composites. Bioresour. Technol.2013;149:575–578.

35. Beach ES, Eckelman MJ, Cui Z, Brentner L, Zimmerman J.B. Preferential technological and life cycle environmental performance of chitosan flocculation for harvesting of the green algae Neochloris oleoabundans. Bioresour. Technol. 2012;121:445–449.

36. Farid MS, Shariati A, Badakhshan A, Anvaripour B. Using nanochitosan for harvesting microalga Nannochloropsis sp. Bioresour. Technol. 2013;131:555–559.

37. Sirin S, Trobajo R, Ibanez C, Salvado J. Harvesting the microalgae Phaeodactylum tricornutum with polyaluminum chloride, aluminium sulphate, chitosan and alkalinity-induced flocculation. J. Appl. Phycol. 2012;24,1067–1080.

38. Kim J, Yoo G, Lee H, Lim J, Kim K, Kim CW et al. Methods of downstream processing for the production of biodiesel from microalgae. Biotechnol. Adv. 2013;31:862–876.

139. Prakash J, Pushparaj B, Carlozzi P, Torzillo G, Montaini E Materassi R. Microalgae drying by a simple solar device. Int. J Solar Energy 1997;18(4):303–11.

140. Desmorieux H, Decaen N. Convective drying of spirulina in thi layer. J. Food Eng. 2006;66(4):497–503.

141. Leach G, Oliveira G, Morais R. Spray-drying of Dunaliell salina to produce a b-carotene rich powder. J. Ind. Microbio Biotechnol. 1998;20(2):82–5.

142. Grima ME, Medina A, Gime´nez A, Pe´ rez SJ, Camacho F Sa´nchez JG. Comparison between extraction of lipids an fatty acids from microalgal biomass. J. Am. Oil Chem. Soc 1994;71(9):955–9.

143. Atabani AE, Silitonga AS, Badruddin IA, Mahlia TMI, Masjuk HH, Mekhilef S. A comprehensive review on biodiesel as a alternative energy resource and its characteristics. Renew Sustain. Energy Rev. 2012;16:2070–93.

144. Vaara M. Agents that increase the permeability of the oute membrane. Micobiol Rev 1992;56:395-411.

145. Young G, Nippen F, Titterbrandt S, Cooney MJ. Direc transesterification of biomass using an ionic liquid cosolven system. Biofuels 2011;2,261-6

146. Amin Hejazi M, Wijffels RH. Milking of microalgae. Trend Biotechnol. 2004;22(4):189–94.

147. Bligh EG, Dyer WJ. A rapid method of total lipid extraction an purification. Can. J. Biochem. Physiol. 1959;37 (8): 911–917.

148. Folch J, Lees M, Sloane-Stanley GH. A simple method for th isolation and purification of total lipides from animal tissues. J Biol. Chem. 1957;226:497–509.

149. U.S. DOE, 2010. National Algal Biofuels Technology Roadma (Technology Roadmap). U.S. Department of Energy, Office c Energy Efficency and Renewable Energy, Biomass Program.

50. Sathish A, Sims RC. Biodiesel from mixed culture algae via a wet lipid extraction procedure. Bioresour. Technol. 2012;118:643–647.

51. Mahanta P, Shrivastava A. Technology development of bio-diesel as an energy alternative. Available from: http://www.newagepublishers.com/samplechapter/001305.pdf; 2011

52. Achten WMJ, Verchit L, Mathijs Franken YJ, Singh E, Aerts VP, Muys RB. Jatropha bio-diesel production and use. Biomass Bioenergy 2008;32(12):1063–84.

53. Rabe ELM. Jatropha oil in compression ignition engines, effects on the engine, environment and Tanzania as supplying country. Eindhoven: Eindhoven University of Technology; 2010.

54. Achten WMJ, Verchit L, Mathijs Franken YJ, Singh E, Aerts VP, Muys RB. Jatropha bio-diesel production and use. Biomass Bioenergy 2008;32(12):1063–84.

55. Beerens P. Screw-pressing of Jatropha seeds for fueling purposes in less developed countries. Eindhoven: Eindhoven University of Technology 2007.

56. Forson FK, Oduro EK, Hammond-Donkoh E. Performance of Jatropha oil blends in a diesel engine. Renew. Energy 2004 ;29(7):1135–45.

57. Tewari DN. Jatropha and biodiesel. India: Ocean Books Ltd.; 2007.

58. Wang H, Gao L, Chen L, Guo F, Liu T. Integration process of biodiesel production from filamentous oleaginous microalgae Tribonema minus. Bioresour. Technol. 2013;142:39–44.

59. Iqbal J, Theegala C. Microwave assisted lipid extraction from microalgae using biodiesel as co-solvent. Algal Research 2013;2:34–42.

60. Santana A, Jesus S, Larrayoz MA, Filho RM. Supercritical carbon dioxide extraction of algal lipids for the biodiesel production. Procedia Engineering 2012;42:1755 – 1761.

161. Macias-Sanchez MD, Mantell C, Rodriguez M, de la Oss EM, Lubian EM, Montero O, Supercritical fluid extractio of carotenoids and chlorophylla from Synechococcus sp. . Supercrit. Fluids 2007;39:323–9.

162. Patil PD, Gude VG, Mannarswamy A, Deng S, Cooke P, Munsor McGee S et al. Optimization of direct conversion of wet alga to biodiesel under supercritical methanol conditions. Bioresou Technol. 2011;102:118–122.

163. Lee JY, Yoo C, Jun SY, Ahn CY, Oh HM. Comparison c several methods for effective lipid extraction from microalga Supplement issue on recent developments of biomass conversio technologies. Bioresour. Technol. 2010;101:75–77.

164. Cravotto G, Boffa L, Mantegna S, Perego P, Avogadro M Cintas P. Improved extraction of vegetable oils under high intensity ultrasound and/or microwaves. Ultrason. Sonochen 2008;15:898–902.

165. Balasubramanian S, Allen JD, Kanitkar K, Boldor D. O extraction from Scenedesmus obliquus using a continuou microwave. Bioresour. Technol. 2011;102:3396–403.

166. Lewis T, Nichols PD, McMeekin TA. Evaluation of extractio methods for recovery of fatty acids from lipid-producin microheterotrophs. J. Microbiol. Methods 2000;43:107–116.

167. Johnson MB, Wen Z. Production of biodiesel fuel from th microalga *Schizochytrium limacinum* by direct transesterificatio of algal biomass. Energy Fuels 2009;23:5179–5183.

168. Vicente G, Bautista LF, Rodríguez R, Gutiérrez FJ, Sádaba I Ruiz-Vázquez RM et al. Biodiesel production from biomass c an oleaginous fungus. Biochem. Eng. J. 2009;48:22–27.

169. Garces R, Mancha M. One-step lipid extraction and fatt acid methyl esters preparation from fresh plant tissues. Ana Biochem. 1993;211(1):139–143.

170. Li L, Du W, Liu D, Wang L, Li Z. Lipase-Catalyze transesterification of rapeseed oils for biodiesel production wit

a novel organic solvent as the reaction medium. J. Mol. Catal. B: Enzym 2006;43:58-62.

71. Shimada Y, Watanabe Y, Sugihara A, Tominaga Y. Enzymatic alcoholysis for biodiesel fuel production and application of the reaction to oil processing. J. Mol. Catal. B: Enzym 2002;17:133-42.

72. Er-Zheng S, Min-Jie ,Jian-Guo Z, Jian-Feng G,Dong-Ziu W. Lipase catalyzed irreversible transesterification of vegetable oils for fetty acid methyl esters production with dimethyl carbonate as the acyl acceptor. Biochem. Eng. J 2007;36:167-73.

73. Pioch D, Lozano P, Graille J. First lipase catalyzed synthesis of fatty carbonate esters. Biotechnol. Lett. 1991;13:633-5.

74. Mangesh KG, Ajay DK, Charan ML, Narayan NS. Transesterification of karanja (Pongamia pinnata) oil by solid basic catalysts. Eur. J. Lipid Sci. Technol. 2006;108:389–97.

75. Du W, Xu Y, Zeng J, Liu D. Novozym 435-catalyzed transesterification of crude soya bean oil for biodiesel production in a solvent-free medium. Biotechnol. Appl. Biochem. 2004;40:187-90.

76. Borugadda VB, Goud VV. Biodiesel production from renewable feedstocks: Status and opportunities. Renew. Sustain. Energy Rev. 2012;16:4763–84.

77. Yin JZ, Xiao M, Song JB. Biodiesel from soybean oil in supercritical methanol with cosolvent. Energy Convers. Manage. 2008;49:908–12.

78. Nath K, Das D. Hydrogen from biomass. Curr. Sci. 2003;85:265–71.

79. Bajaj A, Lohan P, Jha PN, Mehrotra R. Biodiesel production through lipase catalyzed transesterification: a review. J. of Mol. Cata B: Enzym 2010;62:9–14.

80. Ranganathan SV, Narasimhan SL, Muthukumar K. An overview of enzymatic production of biodiesel. Bioresour. Technol.2008;99:3975–3981.

181. Bharathiraja B, et al. Biodiesel production using chemic and biological methods – A review of process, catalyst, acyl. Renew. Sustain. Energy Rev. (2014), http://dx.doi.org/10.1016/ rser.2014.05.084

182. Shahid EM, Jamal J. Production of biodiesel: a technical revie Renew. Sustain. Energy Rev. 2011;15(9):4732–45.

183. Sharma YC, Singh B. Development of biodiesel: current scenari Renew. Sustain. Energy Rev. 2009;13(6–7):1646–51.

184. Jain S, Sharma MP. Prospects of biodiesel from Jatropha i India: a review. Renew. Sustain. Energy Rev. 2010;14(2):763–71

185. Attanatho L, Magmee S, Jenvanitpanjaku P. Factors affecting th synthesis of biodiesel from crude palm kernel oil. In: The join international conference on sustainable energy and environmer (SEE). 2004.

186. Ehimen EA, Sun ZF, Carrington CG. Variables affecting the i situ transesterification of microalgae lipids. Fuel 2010;89(3):677 84.

187. Encinar JM, Gonzalez JF, Rodriguez-Reinares A. Biodiesel fror used frying oil. Variables affecting the yields and characteristic of the biodiesel. Ind. Eng. Chem. Res. 2005;44(15):5491–9.

188. Freedman B, Pryde E, Mounts T. Variables affecting the yields c fatty esters from transesterified vegetable oils. J. Am. Oil Chem Soc. 1984;61(10):1638–43.

189. May CU. Transesterification of palm oil: effect of reactio parameters. J. Oil. Palm Res. 2004;16(2):1–11.

190. Parlak A, Karabas H, Ayhan V, Yasar H, Soyhan HS, Ozsert I Comparison of the variables affecting the yield of tobacco see oil methyl ester for KOH and NaOH catalysts. Energy Fuel 2008;23(4):1818–24.

191. Murugesan A, Umarani C, Chinnusamy TR, Krishnan M Subramanian R, Neduzchezhain N. Production and analysi of bio-diesel from non-edible oils – a review. Renew. Sustair Energy Rev. 2009;13(4):825–34.

92. Khan NA, el Dessouky H. Prospect of biodiesel in Pakistan. Renew. Sustain. Energy Rev. 2009;13(6–7):1576–83.

93. Marchetti J.M. Past, present and future scopes of biodiesel industry. In: WCSET Meeting, Oslo, Norway, 2009.

94. McNeff CV, McNeff LC, Yan B, Nowlan DT, Rasmussen M, Gyberg A et al. A continuous catalytic system for biodiesel production. Appl. Catal. A Gen. 2008;343:39–48.

95. Annoy. Biodeisel-2020. Global market survey, feedstock trends and forecasts. Multi client study, 2nd eds. Multi-Client Study, 2nd Edition (2008 release), 685 pages Published by Emerging Markets Online. ⟨www.emerging- markets.com⟩.

96. EBTP. Biobutanol. European biofuels technology platform (EBTP), http://www.biofuelstp.eu/butanol.html; 2009

97. Jones DT, Woods DR. Acetone-butanol fermentation revisited. Microbiol. Rev. 1986;50(4):484–524.

98. Pfromm PH, Amanor-Boadu V, Nelson R. Bio-butanol vs. bio-ethanol: A technical and economic assessment for corn and switchgrass fermented by yeast or Clostridium acetobutylicum. Biomass Bioenergy 2010;34:515 – 524.

99. Qureshi N, Li XL, Hughes S, Saha BC, Cotta MA. Butanol production from corn fiber xylan using *Clostridium acetobutylicum*. Biotechnol. Prog. 2006;22:673– 680.

100. Adams JM, Gallagher JA, Donnison IS. Fermentation study of *Saccharina latissima* for bioethanol production considering variable pre-treatments. J. Appl. Phycol. 2009;21: 569–574.

101. Powell EE, Hill GA. Economic assessment of an integrated bioethanol–biodiesel–microbial fuel cell facility utilizing yeast and photosynthetic algae. Chem. Eng. Res. Des. 2009;87:1340–1348.

102. Durre P. Fermentative production of butanol- academic perspective. Curr. Opin. Biotechnol. 2011;22:331-336.

203. Garcia V, Päkkilä J, Ojamo H, Muurinen E, Keiski RL. Challenge in biobutanol production: how to improve the efficiency? Renev Sustain. Energy Rev. 2011;15:964-80.

204. Baroghi LG. Enhanced Butanol Production by Free an Immobilized Clostridium sp. Cells using Butyric Acid as Co Substrate: Biotechnol. (Bioprocess Eng. – Biofuels): (2009 THESIS.

205. Lipnizki F, Hausmanns S, Laufenberg G, Field R, Kunz B. Us of Pervaporation-bioreactor Hybrid Process in Biotechnolog Chem. Eng. Technol 2000;23:569-577.

206. Lee SY, Park JH, Jang SH, Nielsen LK, Kim J, Jung KS Fermentative Butanol Production by Clostridia: Biotechno Bioeng. 2008;101:209–228.

207. Ezeji TQN, Blaschek HP. Production of acetone–butanol–etha ol (ABE) in a continuous flow bioreactor using degermed cor and Clostridium beijerinckii. Process Biochem. 2007;42:34–9.

208. Kourkoutas Y, Bekatorou A, Banat IM, Marchant R, Koutina AA. Immobilization technologies and support materials suitabl in alcohol beverages production: a review. Food Microbic 2004;21:377-3.

209. Díaz de Santos, SA Lima, N. e Mota, M. Biotecnologia Fundamentos e aplicações. Biocatálise aplicada Lidel-Ediçõe Técnicas, Lda.(2003)

210. Jones DT, Woods DR. Acetone-butanol fermentation revisitec Microbiol. Rev. 1986;50(4):484–524.

211. van der Wal H, Bram L.H.M. Sperber, Bwee Houweling-Tar Robert R.C. Bakker, Willem Brandenburg. Ana M. López Contreras: Production of acetone, butanol, and ethanol fron biomass of the green seaweed Ulva Lactuca; Bioresour. Techno 2013;128:431–437.

212. Teymouri F, Laureanoperez L, Alizadeh H, Dale E Optimization of the ammonia fiber explosion (AFEX) treatmen parameters for enzymatic hydrolysis of corn stover. Bioresou Technol.2005;96:2014-2018.

13. Alizadeh H, Teymouri F, Gilbert T, Dale B. Pretreatment of switchgrass by ammonia fiber explosion (AFEX). Appl. Biochem. Biotechnol. 2005 ;124:1133-1141.

14. Lee JM, Jameel H, Venditti RA. A comparison of the autohydrolysis and ammonia fiber explosion (AFEX) pretreatments on the subsequent enzymatic hydrolysis of coastal Bermuda grass. Bioresour. Technol. 2010;101:5449-5458.

15. Li B-Z, Balan V, Yuan Y-J, Dale BE. Process optimization to convert forage and sweet sorghum bagasse to ethanol based on ammonia fiber expansion (AFEX) pretreatment. Bioresour. Technol. 2010;101:1285-1292.

16. Noomtim P, Cheirsilp B. Production of butanol from Palm Empty Fruit Bunches Hydrolysate by Clostridium acetobutylicum. Energy. Procedia. 2011;9:140-146.

17. Huang H, Ramaswamy S, Tschirner U, Ramarao B. A review of separation technologies in current and future biorefineries. Sep. Purif. Technol. 2008;62:1-21.

18. Knappert DR, Grethlein HE, Converse AO. Partial acid hydrolysis of poplar wood as a pretreatment for enzymatic hydrolysis. Biotechnol. Bioeng. Symp. 1981;11:67-77.

19. Hill CAS, Abdul Khalil HPS, Hale MD. A study of the potential of acetylation to improve the properties of plant fibers. Ind, Crops. Prod. 1998;8:53–63.

20. Dubois M, Gile KA, Hamilton JK, Rebes PA, Smith F. Colorimetric method for determination of sugars and related substances. Anal. Chem. 1956;28:350–356.

21. Yoza BA, Masutani EM. The analysis of macroalgae biomass found around Hawaii for bioethanol production. Environ. Technol. 2013;34(13–14):1859–1867. http://dx.doi.org/10.1080/09593330.2013.781232.

22. Golueke C, Oswald W. Biological conversion of light energy to the chemical energy of methane. Appl. Environ. Microbiol. 1959;7:219–227.

223. Rösch C, Skarka J, Wegerer N. Materials flow modeling c nutrient recycling in biodiesel production from microalga Bioresour. Technol. 2012;107:191–199.

224. Sialve B, Bernet N, Bernard O. Anaerobic digestion of microalga as a necessary step to make microalgal biodiesel sustainabl Biotechnol. Adv. 2009;27:409–416.

225. Fernandez AV, Vargas G, Alarcon N, Velasco A. Evaluation c marine algae as a source of biogas in a two-stage anaerobi reactor system. Biomass Bioenergy 2008;32:338–44.

226. Salerno, M.,Nurdogan,Y.,Lundquist,T.J.,2009.Biogas productio from Algae Biomass Harvested at Wastewater Treatment Pond In: Bioenergy Engineering Conference, 11–14 October 200 BIO-098023.

227. Yen HW, Brune DE. Anaerobic co-digestion of algal sludg and waste paper to produce methane. Bioresour. Techno 2007;98:130–134.

228. Morken J, Sapci Z, Strømme JET. Modeling of biodies production in algae cultivation with anaerobic digestio (ACAD). Energy Policy (2013), http://dx.doi.org/10.1016/ enpol.2013.04.081i.

229. Ehimen EA, Sun ZF, Carrington CG, Birch EJ, Eaton-Rye J. Anaerobic digestion of microalgae residues resulting from th biodiesel production process. Appl. Energy 2010;88:3454–3463

230. Samson R, LeDuy A. Biogas production from anaerobi digestion of Spirulina maxima algal biomass. Biotechno Bioeng. 1982;24:1919–24.

231. Ras M, Lardon L, Sialve B, Nicolas B, Steyer JP. Experiment study on a coupled process of production and anaerobic digestio of Chlorella vulgaris. Bioresour. Technol. 2011;102:200–206.

232. Chynoweth DP, Turick CE, Owens JM, Jerger DE, Peck MW Biochemical methane potential of biomass and waste feedstock Biomass Bioenergy 1993;5:95-111.

233. Bird KT, Chynoweth DP, Jerger DE. Effects of marine algal proximate composition on methane yields. J. Appl. Phycol.1990;2:207-213.

234. Morand P, Briand X. Anaerobic digestion of Ulva sp. 2. Study of Ulva degradation and methanisation of liqucfaction juices. J. Appl. Phycol. 1999;11:165-77.

235. Briand X, Morand P. Anaerobic digestion of Ulva sp. 1. Relationship between Ulva composition and methanisation. J. Appl. Phycol. 1997;9:511-24.

236. Rasi S, Veijanen A, Rintala J. Trace compounds of biogas from different biogas production plants. Energy 2007;32:1375–1380.

237. Wheeler P, Lindberg A. Biogas Upgrading and Utilisation, IEA Bioenergy— Task 24: Energy from Biological Conversion of Organic Waste 1999;1-19.

238. Clark J, Deswarte F. Introduction to chemicals from biomass. In: Stevens CV, editor. Wiley series in renewable resources. John Wiley & Sons; 2008.

239. Demirbas A. Biomass resource facilities and biomass conversion processing for fuels and chemicals. Energy Convers. Manage. 2001;42(11):1357–78.

240. Hirano A, Hon-Nami K, Kunito S, Hada M, Ogushi Y. Temperature effect on continuous gasification of microalgal biomass: theoretical yield of methanol production and its energy balance. Catal. Today 1998;45(1–4):399–404.

241. Minowa T, Sawayama S. A novel microalgal system for energy production with nitrogen cycling. Fuel 1999;78(10):1213–5.

242. Goyal HB, Seal D, Saxena RC. Bio-fuels from thermochemical conversion of renewable resources: a review. Renew. Sustain. Energy Rev. 2008;12(2):504–17.

243. Raveendran K, Ganesh AK, Khilar C. Influence of mineral matter on biomass pyrolysis characteristics. Fuel 1995;74:1812–22.

244. Demirbas A. Oily products from mosses and algae via pyrolysis Energy Sources Part A—Recovery Utilization and Environmenta Effects 2006;28(10):933–40.

245. Singh SP, Singh D. Biodiesel production through the use c different sources and characterization of oils and their esters a the substitute of diesel: a review. Renew. Sustain. Energy Re 2010;14(1):200–16.

246. Sharma YC, Singh B, Upadhyay SN. Advancements i development and characterization of biodiesel: a review. Fu 2008;87(12):2355–73.

247. Raveendran K, Ganesh A, Khilar KC. Pyrolysis characteristics c biomass and biomass components. Fuel 1996;75:987–98.

248. Li S, Lyons-Hart J, Banyasz J, Shafer K. Real-time evolved ga analysis by FTIR method: an experimental study of cellulos pyrolysis. Fuel 2001;80:1809–17.

249. Bassilakis R, Carangelo RM, Wojtowicz MA. TG-FTIR analysi of biomass pyrolysis. Fuel 2001;80:1765–86.

250. Haiping Yang, Rong Yan, Hanping Chen, Dong Ho Lee, Chuguan Zheng. Characteristics of hemicellulose, cellulose and ligni pyrolysis. Fuel 86 (2007) 1781–1788.

251. Orfao J, Antunes F, Figueiredo J. Pyrolysis kinetics c lignocellulosic materials-three independent reaction mode Fuel 1999;78:349–58.

252. Ghirardi ML, Zhang L, Lee JW, Flynn T, Seibert M, Greenbaur E, et al. Microalgae: a green source of renewable H2. Trend Biotechnol. 2000;18(12):506–11.

253. Yin JZ, XiaoM, SongJB. Biodiesel from soybean oil i supercritical methanol with cosolvent. Energy Convers. Manage 2008;49:908–12.

254. Murugesan A, Umarani C, Chinnusamy TR, Krishnan M Subramanian R, Neduzchezhain N. Production and analysi of bio-diesel from non-edible oils —a review. Renew. Sustair Energy Rev. 2009;13:825–34.

55. Zhang Y, Fan X, Yang Z, Wang H, Yang D, Guo R. Characterization of H_2 photoproduction by a new marine green alga, Platymonas helgolandica var. Tsingtaoensis. Appl. Energy2012;92:38–43.

56. Cantrell KB, Ducey T, Ro KS, Hunt PG. Livestock waste-to-bioenergy generation opportunities. Bioresour. Technol.2008;99(17):7941–53.

57. Akkerman I, Janssen M, Rocha J, Wijffels RH. Photobiological hydrogen production: photochemical efficiency and bioreactor design. Int. J. Hydrogen Energy 2002;27(11):1195–208.

58. Laurinavichene T, Tolstygina I, Tsygankov A. The effect of light intensity on hydrogen production by sulfur-deprived Chlamydomonas reinhardtii. J. Biotechnol. 2004;114:143–51.

59. Melis A, Happe T. Hydrogen production. Green algae as a source of energy. Plant Physiol. 2001;127(3):740–8.

60. Sen U, Shakdwipee M, Banerjee R. Status of biological hydrogen production. J. Sci. Ind. Res. 2008;67:980-993.

61. Philipps G, Happe T, Hemschemeier A. Nitrogen deprivation results in photosynthetic hydrogen production in Chlamydomonas reinhardtii. Planta 2012;235:729–45.

62. Melis A. Green alga hydrogen production: progress, challenges and prospects. Int. J. Hydrogen Energy 2002;27(11–12):1217–28.

63. Skjånes K, Lindblad P, Muller J. BioCO_2 – A multidisciplinary, biological approach using solar energy to capture CO_2 while producing H_2 and high value products. Biomol. Eng. 2007;24(4):405-413.

64. Melis A, Melnicki MR. Integrated biological hydrogen production. Int. J. Hydrogen Energy 2006;31(11):1563-1573.

65. Giannelli L, Torzillo G. Hydrogen production with the microalga Chlamydomonas reinhardtii grown in a compact tubular photobioreactor immersed in a scattering light nanoparticle suspension. Int. J. Hydrogen Energy 2012;37:16951–61.

266. Oncel S, Sabankay M. Microalgal Biohydrogen productio considering light energy and mixing time as the two key feature for scale-up. Bioresour. Technol.2012;121:228–34.

267. Tsygankov A, Kosourov S, Seibert M, Ghirardi ML. Hydroge photoproduction under continuous illumination by sulfu deprived, synchronous Chlamydomonas reinhardtii culture Int. J. Hydrogen Energy 2002;27:1239–44.

268. Fedorov AS, Kosourov S, Ghirardi MI, Seibert M. Continuou hydrogen photoproduction by Chlamydomonas reinhardt using a novel two-stage, sulfate limited chemostat system. App Biochem. Biotechnol. 2005;121:403–12.

269. Tolstygina IV, Antal TK, Kosourov SN, Krendeleva TE, Rubi AB, Tsygankov AA. Hydrogen production by photoautotrophi sulfur-deprived Chlamydomonas reinhardtii pre grown an incubated under highlight. Biotechnol. Bioeng. 2009;102:1055 61.

270. Guan Y, Deng M, Yu X, Zhang W. Two-stage photo-biologica production of hydrogen by marine green alga Platymona subcordiformis. Biochem. Eng. J. 2004;19:69–73.

271. Kojima E, Lin B. Effect of partial shading on photoproductio of hydrogen by Chlorella. J. Biosci. Bioeng. 2004;97:317–21.

272. Guan Y, Zhang W, Deng M, JinYu X. Significant enhancemen of photobiological H_2 evolution by carbonyl cyanidem chlorophenyl hydrazine in the marine green alga Platymona subcordiformis. Biotechnol. Lett. 2004;26:1031–5.

273. Amutha KB, Murugesan AG. Biological hydrogen production b the algal biomass Chlorella vulgaris MSU 01 strain isolated fror pond sediment. Bioresour. Technol.2011;102:194–9.

274. Aoyama K, Uemura I, Miyake J, Asada Y. Fermentativ metabolism to produce hydrogen gas and organic compound in a cyanobacterium, Spirulina platensis. J. Ferment. Bioeng 1997;83:17 20.

275. Maneeruttanarungroj C, Lindblad P, Incharoensakdi A. A newly isolated green alga, Tetrasporasp.CU2551 from Thailan

with efficient hydrogenproduction. Int. J. Hydrogen Energy 2010;35:13193–9.

276. Tsygankov AS, Serebryakova LT, Svesnikov DA, Rao KK, Gogotov IN, Hall DO. Hydrogen photoproduction by three different nitrogenases in whole cells of Anabaena variabilis and the dependence on pH. Int. J. Hydrogen Energy 1997;22:859–67.

277. Markov SA, Thomas AD, Bazin MJ, Hall DO. Photoproduction of hydrogen by cyanobacteria under partial vacuum in batch culture or in a photobioreactor. Int. J. Hydrogen Energy 1997;22:521–4.

278. Yoon JH, Shin JH, Kim MS, Sim SJ, Park TH. Evaluation of conversion efficiency of light to hydrogen energy by Anabaena variabilis. Int. J. Hydrogen Energy 2006;31:721–7.

279. Knothe G. Biodiesel and renewable diesel: acomparison. Prog. Energy Combust. Sci. 2010;36:364–73.

280. Pienkos PT, Darzins A. The promise and challenges of microalgal-derived biofuels. Biofuels, Bioprod. Bioref. 2009;3:431–40.

281. Schenk PM, Hall SRT, Stephens E, Marx UC, Mussgnug JH, Posten C. et al. Second generation biofuels: high-efficiency microalgae for biodiesel production. Bioenergy Res. 2008;1:20–43

282. Scott SA, Davey MP, Dennis JS, Horst I, Howe CJ, LeaSmith DJ, et al. Biodiesel from algae: challenges and prospects. Curr. Opin. Biotechnol. 2010;21:277–86.

283. Tsukahara K, Sawayama S. Liquid fuel production using microalgae. J. Jpn Petroleum Inst. 2005;48(5):251–9.

284. Ramos M, Fernandez C, Casas A, Rodriguez L, Perez A. Influence of fatty acid composition of raw materials on biodiesel properties. Bioresour. Technol.2009;261–268.

285. Shahid EM, Jamal J. Production of biodiesel: a technical review. Renew. Sustain. Energy Rev. 2011;15(9):4732–45.

286. Ahmad AL, Mat Yasin NH, Derek CJC, Lim JK. Microalgae a a sustainable energy source for biodiesel production: a reviev Renew. Sustain. Energy Rev. 2011;15(1):584–93.

287. Karmakar A, Karmakar S, Mukherjee S. Properties of variou plants and animals feedstocks for biodiesel productior Bioresour. Technol. 2010;101(19):7201–10.

288. Kibazohi O, Sangwan RS. Vegetable oil production potential fror Jatropha curcas, Croton megalocarpus, Aleurites moluccana Moringa oleifera and Pachira glabra: assessment of renewabl energy resources for bio-energy production in Africa. Biomas Bioenergy 2011; 35(3):1352–6.

289. Atadashi IM, Aroua MK, Abdul Aziz A. High quality biodies and its diesel engine application: a review. Renew. Sustair Energy Rev. 2010;14(7):1999–2008.

290. Veeken A, Kalyuzhnyi S, Scharff H, Hamelers B. Effect of pl and VFA on hydrolysis of organic solid waste. J. Environ. Eng 2000;126:1076–1081.

291. Sheehan J, Dunahay T, Benemann J, Roessler P. A look bac at the US Department of Energy's aquatic species. Prograr Biodiesel from Algae. Report of National Renewable Energ Laboratory TP-580-24190 (NREL), July 1998.

292. Hirayama S, Ueda R, Ogushi Y, Hirano A, Samejima Y, Hor Nami K et al. Ethanol production from carbon dioxide b fermentative microalgae. Stud. Surf. Sci. Catal. 1998;114:657 60.

293. Bush RA, Hall KM. Process for the production of ethanol fror algae. US Patent 7135308, 2006.

294. Mustaqim D, Ohtaguchi KA. Synthesis of bioreactions for th production of ethanol from CO_2. Energy 1997;22:353–6.

295. Deng MD, Coleman JR. Ethanol synthesis by genetic engineerin in cyano- bacteria. Appl. Environ. Microbiol. 1999;65:523–8.

296. John RP, Anisha GS, Nampoothiri KM, Pandey A. Micro and macroalgal biomass: a renewable source for bioethanol. Bioresour. Technol.2011;102:186–93.

297. Lou D, Hu Z, Choi DG, Thomas VM, Real MJ, Chance RR. Life cycle energy and greenhouse gas emissions for an ethanol production process based on blue- green algae. Environ. Sci. Technol. 2010;44:8670–7.

298. Ma H, Oxley L, Gibson J, Li W. A survey of China's renewable energy economy. Renew. Sustain. Energy Rev. 2010;14:438–45.

299. Jones DT, Woods DR. Acetone-butanol fermentation revisited. Microbiol. Rev. 1986;50(4):484–524.

300. Christi Y. Biodiesel from microalgae. Biotechnol. Adv., 2007, 25, 294-306.

301. Golueke CG, Oswald WJ, Gotaas HB. Anaerobic digestion of algae. Appl. Microbiol. 1957;5:47–55.

302. Mussgnug JH, Klassen V, Schlüter A, Kruse O. Microalgae as substrates for fermentative biogas production in a combined biorefinery concept. J. Biotechnol. 2010;150:51–6.

303. McKendry P. Energy production from biomass (part 3): gasification technologies. Bioresour. Technol. 2002;83(1):55–63.

304. Dote Y, Sawayama S, Inoue S, Minowa T, Yokoyama S-y. Recovery of liquid fuel from hydrocarbon-rich microalgae by thermochemical liquefaction. Fuel 1994;73(12):1855–7.

305. Minowa T, Yokoyama S-y, Kishimoto M, Okakura T. Oil production from algal cells of Dunaliella tertiolecta by direct thermochemical liquefaction. Fuel 1995;74(12):1735–8.

306. Miao X, Wu Q, Yang C. Fast pyrolysis of microalgae to produce renewable fuels. J. Anal. Appl. Pyrolysis 2004;71(2):855–63.

307. Levin D B, Pitt L, Love M. Biohydrogen production: prospect and limitation to practical application. Int. J. Hydrogen Energy 2004;24:173-185.

308. Nakada E, Nishikata S, Asada Y, Miyake J. Photosyntheti bacterial hydrogen production combined with a fuel cell. Int. J Hydrogen Energy 1999;24:1053-1057.

309. Lin CN, Wu SY, Lee KS, Lin PJ, Lin CY, Chang JS. Integration c fermentative hydrogen process and fuel cell for online electricit generation. Int. J. Hydrogen Energy 2007;32:802-808.

310. Hirano A, Ueda R, Hirayama S, Ogushi Y. CO_2 fixation and ethanc production with microalgal photosynthesis and intracellula anaerobic fermentation. Energy 1997;22(2–3):137–42.

311. Dismukes GC, Carrieri D, Bennette N, Ananyev GM, Posewit MC. Aquatic phototrophs: efficient alternatives to land-base crops for biofuels. Curr. Opin. Biotechnol. 2008;19(3):235–40.

312. Qin J. Bio-hydrocarbons from algae—impacts of temperature light and salinity on algae growth. Barton, Australia: Rura Industries Research and Development Corporation; 2005.

3

Macroalgae - An Omnipotent Source for Sustainable Future

8.1 INTRODUCTION

Seaweed is a macroalgae seen in various forms and color. It occurs along the coastal line in marine habitat and India as a peninsular nation has got wide distribution of seaweeds compared to other countries [1, 2]. Several harvesting techniques and cultivation methods for red, brown and green algae were in practice depending upon the species and other environmental factors [4-6]. The biochemical components of seaweeds such as carbohydrates, proteins, vitamins, fat, minerals are numerous and their composition experiences seasonal variation. The various compounds such as sulphated polysaccharides, fucoxanthin, carrageeanans, agarans, terepenoids etc., and organic extracts such as methanolic, ethanolic and butanolic extracts of different seaweeds possess antioxidant, antiviral, antinflammatory, anticancer activity against various pathogens and diseases [85-88]. With the increased domestic, industrial and vehicle uses seaweed compounds extracted and processed are being used as various biofuels. Moreover *Ulva* species has been used in the production of all the four biofuels like bioethanol, biodiesel, biobutanol, biogas while biomass of other macroalgal species such as *Sargassum, Padina, Dictyota, Ascophyllum, Laminaria, Saccharina* were also found to be used in the production of various fuels. On the other hand polysaccharides of seaweeds are used to produce agar, a phycocolloid which has its application in the food industries in the form of stabilizer, thickening agent; other carbohydrates such as fucoidan, laminarin, carrageenan were also analysed to have varied applications in the food and drugs sector. Seaweed extracts from *Laminaria, Undaria, Saccharina* species were known to be employed in cancer treatment whereas other mineral rich extracts treats arthritis and hypothyroidism.

Current world which approaches the trends of nanotechnology, silve and gold nanoparticles from potential seaweed species were found t have intense antimicrobial properties that aids effluent treatment. Apai from these, seaweeds such as *Laminaria, Sargassum,* and *Ulva* specie were being used in treating waste water; *Ascophyllum* species used a beauty enhancers and nutritious animal feed while their alkaline extract as biofertiliser have provided good quality plants.

8.1.1 Morphology and Diversity

Algae have a great variety of forms, sizes and colors in which some ar simple colonies with many cells; some are filamentous, tubular, meshec membranous or saccate algae. Some are more delicate and complex such as, *Sargassum* having structures similar to roots, stems and leaves c plants as well as differentiated air bladders which help the algae floating t water surface to be able to absorb more sunlight. Nevertheless, althoug algae may have different forms, their internal structures are compose of similar cells with simple differentiation instead of true roots, stem or leaves. Seaweeds are grouped into four based on colour: Green alga (Chlorophyta), brown algae (Phaeophyta), red algae (Rhodophyta), an blue-green algae (Cyanophyta) which are being harvested and utilize for several purposes [1]. The thallus (the algal body), lamina or blade (flattened leaf-like structure), stipe (a stem-like structure), and holdfast (basal structure that provides attachment to a surface) form the parts c a seaweed which are shown in the Figure 1.

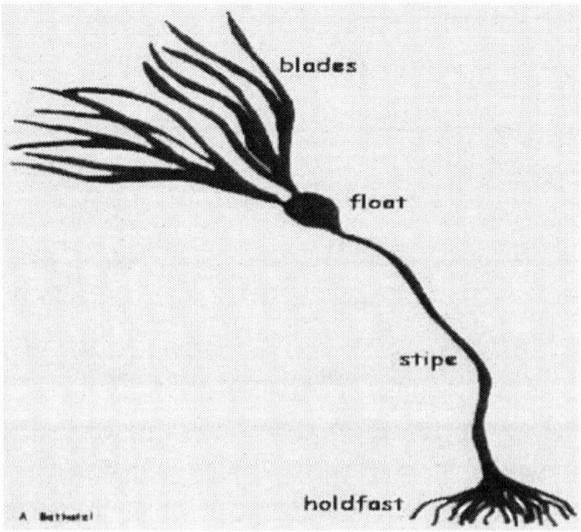

Figure 1 General structure of a seaweed

.1.2 Occurrence and Distribution

eaweed is particularly available in abundant in three areas: the warm orthern waters around the Kermadec and three king island, the Cook trait Kaikoura coast region in central New Zealand and the south in n area encompassing Fiordland, Stewart island and the Otago coast. lso it was found that innumerable amount of varied species belonging o different genus are found along the Indian coast [2]. The figures & 3 show the different kinds of species distribution that includes hlorophyta, Phaeophyta, Rhodophyta and Cyanophyta alongside ountries of Indian Ocean and areas of Indian coastal line [3] whereas igure 4 picturises the seaweed strength along the coastal areas of ther countries like China, Korea, Japan, USA, Ireland, Iceland, and cotland [242].

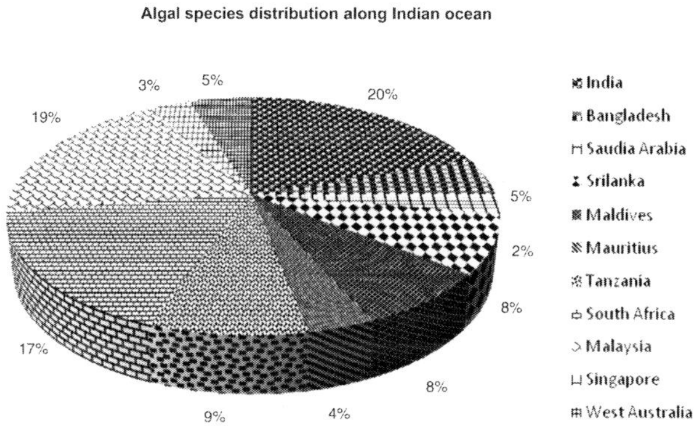

Figure 2 Distribution of species along Indian Ocean

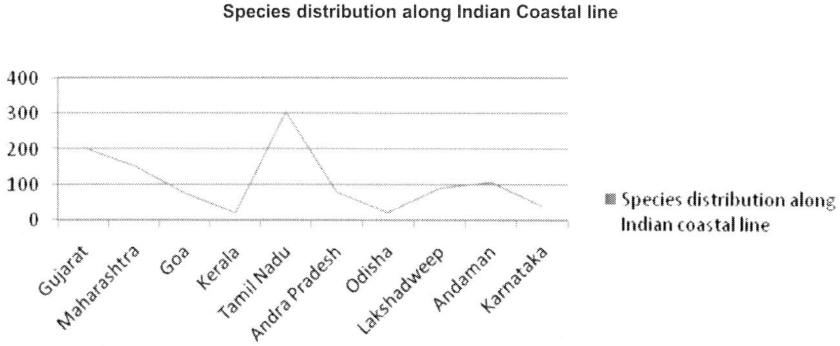

Figure 3 Graphical representation of species distribution

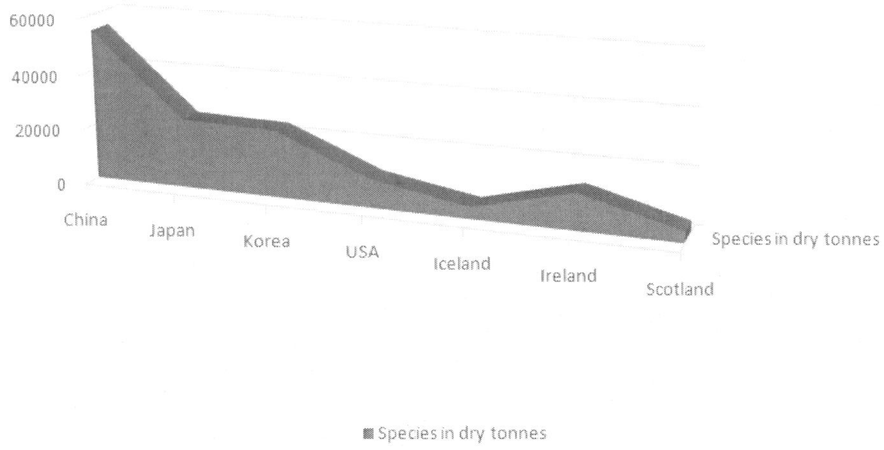

Figure 4 Macroalgal species weightage in other countries

8.1.3 Cultivation and Harvesting

With the advent of various seaweed species, cultivation of seawee attracts more attention to put them in varied uses. The physica chemical and biological parameters should be investigated befor cultivation. Apart from this, the general environmental factors tha influence the cultivation and harvesting processes are location, wate and land quality, fauna, climatic factor, marketing accessibility, farr management operations, industrial and domestic pollution. Dependin upon the properties and natural environment of individual seawee species or species in groups, the harvesting methods are selected. Th seaweed culture methods that are in practice are off-bottom methoc raft method, horizontal culture method, mixed culture method, an long line method. It is noted that the major problem faced by th seaweed industry is the lack of efficient harvesting technique. Le us see the viable harvesting that has been used from the past. Th culturing method varies between species to species and for instanc the red algae are generally seen in deep waters and are much smalle in size compared to the brown algae. In addition, the harvesting of re algae is expensive and involves complexity and it was reviewed tha two planting techniques were commonly used for red algal farming One is the direct method that involves the direct burial of the thal into the sandy bottom using different types of tools and the other is th Plastic Tube Method which consists of fastening bundles of thalli t plastic tubes filled with sand, which anchor the algae to the sea bottom [4]. It was estimated from the experience gained from subtidal farms that hand-pulling of thalli has given more production rather than the use of tools [5]. Mechanized harvesting that uses power-driven barges equipped with reciprocating underwater mowers (cropping vessels) were used for cultivating *Macrocystis* that grows in large beds. Similar cropping equipments were used to harvest *Ascophyllum* species which grows at high tide (Booth, E., In Firth, 1969) [6]. The predominant species *Porphyra yezoensis* and *Porphyra tenera* grow in the inner parts of estuaries and bays and can survive in high salt conditions whereas *Porphyra pseudolinearis* grows in the deeper waters. The conchocelis (spore) culture technique was used to cultivate these *Porphyra* (nori) species [7]. The rope cultivation and stone techniques were used in Japan to cultivate the brown alga *Undaria* or "wakame". The "collector strings" are hung in the water for the sporophytes of *Undaria* to grow and tanks were used for seedling. As an expansion of past technique, along with natural beds a new substrate of rocks exploded with dynamite was also used to grow *Undaria* [8]. Production of hybrid varieties of Undaria was also made possible by crossing closely related species [9]. To harvest *Laminaria* that grows only on hard, rocky ocean bottoms, reciprocating cutters mounted on dredges or a system of continuous grapnels were used and the development of techniques for cutting the normal 2-yr growth period to harvest down to 1 yr was researched to increase the production [10]. It was reported that in Norwegain coastal area the sugar kelp *Saccharina latissima was cultivated by integrating with salmon* (*Salmo salar*) aquaculture and the growth seemed to be good in late autumn and in spring [11]. Green algae may be found in marine or freshwater habitats, and some even thrive in moist soil. The green macro algae were cultivated in outdoor tanks and the biocrude was obtained through hydrothermal liquefaction in batch reactor [12]. Also, the green algae were cultivated in ponds, open lagoons and in cages. It was reported that the *Oedogonium* species has given the maximum yield followed by Derbesia and Ulva species [13].

3.2 BIOCHEMICAL COMPOSITION

The composition of seaweeds varies depending upon season, density and other environmental factors. Compared to vegetables, fruits, pulses and cereals, seaweed records the maximum content of carbohydrates, proteins, vitamins, minerals, fat, fibre, ash, and moisture. Earlier wor has revealed that the different species gathered from similar are family, environmental factors showed greater variations with respec to their elemental composition [22]. Also it has been observed that re seaweed contains thirty times more potassium than bananas and 20 times more iron than beetroot. The Nori seaweed constitutes twice th protein than meat and the Hijiki seaweed contains twice the amour of calcium compared to full-cream milk. Further, it was noted tha 15,000 novel compounds were chemically determined and algae ar considered to be the predominant source for novel biologically activ compounds that is required for human nutrition [57]. While seasona variations are of great concern, it was showed that the phaeophyt and rhodophyta recorded the maximum biomass during summer an cholorophyta was maximum during autumn season [20]. The Table shows the composition of some commercially important seaweeds.

Table 1 Biochemical composition of some seaweed species

Species	Biochemical compound	Yield	Reference
Gracillaria species	Vitamin A	865μg RE /100g	[14]
	Vitamin C	28.5± 0.1mg/100g	-
	Fiber	(24.7±0.7)%wet wt	-
	Lipid	0.43-(3.3±0.2)%wet wt	[14,46]
	PUFA	2.43-51.77% w/w	[76]
	Protein	(6.9±0.1)%wet wt	-
	Minerals	153.3±2.64mg/100g	-
	Ash	(22.7 ±0.6)%	[45]
	Amino acid	116.37mg/g	-
Padina species	Carbohydrate	50.9±0.52mg/g dry wt	[15]
	Lipid	58.38-184.0 mg/g dry wt	[15,76]
	Protein	17.1% d.wt	[58]
	Vit B6	115±0.577 μg/g	[15]
	Vit B2	75±1.732μg/g	-
	Vit B1	46±2.309μg/g	-
	Minerals	90.93±0.257 μg/g	[15]
	Trace elements	30-110 ppm	[65]
Enteromor-pha species	Omega -3 FA	10.38%	[19]
	Omega -6FA	10.9g/100g	[16]
	Fiber	33.4%dry wt	[18]
	Protein	19.5%	[17]
	Carbohydrate	64.9%	[17]
	Ash	15.2%	[17]
	Lipid	0.3%	[17]
	Moisture	9g/kg	[3]

Species	Biochemical compound	Yield	Reference
Ulva species	Carbohydrates	(17.8±2.3)-70.1% d.wt	[23,25]
	Protein	14.7%	[25]
	Minerals	247.54mg/100g d.wt	[24]
	Lipids	0.51-1.18 %	[26,27]
	SFA	65.40±1.65%	[57]
	Aminoacids	829.48±6.51mg/100g	[28]
	Fibre	51.3-62.2%d.wt	[80]
	Ash	25.9-62.2% d.wt	[80]
	Vitamin E	19.70 µg/g d.wt	[68]
Laminaria species	Carbohydrates	23.61%	[29,31]
	Protein	16±3.8g/100g	[31,32]
	Lipids	26.05±0.3g/10g	[30,31]
	Aminoacids	86.5±14.1g/100gN	[30]
	Fibre	36.0±5.7g/100g	[30]
	Ash	24.9-36.4%	[32]
	Minerals	1929.57mg/100g d.wt	[34]
	Vitamin E	34.38 µg/g d.wt	[69]
Porphyra species	Carbohydrates	45.1%	[41]
	Protein	47% d.wt	[37]
	Lipids	2.1±1.2g/100g	[30]
	Fibre	47.7±4.25g/100g	[30]
	Ash	20.59±0.16g/100g d.wt	[35]
	Minerals	106.07±3.44	[64]
Palmaria species	Protein	28.25% d.wt	[55,37]
	Minerals	1686.78mg/100gd.wt	[34]
	Fatty acids	25.53±2.37%	[79]
	Moisture	84%	[21]
	Ash	12-37%	[21]
Undaria species	Ash	39.26±0.24g/100g d.wt	[35]
	Minerals	9992.42±383mg/100g	[35,34]
	Fibre	d.wt	[34]
	Fatty acids	35.3%	[79]
		33.36±3.91g/100g d.wt	
Sargassum species	Carbohydrate	29.37-47.04±0.7% d.wt	[71,56,59,60,62]
	Protein	11.52-16.59**±0.86%**	[36,63]
	Fattyacid	0.16±0.006-20.96%	[59,71]
	Fibre	77.40±16.14g/kg	[46]
	Ash	194.30±48.65g/kg	[46]
	Mineral	1901.74mg/100g d.wt	[59]
	Moisture	14.33±1g/kg	[46]

Species	Biochemical compound	Yield	Reference
Fucus species	Carbohydrate	65.7**±0.4%**	[77]
	Lipid	1.4±0.1%	[77]
	Protein	7%	[37]
	Mineral	11736±127mg/100g d.wt	[35]
	Ash	0.11-22.71±3.1%	[70,73]
	vitamin	820 µg/gd.wt	[67]
Hypnea species	Carbohydrate	13.69%	[48,49]
	Protein	28.63**±0.39%**	[41,48,49]
	Lipid	1.43%	[48,49]
	Fibre	47.0025%	[48,49]
	Ash	21.28%	]48,49]
	Moisture	11.18%	[48,49]
	Aminoacid	77.1g/100g	[41]
Gelidium species	Protein	11.55**±1.02%**	-
	Carbohydrate	41.87±2.21%	[47,51]
	Lipid	1.53±0.61%	-
	Fibre	24.74±1.05%	-
	Ash	17.57±0.74%	-
	Moisture	10.85±0.98%	-
Turbinaria species	Protein	14.68%	[42,52,58]
	Carbohydrate	18.46**±0.685%**	[42,52,58]
	Lipid	2.53±0.14%	[52,58]
	Moisture	83.79**±0.32%**	[20]
	Ash	21.37**±0.60%**	[20]
	Fibre	29.61**±1.59% d.wt**	[33]
Acanthopora species	Carbohydrate	23.54**±0.10%**	-
	Lipid	55.33**±5.1g/mg**	[43,44]
	Fattyacid	64%	-
	Ash	16.4**±2.4g/mg**	-
	Protein	24%	[53]
Colponemia species	Carbohydrate	32.1**±1.75%**	[56]
	Protein	9.2**±1.78%**	-
	Lipid	1.5**±0.29%**	-
	Moisture	11.5±0.20%	-
	Fatty acid	1.0%	-
	Ash	28.1±0.86%	-
Ascophyllum species	Carbohydrate	69.6**±0.2%**	[77]
	Protein	(3-15)% d.wt	[37]
	Lipid	1.2**±0.1** %	[77]
	Oil	2.85 %	[23]
	Ash	22.5**±0.1** %	[77]
	Vitamin C	81.75 µg/g d.wt.	[69]
	Vitamin E	3.63 µg/g d.wt	[69]

Species	Biochemical compound	Yield	Reference
Dictyota species	Carbohydrate	10.63%	[42]
	Lipid	16.1-20.2 % d.wt	[82]
	Protein	9.8%	[42]
	Minerals	0.559**±0.009 ppm**	[78]
Caulerpa species	Carbohydrate	9.7±0.477 %	[36]
	Protein	24.55±0.84 %	[36]
	Lipid	0.9±0.38%	[36]
	Fiber	1.36 g/kg	[72]
	Ash	24.2 % d.wt	[81]
Jania species	Protein	2.53±0.15% wet wt	[74]
	Lipid	0.22±0.035% wet wt	[74]
	Moisture	40.13±1.54 % wet wt	[74]
	Ash	48.14±1.88 % wet wt	[74]
	Fatty acid	11.319 % wet wt	[74]

.3 PROPERTIES

t was reviewed that seaweeds are rich in bioactive compounds and some of the species to say are *Laminaria* species, *Fucus* species, *Ascophyllum odosum*, *Chondrus crispus*, *Porphyra* species, *Ulva* species, *Sargassum* pecies, *Gracilaria* species and *Palmaria palmate* [83]. Since chemical reservatives were proved to cause deleterious health hazards like ancer, asthma, etc., seaweeds have been known to be the safe and romising replace as food additives that have antimicrobial and ntioxidant properties [84]. Also many secondary metabolites from he marine source with effective antibacterial, antifungal and antiviral ctivities which are being used as antibiotics and drugs to treat various nfectious diseases were also identified [85]. Bioactive compounds of narine macro algae are known to have diverse mechanism of action gainst diseases. In addition to this, red, brown and green macro lgae were detected to contain compounds with cytostatic, antiviral, ntihelminthic, antifungal and antibacterial activities [86].

.3.1 Antioxidant Activity

lgae are antioxidant in nature because of their non-enzymatic ntioxidant components like reduced glutathione (GSH), ascorbic acid, lpha tocopherol, betacarotenoids, flavonoids, hydroines, pycocyanin, roline, mannitol, myoinositol, phenolics and polyamines [87]. The antioxidants can inhibit the intiation of the oxidative chain reactio and thus prevent cell damage caused by Reactive Oxygen Species (ROS [88]. The antioxidant molecules can destroy free radicals by donatin hydrogen atoms or by electron donation. In most of the cases DPPH had been used as a free radical to evaluate reducing substances rathe than nitric oxide, deoxyribose, hydrogen peroxide, ABTS. The DPPH radical scavenging activity was evaluated using an ESR (Electron Spi Resonance) spectrometer [89]. From the earlier works done it wa reported that the extracts of species such as *Hijik fusiformis*, *Cladosipho okamuranus*, *Undaria pinnatifida*, and *Sargassum fulvellum* were know to possess effective DPPH radical scavenging activity [91, 92]. Als carotenoids have radical scavenging which helps to keep up healt and in prevention of disease whereas fucoxanthin has been reporte to effectively scavenge chemically-generated free radicals like DPPH [93]. Based on quenching rate constants it was shown that the radica scavenging activity of fucoxanthin and its metabolite fucoxanthino were higher than that of α-tocopherol and lower than β-caroten [94]. Fucoxanthin (FX), an orange colored carotenoid belonging t non provitamin carotenoids which is also known as xanthophylls ar present in edible brown seaweeds such as *Undaria pinnatifida*, *Hijiki fusiformis*, *Laminaria japonica* and *Sargassum fulvellum* that posses significant antioxidant activity [99, 100]. Sulfated polysaccharides (SP are the compounds found in the extracellular matrix of seaweeds tha has got antioxidant property and the well-known sp. are carragenans agarans, xylase, galactose, mannose, fucan, fucoidan [96]. Based o the rheological behavior, the polysaccharide content and the methano extract of red algae *Gracilluria biridae* and *Graciluria verrucosa* wer reported to possess antioxidant property [97, 98]. Further it was note that the water extract of Laminaria species had higher antioxidan activity than its ethanol extract [90]. Thus high levels of oxidative stres leads to many harmful diseases such as atherosclerosis, Parkinson disease, Alzheimer's disease, acute myocardial infarction, chroni fatigue syndrome and fibromyalgia. FX was estimated as an effectiv tool to prevent and treat these diseases [95].

8.3.2 ANTICANCER ACTIVITY

Cancer being a fatal disease has become a major health problen worldwide mainly because of bad food habits. The conventiona chemotheraphy or radiotheraphy treatment to control tumors an educe the risk of mortality rates has resulted in other ill-effects such as ong term side effects, destruction of healthy tissues, etc. Several types f carcinomas such as prostate cancer, leukemia, colon cancer, breast ancer, liver cancer, melanoma and lymphoma are in existence [101]. n a quest to cure and uproot cancer, seaweeds were identified as the est source that has antitumor properties. The seaweed compounds r extracts undergo several types of mechanisms against the cancer ells and some of them are apoptosis induction, inhibition of tumor nvasion, hyaluronidase activity inhibition, anti-angiogenic activity nd regulation of mammary gland integrity [102]. Polysaccharides resent in the cell wall of macroalgae contain immunomodulatory and nticancer effects and are most considered in the medical areas of study 103]. Sulfated polysaccharides from brown species viz., *Sargassum, aminaria, Ecklonia* inhibited growth of Sarcoma-180 cells and acted s antitumor against L-1210 leukemia [104-106]. Further fucoidan r fucose containing sulfated polysaccharides (FCSP) and MAPK mitogen activated protein kinase) along with FCSP in brown seaweeds vas found to enhance and augment macrophage mediated immune ignaling molecules production and thereby induced apoptosis [107]. Breast cancer ranks the second most common cancer in the world and s the major cause for mortality in women [108]. As a remedy for this, the nethanol extract of *Sargassum muticum* activity against proliferation of breast cancer cell lines were evaluated for its apoptosis property 109]. Also the combination of seaweed *Porphyra dentate*, β-sitosterol nd campesterol was known to reduce the tumor size considerably 110]. Next the colorectal cancer which affects both men and women, heir cancer cell lines proliferation could be inhibited by the extracts of *aminarian species* and *Ulva faciata* by apoptosis induction mechanism 111-113]. Furthermore based on dosage, the anti-tumor and anti-netastatic activities of fucoidan isolated from *Fucus evanescens* were studied. Alginates and palmitic acid from the species *Sargasum vulgare* brown algae) and *Amphiroa zonata* (red algae) were also reported to possess antitumor property [114,115].

3.3.3 Antimicrobial Activity

Algal biomass has been analysed to have antimicrobial activity and it nainly depends upon the algal species taken and the extraction method ollowed. The antimicrobial compound extracted from a biological source is generally by means of attacking the cell wall and cell membrar of the target organism. Further it disrupts the electron transport chai coagulates protein and nucleic acid synthesis [116]. Among seawe species the ethyl acetate extracts of *Sargassum* species have stro antimicrobial activity against bacteria and fungi than the water extra and the activity is due to the presence of meroterepenoids [117]. Th brown alga *Stoechospermum marginatum* was active against bacteri strains *Klebsiella* and *Vibrio cholerae* whereas the green alga *Cladopho prolifera* was bacteriocidal against *Saccharomyces aureus* and *Vibr cholera* [118,119]. It was found that the extracts of red algae *Gracilar fisheri* inhibits the pathogen *Vibrio harveyi* which affects the shrim population [120]. Also the butanolic extracts of the seaweeds *Ulva lactu* and *Sargassum wightii* exhibited considerable inhibition zone against t shrimp pathogen *Vibrio parahaemolyticus* [121]. The phlorotannins *Ascophllum nodosum* was more active against the *Escherichia coli* strai than the condensed or terrestrial tannins [122]. The dichloromethar extracts of several seaweeds have shown significant antibacterial actic against fish pathogens such as *Asparagopsis armata* and *Falkenberg rufolanosa* [123]. The dimethyl sulfoxide (DMSO) extracts of seawe species and their antiprotozoal activity against *Plasmodium* species hav been studied and the selectivity index was the parameter used to evalua the activity [124]. The same DMSO extract of *Sargassum longifoliu* showed inhibitory activity against various bacterial strains where the acetic acid extract showed maximum inhibition against *Prote* species and minimum activity against Streptocccus species [125]. Th antifungal activity against *Phythium phanidermatam* and *Colletotrichu capsici* was maximum in *Ulva fasciata* [127]. Ultimately nanopartick which is the most welcomed compound because of its size and viabili was found to be extracted as silver nanoparticles from *Ulva lactuca* b agar well diffusion method and it was analysed for its antibacteri activity [126].

8.3.4 Antiviral Activity

Promising feature of seaweed is its antiviral potential by which mar fatal diseases can be treated. AIDS is a fatal disease caused by Huma Immunodeficiency Virus (HIV) which belongs to retro virus family an has no effective treatment till date. The antiviral activity of seawee depends on the dosage and time. Apart from other causes, the Herp implex Virus (HSV-1) and HSV-2 which causes infection in mouth, ace and genital area was also identified as a major risk factor for Iuman immunodeficiency Virus (HIV) [128,129] and the antiviral ctivity against these viruses has been conferred with SQDS (Sulfo uinovosyl diacyl glycerol) fractions extracted from *Sargassum vulgare* 130]. Also the bioactive alginates from Sargassum species has antiviral roperty and it was studied that the extracts of this species acted gainst the viruses Human T cell Lymphotropic Virus Type 1 (HTLV1) nd Human Immunodeficiency Virus Type 1 (HIV-1) [131-133]. The ucoidan polysaccharide was against HIV and human cytomegalovirus HCMV) while its derivative galactofuran extracted from the seaweed *denocystis utricularis* showed inhibitory action towards the retro iruses (HSV) 1 and 2 [134]. Further the diterpenes isolated from he Dictyota species exerted antiviral action and galactofucan sulfate xtract from *Undaria pinnatifida* worked against HSV-1, HSV-2 and ICMV (Human Cyto Megalovirus) [135,136]. Considering plant viruses he seaweed polysaccharides such as fucans, laminarin, alginates, lvans obtained from *Fucus vesiculosus, Laminaria digitatum, Lessonia pecies, Ulva species* respectively, when injected into tobacco plants, rotected against Tobacco Mosaic Virus (TMV) by inducing Jasmonic cid (JA) and Salicylic acid(SA) signaling pathways [137].

.3.5 Anti-Inflammatory Activity

n general inflammatory response is an auto defensive mechanism that s met with huge leukocyte production and the inflammatory reactions re generally due to the presence of ROS, nitric oxide and other factors vhich results in tissue damage [139]. Caulerpin, an alkaloid extracted rom *Caulerpa racemosa* was found to exhibit anti-inflammatory roperties [140] and this compound acts by suppressing the antigen, istamine secretion, lymphocyte and natural killer cell proliferation 138]. The sulphated polysaccharides play a major role in treating nflammation. Also it was demonstrated that the fucans of *Fucus* species nd *Laminarian* species on injection into rats reduced the peritoneal nflammation by leukocyte inhibition [141]. The ulvan polysaccharides f *Ulva rigida* also exerted anti inflammation by reducing immune timulation [142]. The fucoxanthin from seaweeds was active against nflammation and allergic reactions by degranulating the mast cells which secretes histamine [143]. It is noted that the carrageena produced from *Eucheuma* or *Chondrus* or *Hypnea* species was used analyse anti-inflammatory activity [144].

8.4 APPLICATIONS OF SEAWEED

Macroalgal species are being used as a major viable source in variou fields of applicatons. The carbohydrates content of seaweed has bee the major area of interest in several fuel, food and medicinal industrie Many food additives like agar, carrageenan, laminarin, and fucoida extract from seaweeds are found to be implemented in food industrie Since the nanoparticles have gathered interest among researchers i the current world, seaweed nanoparticles being a natural source hav potent application in biomedical industries. Apart from several huma uses, seaweeds are used as feed for cattle and aquatic animals an as fertilizer for various food and cash crops. The various species c seaweed and its targeted content of application have been discussed i the following sections.

8.4.1 Biofuels

As there is a tremendous increase in fuel consumption these day mankind has been in pursuit of natural source for fuel recovery whic ensures pollution-free environment. And now scientists have foun certain seaweeds can be the raw material to produce fuel. Thes macroalgal biomass must be pre-treated for most biofuel application The first step of pre-treatment is to remove foreign objects and debri such as stones, sand, snails, or other litter that may be caught in th biomass either manually or by washing in many cases, chopping o milling is then required to increase the surface area/volume ratio [156 Finally, the biomass should be dewatered to 20%–30% to increase she life and reduce transportation costs in situations where it must b stored for long periods or transported over long distances before furthe processing. The principal energy process considered for seaweed i fermentation, either anaerobic digestion, to create biogas, or ethan fermentation. Other thermochemical options for macroalgae utilizatio include direct combustion, gasification, pyrolysis and liquefactio

161]. Figure 5 explains about the various methods of production of iofuel from the algal biomass for better understanding.

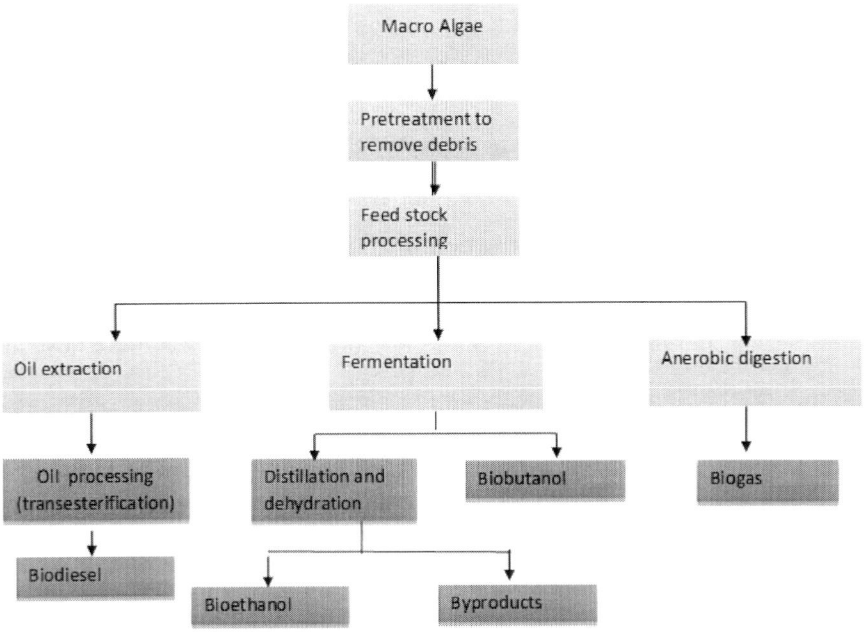

igure 5 Flowchart showing various biofuel production from algal biomass

.4.1.1 ***Bioethanol*** Generally, bioethanol is produced from wood, rasses, and other inedible parts of plant but it is a tedious process to nake sugar monomers. In order to overcome this, marine algae can be ısed as source for bioethanol production. Algae contain large quantities of carbohydrate biomass and high photon conversion efficiency for oioethanol production [148,149]. In addition, marine algae have buoyant oroperty which can simplify the process of bioethanol production by ıeglecting the pre-treatment steps [150]. It is noted that in Japan they ıse 4.47 million sq.km for harvesting *Sargassum horneri* for bioethanol oroduction [151]. *Ulva reticulata* macroalgae which can grow quickly vas said to possess the potential to produce bioethanol [146]. It is eported that other species such as *Padina japonica, Sphacelaria rigidula, Dictyosphaeri acavernosa,* and *Sargassum polyphyllum* have appreciable lry weight content [147]. Brown seaweed *Laminaria hyperborea* has high amount of soluble carbohydrate [152]. Bioethanol from algae wa known to reduce greenhouse gas emission by 85% over reformulate gasoline [145].

8.4.1.2 *Biodiesel* Since diesel price is on hike irrespective of th demand, biodiesel which is a natural fuel can be made from oi available within the algae. It is significantly noted that the macroalga contains oil within its cell and biodiesel obtained with this oil is engir compatible. There are different varieties of macroalgae which contai oil. Once grown, the oil is removed from the macroalgae using chemica or by squeezing oil out of the cells using scientific equipment. Then th oil is used as an ingredient in biodiesel. This oil is changed chemical from plant oil to biodiesel. The finished product can be used on its ow as pure biodiesel but is normally mixed with ordinary diesel and use directly in cars. Biodiesel is an alternative biodegradable energy sourc which has less CO_2 and no emission [153]. Oil extraction from algae an transesterification process are the two major steps involved in biodies production and the brown seaweed *Ascophyllum nodosum* has high o content compared to the species such as *Sargassum, Codium, Ulva,* an *Enteromorpha* [154].

8.4.1.3 *Biogas* For use in domestic and industrial gas fuel, seawee can be subjected to anaerobic digestion for the production of methan gas. It was estimated that the methane yield from anaerobic digestio of seaweed was 0.12 N/CH_4/g [155]. *Saccharina latissima* is a brow seaweed rich in carbohydrate and was anaerobically digested for bioga production. It was reported that two process parameters such as stear explosion and thermal pretreatment method for biomass degradatio seemed to affect the biogas yield. In general, the brown algae are mor easily degraded than the green algae, and the green are more easil degraded than the red [156]. Anaerobic digestion has been the mos efficient method for the production of biogas rather than fermentatio and thermal treatment. The biogas yield from certain species vi *Saccorhiza polyschides, Ulva* species, *Laminaria digitata, Fucus serratu* and *Saccharina latissima*are is discussed in the following table 3 [159 Despite the variations in the quantity and material ratios, the rati etween the methane produced and the input chemical oxygen demand 'as reported to be stable [160].

.4.1.4 *Biobutanol* It is important to note that the biobutanol has ecome a very competitive renewable biofuel for use in internal ombustion engines and has been a boon to mankind. While comparing ıel properties it indicates that n-butanol is potent enough to remove ıe drawbacks brought by low-carbon alcohols or biodiesel. The pplications of butanol as a biofuel are considered as three aspects, nd they are as combustion experiments in some well-defined burning eactors, as gasoline in spark ignition engine, and as diesel fuel in ompression ignition engine. From these demonstrations, butanol ; estimated to be effective as a second generation biofuel, from the iewpoints of combustion characteristics, engine performance, and xhaust emissions [162]. *Ulva lactuca* was considered as the reserve pecies for acetone butanol ethanol (ABE) fermentation [157]. The ermentation done with the bacterial strains such as *Clostridium eijerinkii*, *Clostridium saccharoperbutylacetonium* and the algal arbohydrate resulted in the production of biobutanol. The yield and he concentration of biobutanol obtained from media were 0.29g utanol/g sugar and 4g/litre respectively [158].

.4.2 Polysaccharides of Seaweed

eaweed polysaccharide is known for their varied functions and tructures. They constitute natural sugars and sugar acids similar to and plants and in animal polysaccharides. As these polysaccharides ontain hexose sugar, glucose, galactose and mannose, they have dentical chemical formula, shape, properties and specific atomic orientation [174].

.4.2.1 *Agar* Seaweed agar is the most ancient phycocolloid found n Japan and discovered by Minoya Tarozaemon in 1658 and first time nanufactured in monument [165]. Agar is the major component of he cell-wall of certain red algae, which are the members of families,

Gelidiaceae, Gelidiellaceae and Gracilariceae [163]. Agar consists of chain of 9-p galactopyranose units linked in 1,4 bonds with a sulphate galactose. In order to increase the yield and gel strength of agar, a alkaline treatment was done with sodium hydroxide for nearly one hou at the rate of 2 to 3% alkali solution of 20,000 l/tonne at 90°C [164]. Als the sun-bleached seaweed was washed well in water and soaked for 24 and then ground to a pulp and rinsed again in water and then the pulp wa then extracted with water under pressure for 2 h after bringing the pH t 6 by adding of acetic acid. The agar gel was subjected to freeze thawin and bleached with NaClO before drying in a current of hot air [167]. is insoluble in cold water but soluble in boiling water. When agar wa cooled to 34-43°C it forms a firm gel and does not melt further belo 85°C [166]. In food technology agar is used as gelling and thickenin agent in the confectionary and bakery industries; and as stabilizer in th preparation of cheese and for salad dressings. In fish and meat industr agar is applied for canned products, as a protective coating to avoi shaking during transport of these products. The agarose polycolloi plays a prominent role in the DNA research and gel electrophoresis Further agar is widely used in pharmaceutical industry as a laxative drug vehicle and as a medium for bacterial and fungal cultures and als used as an ion exchanger in the ion exchange resins [168].

8.4.2.2 ***Alginate*** Algin or alginic acid is a membrane mucilage an a major constituent of all alginates and the trade name is sodiur alginate. Alginic acid is obtained from brown seaweed species suc as *Ecklonia, Macrocysiis, Undaria, Laminaria* and *Duruillea* fror temperate area and *Turbinaria, Sargassum, Cystoseira* and *Harmophys* from the tropical areas. Alginic acid which is the major polysaccharid of the brown seaweeds consists of unbranched chains comprising c contiguous fl-l,4-linked d-mannuronic acid and blocks of contiguou e-l,4-linked e-guluronic acid [175]. The proportions of D-mannuroni acid and l-guluronic acid varies between different species and fror different parts of the same weed [176]. Alginates are found in both th intercellular region and the cell walls and it does biological function of structural and ion exchange type. Also it was extracted from th species *Laminaria digitata by alkaline extraction protocol* [173,177]. Th molecular weight of alginate ranges generally between 500 and 100 kda [169]. As discussed earlier, the alginate composition of differen eaweed species are *Ascophylum nodosum* (22–30%); *Laminaria igitata* fronds (25–44%); *Laminaria digitata* stipes (35–47%); *Laminaria yperborea* fronds (17–33%); *Laminaria hyperborea* stipes (25–38%) 170]. Alginate content ranging between 17 and 45% are extracted 1 *Sargassum* species [171]. In industries the alginates are extracted rom brown seaweed *Sargassum turbinarioides* by cutting the thallus vith a knife near the rhizoid and the algae were washed and sun-dried t ambient temperature and stored in aerated bags [172]. Japanese vork on the brown seaweed has revealed that alginate is subjected o polymerization in the cytoplasm and then transported to the cell urface [178]. It is observed as a d-mannuronic acid precursor of olymannuronic acid while separating from the brown seaweed, *Fucus ardneri* in young tissue. They identified the presence of trace quantity f a GDP-guluronic acid [179]. Further on extracting epimerase from *'elvetia canaliculata* the conversion of polymannuronic acid into a nixed poly-d-mannuronic-l-guluronic polymer has been witnessed by ritium incorporation [180]. In beverages, alginates acts as clarifying gents for making wines and liquor whereas it acts as foam stabilizer n lager beer and malt beer [181]. Artificial casings are made with lginates as it ensures longer shelf-life in sausage industries and lginates are used in the form of gel for deep freezing of fish, meat and oultry products in western countries [182].

.4.2.3 *Carrageenan* Macroalgal carrageenan is a sulphated olymer obtained from various red seaweeds and it differs from agar n its high sulphate and ash content. The major difference between the gars and carrageenan is that the former contains d- and l-galactose nits whereas the latter consists entirely of the d-sugar [174]. They are ommercially important hydrophilic colloids present in the matrix of ed seaweeds (rhodophyta) and does structural functions also. They are onsidered as high sulfated galactans and as strong anionic polymers 183]. Seaweed species *Kappaphycus, Eucheuma, Chondrus crispus, Gigartina stellata, Iridaea,* and *Hypnea* species have high content of arrageenan [184]. Carrageenan can be recovered either by direct lrying on steam-heated rolls or by precipitation of the carrageenan rom solution by 2-propanol or other alcohols. It is to be noted that in he past alcohol precipitation method was used to recover carrageenan rom irish moss [185]. Commercially three types of carrageenan are available. They are kappa-, iota-, and lambda-carrageenan. Furcellara which is similar to kappa carrageenan is a preferred product for us in milk pudding powders. The iota-carrageenan is used in dessert g formulations, whereas lambda carrageenan is being used to thicke dairy product. In food industry, carrageenan is also being used i bakery, confectionery and in culinary works. Carrageenan is also calle as "painters" because of its wide range of use in paint manufacturir and in stabilising pigments. Carrageenan gel beads are the exceller media for enzyme entrapment which in turn catalysis the synthes and conversion processes [186].

8.4.2.4 ***Fucoidan*** Fucoidan is a sulfated polysaccharide found mainl in various species of brown seaweed and fucoidan is used as dietar supplements [188]. Fucoidan is a class of sulphated, fucose-ricl polysaccharides found in the fibrillar cell walls and intercellular space of brown seaweeds. Fucose-containing sulphated polysaccharide (fcsps) consist of a backbone of (1→3)- and (1→4)-linked α-fucopyranose residues, that may be organized in stretches of(1→3 α-fucan or of alternating α(1→3)- and α(1→4)-bonded l-fucopyranos residues [187]. Apart from fucose and sulphate groups fucoidans als contain galactose, xylose, mannose and other uronic acids [186,187,188 Fucoidan was extracted using dilute acetic acid from various specie of laminaria and fucus [201]. Fucoidans were reported to posses anti-inflammatory, anticoagulant, antithrombotic [195,196], antivira including anti-HIV [194,193], immunomodulatory [196], antioxidar [197], and antitumor properties [198]. Fucoidan from laminaria specie was found to inhibit a variety of DNA and RNA enveloped viruses an also was found to be useful in elucidation of mammalian sperm an egg [199]. The "fucan" extracted from *Pelvetia canaliculata* had a ver strong affinity for magnesium which in turn can assist the contact c their fronds with seawater. It was also studied that *Pelvetia canaliculat* that grows on the higher part of the shores have a high "fucan" conter [200].

.4.2.5 ***Laminarin*** Laminarin is a storage glucan found in brown lgae and is used as a carbohydrate food reserve similar to diatoms 202]. Laminarin is a β-glucan and it is known to induce anti-apoptotic nd anti-tumoral activities [203]. This is a water-soluble polysaccharide ontaining approximately 20-25 glucose units. It has two types of hains namely, g-chains terminated at the reducing end with glucose nd m-chains terminated by mannitol [204]. In addition, the laminarin, rom *Eisenia bicyclis,* may contain 1,6-linked units in the chains, or the hains may be branched at C-6 [205]. It was found that the mannitol nd laminarin are active metabolites which can be interconverted 206]. Also it was shown that laminarin isolated from the cytoplasm of eveloping zygotes of fucus species decreased during the first 7 hours f wall assembly while the content in cellulose in the wall increased aminarin structure and composition vary according to algae species 207]. Based on the degree of polymerization, the molecular weight of aminaran has been found as 5000 Da approximately [208].

.5 PHARMACEUTICAL APPLICATION OF SEAWEED

Most inspired bioresource as far as pharmaceuticals are concerned is eaweed because of the high nutrient content. Seaweed has been used as ood throughout Asia, Japan, China, Rome, etc., to treat various health isorders. The Romans used seaweed in the treatment of wounds, burns, nd rashes [209] whereas in Scotland physicians used dried seaweed tem to drain abdominal wall abscesses and they also inserted seaweed nto the cervix to treat dysmenorrhea and seaweed was also employed ntravaginally for vaginal atresia for which it was used urethrally and ectally for strictures [210,211]. Traditional Chinese medicine includes se of the brown alga *Laminaria* in the treatment of cancer. The ncient Egyptians used seaweed to treat breast cancer. Seaweeds are eing extensively used in cardiovascular conditions as it can reduce holestrol level. In general alginates from seaweeds has been used in vound dressings and as fillers in tablets, pills and as ointment base vhereas seaweed carrageenan acts as good emulsifiers in mineral oil nd drug preparations. *Saccharina japonica* and *Undaria pinnatifida* vere analysed to contain fucoidan which can destruct cancer cells. Also complete wipe out of cancer cells by fucoidan treatment has been demonstrated in Japan. It is noted that cancer mortality rates an breast cancer rates are considerably low in Japan because of seawee consumption [212,213]. Seaweed extracts being a source of calcium magnesium, selenium and other minerals has been evaluated to tre osteoarthritis [214,215]. And it was reported that intake of seawee powder 5g/day, 12g/day and 4-6g/day in diet controls cholestero hypertension, and metabolic syndrome respectively [216,217]. Iodine rich seaweeds like *Asparagopsis tcudjirmis* and *Sarconema* species wer reported to cure hypothyroidism (goitre). Also in the earlier researc works seaweed extracts were known to stimulate B lymphocytes an macrophages that in turn modulates immune response [221,222].

8.6 APPLICATION OF SEAWEED IN OTHER AREAS

On pursuing the trends of biotechnology and its advancements, seawee as a single macroalgal source with varied types, characteristics an multipotent properties was known to be employed in various secto as from animal feed to the latest emerging field of nanotechnolog From the earlier literature works done the other areas of effective use of seaweeds and their distinct types are summarised in the followin sections.

8.6.1 Seaweed Nanoparticles

Increasing trend in nanotechnology has gathered interest among th researchers, thereby algal nano particles are known as bio nano factorie as they are highly stable, easy to handle and avoids cell maintenance Added to this, metal nano particles from seaweed have excellent potentia in biomedical applications [223]. Benign nanoparticle synthesis which i nontoxic has been an emerging trend in today's world [220]. Ecofriendl gold nanoparticles synthesized from *Turbinaria conoides* was confirme to be associated with carboxylic, amine, and polyphenolic groups b fourier transform-infrared spectroscopy [219]. Green seaweed *Caulerp peltata*, red *Hypnea valencia*, and brown *Sargassum myriocystum* seaweed were used to synthesize zinc oxide nanoparticles and they can be used i effluent treatment process to reduce microbes [224]. Silver nanoparticle from *Sargassum tenerrimum* along with the presence of phytochemical as reducing agents was found to have excellent antimicrobial activit

225]. Further report identified that Fe_3O_4 nanoparticles obtained from *argassum muticum* which contain sulphated polysaccharides as the educing agent were determined to have antimicrobial potential and tabilizing capacity and this on fabrication came out with other metal xides [226].

.6.2 Seaweed in Wastewater Treatment

'he two major areas of waste water treatment in which seaweed has s prominent role to play are one is to treat sewage and agricultural vastes to exploit nitrogen-phosphorous wastes and the other is to emove toxic metals from industrial effluents and it was suggested to e a boon in coastal areas [227]. Alginates extracted from the seaweed *'argassum sinicola* was used to co-immobilize the microalgae *Chlorella orokiniana* and the bacterium *Azospirillum brasilense which promotes rowth when employed in waste water treatment*. Seaweeds were used n wastewater treatment because of its ability to absorb nutrients nd heavy metal ions that are toxic. Also seaweeds were suggested s biological indicators of marine pollution like eutrophication on onsidering its capacity to take up ammonia in the nitrogen form nd phosphorous. Despite varying concentration and type seaweeds uch as *Sargassum*, *Laminaria*, *Ecklonia*, *Ulva* and *Enteromorpha* were dentified as indicators of heavy metal pollution [228].

.6.3 Seaweed as Beauty Promoters

Seaweeds possess potentials to stimulate blood circulation and evitalize, nourish and eliminate toxins of the skin. Seaweed bath is one uch instance that has been practiced in Ireland with *Fucus serratus* pecies promoted to treat rheumatism and arthritis. Also all the required utrients, aminoacids and oils were found to be absorbed by the skin in eaweed bath [230]. An Irish company has been established to produce eaweed powder from *Ascophyllum nodosum* for the cosmetic and lgotherapy and it was said to improve damaged hair by means of ionic nteractions with the proteins of hair [232]. Also seaweed enzymes were ound to heal dandruff and stimulate hair follicles for hair growth [229]. Silicon from seaweeds was analysed to have anti-wrinkle effect on facial skin and other anticellulite preparations from seaweeds have been used n the form of creams and lotions for hip, thigh and neck [230].

8.6.4 Seaweed as Animal Feed

Seaweed meal rich in proteins and carbohydrates have been used a animal feed after getting the raw biomass cleaned, dried and milled. Als seaweeds were reported to be used as domestic animal feed in Norwa Iceland and France for years. Seaweed feed is considered to have 3 percent of the nutrition of the grains. It was being added to poult diets in a ratio of 5-15 percent of the diet depending upon seawee species and the targeted animal age. In addition, feeding seaweed me and sardine oil to chickens has resulted in the reduction of choleste level and increase in omega-3 fatty acid content. Seaweed extra from *Ascophyllum nodosum* on consumption by pigs was known t promote prebiotic activity as it has ascorbic acid, tocopherols, vitami B, minerals and it was found to provide iodine for hogs, chicken sheep and dairy cows [235]. The wet seaweed was being processed b passing through hammer mills with small screens to get fine particle followed by passing through drum dryer at 700-800°C. It was reporte that *Ascophyllum* which is a dark brown seaweed has high content o phenolic compounds whereas *Alaria esculenta* which is light brow seaweed was found to be more effective as animal feed than the forme *Macrocystis pyrifera* and *Sargassum* species were evaluated as fodde supplements in goat and sheep diets. Also Sargassum species was fe to Nubian goats which replaced 25% flour feed [236].

8.6.5 Seaweed as a Biofertilizer

Seaweeds have been used to nourish worn out soil around coastal area Use of seaweed fertilizer stimulated root volume, plant growth and eve promoted fruit development thus resulting in the production of hig quality agricultural products [232]. Further it was revealed that th use of seaweed fertilizer improved germination and disease resistar capacity in plants [233]. Water and alkaline extracts *of Ascophyllur nodosum* have given tomatoes of appreciable mass and also yielde good quality fruits [234]. Seaweed extracts and fertilizers have give promising turn-outs in agriculture for which it is a mainstay in th catalogue of commonly preferred soil nourishers.

.7 INDIAN SCENARIO

ı the earlier works it has been reported that there are 271 genus and 153 species of marine algae found along the Indian coast. It was stimated that the southern coast of India has got more than 200 species f seaweeds which stands a prominent source for agar and sodium lginate production in India. Also it was found that of the 75,373 ›nnes of seaweeds the maximum mass are scattered along the Tamil Jadu coastal line covering an area of 1863 sq km from Rameshwaram › Kanyakumari. It was reported that there are about 25 seaweed rocessing small-scale industries of which 20 are agar producers owever 110-132 tonnes of dry agar are being produced annually based n the harvest of 880-1,100 tonnes (dry weight) in Tamil Nadu and Kerala [150,151]. Seaweeds like *Gracilaria, Gelidium, Kappaphycus* re being cultivated on a large-scale in India for the production of hycocolloids [152]. Though India has got a longer coastal line and ich availability of seaweed resource, the awareness in India about eaweed culture and its commercial applications is not fair enough. So the common people need more awareness about seaweeds to romote a better and healthy future.

.8 ECONOMIC IMPORTANCE

From all the seaweed potentials studied earlier it is evident that the lobalization of seaweed cultivation and seaweed derived products an improve the economic status of various countries all over the vorld. The European, Eastern and Southeast Asian countries are he major producers and consumers of seaweeds [240]. According o FAO 2014, Chile tops the global natural seaweed producers and it vas estimated as one of the first agar producers in the world [239]. Various seaweeds are being harvested for human consumption and ndustrially used in the production of a wide range of nutraceuticals. Patagonia is an outstanding place where all the seaweeds are harvested and commercialized as such biomass and bioproducts [237]. The red seaweeds are harvested for agar and carrageenan industries whereas he brown species are utilized by alginate industries [238]. It is to be ıoted that rapid growth seen in Japan and Korea is mainly because of seaweed production. The 2010 estimate has shown that the glob. seaweed production was 19 million tons and it accounts for US $5. billion [241]. The increasing demand for seaweed-based edible produc puts its cultivation in a profiting trend.

8.9 CONCLUSION

Seaweed thus serves as a sustainable feedstock and an ecofriendl resource for various purposes. Also many bioactive compounc and pharmacologically active substances have been isolated froı macroalgae and put into use in various forms. Now-a-day's seaweec are under threat in developing nations because of human settlement natural barriers and lack of awareness regarding its potential us. Awareness about the importance of harvesting and commercializatio of seaweeds and products should be effectively promoted i developing nations. The cultivators should be well trained to mak them familiar with various harvesting techniques, the parameters an specificity involved in growing various classes of species for successfı commercialization of seaweed farming. India, since it has got lon coastal line with rich availability of seaweed species, if encourage and promoted with financial aids by the government, malnutrition an poverty can be uprooted to the maximum extent.

8.10 REFERENCES

1. Arosa, N. W. Spain et al., Community structure, succession, an production of seaweeds associated with mussel-rafts in the Ri de Arosa, N.W.Spain. Marine Ecology - Progress Series 5 (1981 243-253.

2. P.V. Subba Rao, Vaibhav A.Mantri, Indian seaweed resources an sustainable utilization: Scenario at the dawn of a new centur Current Science 91 (2006) 2.

3. Vinod K.Dhargalkar, Uses of seaweeds in the Indian diet fo sustenance and well-being, Science and culture 80 (2014)7-8.

4. A.H. Buschmann, R.Westermeier, C. Retamales, Cultivation c *Gracilaria* in the sea-bottom in southern Chile: a review, J. App Phycol. 7(1995) 291–301.

. R. Westermeier, P. J. Rivera, L. Go´mez, Cultivo de *Gracilaria chilensis* Bird, McLachlan y Oliveira, en la zona intermareal y submareal del estuario Cariquilda, Maull´ın, Chile. Rev. Chil. Hist. Nat 64(1991) 307–321.

. E. Booth, In Firth Production trade and utilization of seaweeds and seaweed products.Fisheries and aquaculture department(1969).

. S.Suto, Variation in species characters of *Porphyra* under culture condition. Seminar on the Contributions of Culture, Laboratory, Field and Life History Studies to the Systematics of Benthic Marine Algae of the Pacific Abstracts, Jap.-U.S. Comm. on Sci. Coop (1971) 28-29.

. T. Tamura,Marine aquaculture(1966). 2nd ed. (Translated from the Japanese by M. I. Watanabe, p. C-1--A2-30; available U.S. Dep. Commer., Natl. Tech Inf. Serv., Springfield, VA, as PB194051T.)

. Saito. On the effects of environmental factors on morphological characteristics of *Undaria pinnatifida* and the breeding of hybrids among the genus Undaria, Jap.-U.S. Comm. on Sci. Coop. (1971) 18-19.

0. Y. Hasegawa. Forced cultivation of *Laminaria. In* Proc. Seventh Internatl. Seaweed Symp., Sapporo, Japan (1971) 391-393.

1. Aleksander Handaet al.,Seasonal- and depth-dependent growth of cultivated kelp (*Saccharina latissima*) in close proximity to salmon (*Salmo salar*) aquaculture in Norway,414 (2013) 191–201.

2. N. Neveux etal., Biocrude yield and productivity from the hydrothermal liquefaction of marine and freshwater green macroalgae.Bioresource Technology 155 (2014)334–341.

3. Lars Nikolaisen etal., Energy Production from Marine Biomass (Ulva lactuca), Renewable Energy and Transport (Biomass section) PSO Project No. 2008-1-0050.

4. M. H. Norziah, C. Y. Ching,Nutritional composition of edible seaweed Gracilaria changgi. Food Chemistry (2000) 69-76.

15. P. Sethi, Biochemical composition of marine brown algae *Padir tetrastromaticahauck*. ISSN-0975-7006 4 (2012)117-118.

16. Aguilera-Moraleset al., Chemical composition ar microbiological assays of marine algae Enteromorpha sp. as potentialfood source, Journal of Food Composition and Analys 18 (2005) 79–88.

17. V. K. Dhargalkar, N.Pereira,Seaweed: Promising plant of tl millennium Science and culture 71 (2005) 3-4.

18. Patricia Burtin,Nutritional value of seaweeds,Patricia Burti EJEAFChe, 2(2003)ISSN: 1579-4377.

19. Katarzyna Chojnacka et al.,Biologically Active Compounds i Seaweed Extracts - the Prospects for the Application,The Ope Conference Proceedings Journal (2012) 50-372.

20. A Dadolahi-Sohrabetal., Seasonal variations in biomass an species compositionof seaweeds along the northern coasts c Persian Gulf(Bushehr Province), J. Earth Syst. Sci.(2012), 241 250.

21. S. L. Holdt, S. Kraan, Bioactive compounds in seawee functional foodapplications and legislation, J Appl Phyco (2010),DOI 10.1007/s10811-010-9632-5.

22. K. Murugaiyan,S. Narasimman, Elemental composition c *Sargassum longifolium* and *Turbinaria conoides* from Pamba Coast,International Journal of Research in Biological Science 2 (2012) 137-140.

23. B. A. Yoza, E. M. Masutani,The analysis of macroalgae biomas found aroundHawaii for bioethanol production,Environment Technology 34 (2013)13-14.

24. P. Ratana-arporn, A. C. Kasetsart, Nutritional Evaluation c Tropical Green Seaweeds *Caulerpa lentillifera and Ulva reticulat* Nat. Sci. 40 (2006) 75 – 83.

25. S. Ramesh kumar,C. M. Ramakritinan, M. Yokeshbabu Proximate composition of some selected seaweeds from Pal

bay and Gulf of Mannar, Tamilnadu, India,Asian Journal of Biomedical and Pharmaceutical Sciences 3 (2012)1-5.

6. G. Kokilam, S. Vasuki,Biochemical and phytochemical analysis on *Ulva fasciata andCaulerpa taxifolia*,ISSN: 2249-0337 (2013).

7. K. Banerjeeetal,Biochemical Composition of Marine Macroalgaefrom Gangetic Delta at the Apex of Bay of Bengal, African Journal of Basic & Applied Sciences 5-6 (2009) 96-104.

8. J. Ortiz et al, Dietary fiber, amino acid, fatty acid and tocopherol contents of the edible seaweeds *Ulva lactuca and Durvillaea Antarctica.*

9. N. Tatiana, T. Zvyagintseva etal., Water-soluble polysaccharides of some brown algae of the Russian Far-East. Structure and biological action of low-molecular mass polyuronans;Journal of Experimental Marine Biology and Ecology 320 (2005) 123–131.

0. C. Dawczynki, R. Schubert, G. Jahreis, Amino acids, fatty acids, and dietary fibre in edible seaweed products, Food Chemistry 103 (2007) 891–899.

1. N. M. Shevchenkoet al., Polysaccharide and Lipid Compositionof the Brown Seaweed*Laminaria gurjanovae*. Russian Journal of Bioorganic Chemistry 33(2007) 88–98.

2. Berna kılınç etal., Seaweeds for food and industrial applications

3. Fisal ahmad etal., Proximate compositions and total phenolic contents of selected edible seaweed from semporna, sabah, Malaysia borneo science 31 (2012)

4. S. Mabeau and J. Fleurence, Seaweed in food products: Biochemical and nutritional aspects 4 (1993)

5. P. Rupe´rez, Mineral content of edible marine seaweeds, Food chemistry 79 (2002) 23–26.

6. S. Dharka etal.,Proximate composition of marine macro algae, International journal of research in marine sciences 1 (2012) 1-3.

37. J. Fleurence,Seaweed proteins: Biochemical nutritional aspec and potential uses, Trends in food science and technology 1 (1999) 25–28.

38. L. Rioux etal., Effect of season on the composition of bioacti polysaccharides from the brown seaweed *Saccharina longicruri* Phytochemistry 70 (2009) 1069–1075.

39. C. Denis etal., Study of the chemical composition of edible re macroalgae *Grateloupia turuturu* from brittany (france), Foo chemistry 119 (2010) 913–917.

40. A. Vanessa Gressler etal., Analytical methods:Lipid, fatty aci protein, amino acid and ash contents in four brazilian red alga species, Food chemistry 120 (2010) 585–590.

41. K.H. Wong and Peter C.K. Cheung,Nutritional evaluation c some subtropical red and green seaweeds part ii. in vitro protei digestibility and amino acid pro®les of protein concentrate Food chemistry 72 (2001) p11-17.

42. C. Parthiban etal.,Biochemical composition of some selecte seaweeds from tuticorin coast.Advances in applied scienc research 4(2013)362-366.

43. K. Pazdro, A. Kosakowska,J. Pempkowiak, Fatty acid content i seaweeds from the baltic sea and the Indian ocean, Oceanologi 39 (1997) 279–287.

44. A. Arun kumar, A. Palanivelu, A. Darsis, Proximate compositio nutraceutical constituents and fatty acid profile on gcmsc seaweeds collected from balk bay (thondi), India int j curr sci 1 (2014) 57-71.

45. O. Benjama and P. Masniyom, Biochemical composition an physicochemical properties of two red seaweeds (*gracilari fisheri* and *g. tenuistipitata*) from the pattani bay in souther Thailand, Songklanakarin j. sci. technol., 34 (2012) 223-230.

46. Marinho-soriano etal.,Seasonal variation in the chemic composition of two tropical seaweeds, Bioresource technolog 97 (2006) 2402–2406.

7. M.A.M.Siddique, M.S.K. Khan, M.K.A.Bhuiyan, Nutritional composition and amino acid profile of a sub-tropical red seaweed *Gelidium pusillum* collected from st. martin's island, Bangladesh, Middle-east journal of scientific research 4 (2009)72-77.

8. M.A.M. Siddique, M. Aktar, M.A. Mohd khatib, Proximate chemical composition and amino acid profile of two red seaweeds (*hypnea pannosa* and *hypnea musciformis*) collected from St. martin's island, Bangladesh, Journal of fisheries sciences 7 (2013): 178-186.

9. K.H. Wong and C.K. Cheung, Nutritional evaluation of some subtropical red and green seaweeds part i: Proximate composition, amino acid profiles and some physicochemical properties, Food chemistry 71 (2000) 475-482.

0. P. Matanjun etal., Nutrient content of tropical edible seaweeds, *Eucheuma cottonii, Caulerpa lentillifera* and *Sargassum polycystum*. Journal of applied phycology 21(2008) 75-80.

1. G.G. Foster and A.N. Hodgson, Consummation and apparent dry matter digestibility of six intertidal macro-algae by *Turbo sarmaticus,*Aquaculture 167 (1998) 211-227.

2. K. Manivannan etal., Proximate composition of different group of seaweedsfrom vedalai coastal waters (gulf of mannar): southeast coast of India, India middle-east journal of scientific research 4 (2009) 72-77.

3. V. Sitakara rao and U.K. Tipnis, Protein content of marine algae from Gujarat coast. Current science, 33(1964) 16-17.

4. V. Sitakara rao and U.K. Tipnis, Chemical constituents of marine algae from Gujarat coast. Proceedings and seminar an sea salt and plants, Csmcri, bhavanagar, (1967) 277-288

5. K.C. Morgan, J.L.C. Wright, F.J. Simpson., Review of chemical constituents of the red algae *Palmaria palmate*, Economic botany 34 (1980) 27-50.

6. K. Rohani-Ghadikolaei, E. Abdulalian, N.G. Wing-keong, Evaluation of the proximate, fatty acid and mineral composition of representative green, brown and red seaweedsfrom the

persian gulf of iran as potential food and feed resources, J foc sci technol 49 (2012) 774–780.

57. V. Ivanova, M. stancheva, D. petrova, Fatty acid composition (black sea *ulva rigida*and*cystoseira crinite,*Bulgarian journal (agricultural science19 (2013) 42–47.

58. K. Manivannan, G. Thirumaran, G. Karthikai devi,Biochemic composition of seaweedsfrom mandapam coastal regions alor southeast coast of India. American-Eurasian journal of botany (2008) 32-37.

59. Y. Peng et al.,Nutritional and chemical composition and antivir activity of cultivated seaweed *sargassumnaozhouense*

60. B. Wang etal., Analysis on nutrition constituents (*Sargassumnaozhouense* sp., nov. food res. dev. 31 (2010) 19 197.

61. M.J. Zhao, Nutrition evaluation of the edible seaweed,fish. sc 1990:28–31.

62. Z. Daietal.,Evaluation on nutritional components (*sargassumfusiforme*, J. fish. china. 26 (2002) 382–384.

63.1 S Chem etal., Chemical constituents i *sargassumhenslowianum*and its nutrition evaluation, Food re dev. 31 (2010)154–156.

64. P.V. Subba rao etal., Mineral composition of edible seawee *Porphyra vietnamensis,* Food chemistry 102 (2007) 215–218.

65. T. Nedumaran, Mineral composition and plant growt regulators in brown seaweed, *Padina boergeseniiint.* J. cur.tr. re 1 (2012)137-139.

66. M. Francavilla et al, The red seaweed *gracilaria gracilis* as a mul products source,Mar. drugs 11 (2013) 3754-3776.

67. J.A. Haugan, S. Liaaen-jensen, Algal carotenoids 54. carotenoid of brown algae (phaeophyceae),Biochem. syst. ecol. 22 (1994 31–41.

8. C. Taboada, R. Millán,I. Míguez, Composition, Nutritional aspects and effect on serum parameters of marine algae *ulva rigida*,J. sci. food agric. 90(2010) 445–449.

9. P. Macartain et al, Nutritional value of edible seaweeds,Nutr. rev. 65 (2007) 535–543.

0. A.B. Ross et al,Classification of macroalgae as fuel and its thermochemical behavior, Bioresource technology 99 (2008) 6494–6504.

1. K. Ganapathi, V. Subramanian, S. Mathan, Bioactive potentials of brown seaweeds, *Sargassum myriocystum ,S.plagiophyllum* and *S.ilicifolium,* Int. res j pharm. app sci. 3(2013) 105-111.

2. D. Robledo, Y. Freile-pelegrin, Chemical and mineral compositionof six potentially edible seaweed species of yucata´n, Botanica Marina 40 (1997) 301–306.

3. M. Greenwell, C.J. Bird, J. Mclachlan, Depth-related variation in gross chemical composition of several seaweeds, Aquatic botany, 20 (1984) 297-305.

4. S. Polat, Y. Ozogul, Seasonal proximate andfatty acid variationsof some seaweedsfrom the northeastern mediterranean coast, Oceanologia 55(2013) 375–391.

5. G. F. El-said and A. el-sikaily, Chemical composition of some seaweed from Mediterranean sea coast, Environ monit assess 185 (2013) 6089–6099.

6. M. Dahlia, El maghraby, E.M. Fakhry, Lipid content and fatty acid composition of mediterranean macro-algae as dynamic factors for biodiesel production http://dx.doi.org/10.1016/j.oceano.2014.08.001 0078-3234/# 2014

7. E. Rioux, S.L. Turgeon, M. Beaulieu, Characterization of polysaccharides extracted from brown seaweeds, Carbohydrate polymers 69 (2007) 530–537.

8. P. Anantharaman, etal.,Mineral composition of marine macroalgae from mandapam coastal regions; southeast coast of India, Recent research in science and technology 2 (2010) 66-71

79. D.I. Sa´nchez-machado et al., Fatty acids, total lipid, prote and ash contents of processed edible seaweeds, Food chemist 85 (2004) 439–444.

80. O. Benjama, P. Masniyom; Nutritional composition ar physicochemical properties of two green seaweeds (*Ulva pertus* and *U. intestinalis*) from the pattani bay in southern Thailan Songklanakarin j. sci. technol. 33 (2011) 575-583.

81. P. Ratana-arporn and A. Chirapart, Nutritional evaluation tropical green seaweeds *Caulerpa lentillifera*and*Ulva reticulat* Kasetsart journal (natural science) 40 (2006) 75–83.

82. K.J. Mcdermid and B. Stuercke, Nutritional composition edible hawaiian seaweeds, Journal of applied phycology1 (2003) 513–524.

83. S.L. Holdt and S. Kraan, Bioactive compounds in seawee functional foodapplications and legislation. J appl phycol(2010 doi 10.1007/s10811-010-9632-5.

84. K.P. Devi et al,Bioprotective properties of seaweeds: *In vitr* evaluation of antioxidant activity and antimicrobial activit against food borne bacteria in relation to polyphenolic conten *BMC* Complementary and Alternative Medicine 2008, 8:3 doi:10.1186/1472-6882-8-38.

85. M. J. Abad, L. M. Bedoya, P. Bermejo, Marine Compounc and their Antimicrobial Activities, FORMATEX 2011, Scienc against microbial pathogens: communicating current researc and technological advances.

86. D.J. Newman, G.M. Cragg, K.M. Snader. Natural products a sources of new drugs over the period 1981-2002,Journal c naturalproducts 66(2003) 1022-1037.

87. Fu-jin wang et al, Composition of two seaweed-oligosaccharid lysates derived from agarase as-ii and their ferrous-chelatin antioxidant power, journal of marine science and technology 1 (2011)557-564.

8. Sung, et al, Evaluation of antioxidant properties of a new compound, pyrogallol-phloroglucinol-6,6'-bieckol isolated from brown algae, *Ecklonia cava*

9. H. Indu,R. Seenivasan, Invitro antioxidant activity of selected seaweeds from southeast coast of India 5 (2013), issn-0975-1491

0. A. Ismail and T.S. Hong,Antioxidant activity of selected commercial seaweed,Mal j nutr. 8(2002) 167-177.

1. T. Mise,M. Ueda, T. Yasumoto, Production of fucoxanthin-rich powder from *Cladosiphon okamuranus*,Adv. j. food sci. technol. 3 (2011) 73–76.

2. X. Yan et al, Fucoxanthin as the major antioxidant in *Hijikia fusiformis*, a common edible seaweed,Biosci. biotechnol. biochem. 63 (1999) 605–607.

3. W. Stahl, H. Sies, Photoprotection by dietary carotenoids: concept, mechanism, evidence and future development,Mol. nutr. food res. 56 (2012) 287–295.

4. K. Mikami and M. Hosokawa, Biosynthetic pathway and health benefits of fucoxanthin,an algae-specific xanthophyll in brown seaweeds, Int. j. mol. sci. 14 (2013) 13763-13781

5. N. D'orazio et al, Fucoxantin: a treasure from the sea, 10(2012) 604-616.

6. N. Dantas-santos et al,Freshwater plants synthesize sulfated polysaccharides: Heterogalactans from water hyacinth (*Eicchornia crassipes*), Int. j. mol. sci. 13 (2012) 961-976.

7. C.L.F. De almeida, H.D.S. Falcao, G.R. De-m. lima, Bioactivities from marine algae of the genus gracilaria, Int j . mol.sci. 12 (2011) 4550-4573.

8. B.W.S. Souz, Chemical characterization and antioxidant activity of sulfated polysaccharide from the red seaweed*Gracilaria biridae*, Food hydrocolloids 27(2012) 287-292

9. G. Britton, S. liaaen-jensen, H. Pfander et al, Carotenoids with a c40 skeleton incarotenoids—handbook: basel, switzerland (2004) 563.

100. T. Hu, D. liu, Y. Chen, J. Wu, S. Wang, Antioxidant activity o sulfated polysaccharide fractions extracted from *Undari pinnitafida* in vitro.Int. j. biol. macromol. 46 (2010) 193–198.

101. T. Rengarajan et al, Cancer preventive efficacy of marir carotenoid fucoxanthin: cell cycle arrest and apoptosis,Nutrients (2013)4978-4989.

102. S. Mohamed, S.N. Hashim, H.A. Rahman, Seaweeds: sustainable functional food for complementary and alternati therapy, Trends in food science & technology 23 (2012)

103. S. Gupta and N. Abu-ghannam, Bioactive potential and possibl healtheffects of edible brown seaweeds, Trends in food scienc & technology 22 (2011),

104. L. Yamamoto et al, Antitumor effect of seaweeds, antitumc effect of extract from *Sargasum* and *Laminaria*,jpn. j. exp. mec 44 (1974) 543–546.

105. L. Yamamoto et al, Antitumor effect of seaweeds, 3. antitumc effect of an extract from *Sargassum kjellmanianum*,jpn. j. exp med.51 (1981) 187–189.

106. L. Yamamoto et al, Antitumor activity of edible marine alga effect of crude fucoidan fractions prepared from edible brow seaweed against l-1210 leukemia,Hydrobiology (116/117) 198 145–148

107. M.Tutor Ale et al, Important determinants for fucoidan bioactivit a critical review of structure-function relations and extractio methods for fucose-containing sulfated polysaccharides fror brown seaweeds, Mar. drugs 9 (2011) 2106-2130.

108. L. Giacinti et al, Epigenetic information and estrogen recepto alpha expression in breast cancer,Oncologist 11 (2006) 1–8.

109. F. Namvar et al,Antioxidant, antiproliferative an antiangiogenesis effects of polyphenol-rich seaweed (*Sargassur muticum*), Biomed res. int. (2013) 604-787.

110. A.O. Maranguy et al, Anticancer effects of different seaweeds o human colon and breast cancers,Mar. drugs 12 (2014) 4898-491

11. M.J. Ryu et al,The green algae *Ulva fasciata* delile extract induces apoptotic cell death in human colon cancer cells,In vitro cell. dev. biol. anim. 49 (2013) 74–81.

12. H.K. Park et al, Induction of apoptosis by laminarin, regulating the insulin-like growth factor-ir signaling pathways in ht-29 human colon cells,Int. j. mol. Med.30 (2012) 734–738.

13. H.K. Park, J. Kim,T.J. Nam, Induction of apoptosis and regulation of erbb signaling by laminarin in ht-29 human colon cancer cells,Int. j. mol. med. 32 (2013) 291–295.

14. Alekseyenko et al,Antitumor and antimetastatic activity offucoidan, a sulfated polysaccharide isolated from the okhotsk sea *fucus evanescens* brown alga. Bulletin of experimental biology and medicine, 143, 730-732.

15. H. Harada et al,Antitumor activity of palmitic acid found as a selective cytotoxic substance in a marine red alga.Anticancer res. 22 (2002) 2587–2590.

16. S. Gupta and N. Abu-ghannam, Recent developments in the application of seaweeds or seaweed extracts as a means for enhancing the safety and quality attributes of foods, Innovative food science and emerging technologies 12 (2011) 600–609.

17. S. Horie et al, Antibacterial quinine metabolites from the brownalga, *Sargassum sagamianum*, Bullet in of the chemical society of japan 81,1125–1130

18. N.B. Muruleedharan et al, Growth and survival of escherichia coli and enterococci populations in the macro-alga cladophora (chlorophyta), Fems microbiol. Ecol. (2003), 203–211.

19. R. Ely, T. Supriya, C.G. Naik, Antimicrobial activity of marine organisms collected off the coast of south east India, Journal of experimental marine biology and ecology 309 (2004) 121– 127.

20. K. Kanjana et al,Solvent extracts of the red seaweed *gracilaria* fisheri prevent vibrio harveyi infections in the black tiger shrimp penaeus monodon. Fish& shellfish immunology, 30, 389–396.

121. G. Immanuel et al, Effect of butanolic extracts from terrestri herbs and seaweeds on the survival, growth and pathogen (vibri parahaemolyticus) load on shrimp *Penaeus indicus* juvenile Aquaculture 236 (2004) 53–65.

122. Y. Wang et al, Sensitivity of *Escherichia coli* toseawee (*Ascophyllum nodosum*) phlorotannins and terrestrial tannin Asian- australian journal of animal science 22, 238–245.

123. A. Bansemir et al, Screening of cultivated seaweeds fo antibacterial activity against fish pathogenic bacteri Aquaculture 252 (2006) 79- 84.

124. Catherine vonthron-sénécheau , etal., Antiprotozoal activities o organic extracts from french marine seaweeds, *mar. drugs*201 9, 922-933.

125. P. Kajamideen et al, Explore the antimicrobial potenti from organic solvents of brown seaweed extracts(*Sargassu longifolium*) alleviating to pharmaceuticals, Ijp 1 (2014) 1-4.

126. S. Bharathiraja et al,Biomimetic of silver nanoparticles by *Ulv lactuca* seaweed and evaluation of its antibacterial activity (2012), issn- 0975-1491.

127. K.Sivakumar et al, Antifungal activity of certain seaweeds fror puthumadam coast,Ijrrpas 3(2013) 341-350.

128. T. Vo, D. Ngo, Q. V. Ta, S. Kim, Marine organisms as a therapeuti source against herpes simplex virus infection, Eur.j.pharm.sc 44 (2011) 11-20.

129. C.I. Celum, The interaction between herpes simplex virus an human immunodeficiency virus, Herpes 11 (2004)36a–45a.

130. E. Plouguerné et al, Antiviral sulfoquinovosyldiacylglycerol (sqdgs) from the brazilian brown seaweed *Sargassu vulgare*,Mar. drugs11 (2013) 4628-4640.

131. A.N. Lei liu et al, Towards a better understanding of medicin uses of the brown seaweed *Sargassum* in traditional Chines medicine: A phytochemicaland pharmacological review journa of ethnopharmacology 142 (2012) 591–619.

32. M. Romanos et al, Inhibitory effect of extract Brazilian marine algae onhuman T-cell lymphotropic virus type1 (htlv-1)-induced syncytium formation in vitro, Cancer investigation 20 (2002) 46–54.

33. M.J. Ahn et al, Inhibition of HIV-1 reverse transcriptase and HIV-1 integrase and antiviral activity of korean seaweed extracts, Journal of applied phycology 14, 325–329.

34. N.M. Ponceet al, Fucoidans from the brown seaweed adenocystis utricularis: Extraction methods, antiviral activity and structural studies,Carbohydrateresearch 338 153-165.

35. P. Siamopoulou et al, Diterpenes from the brown algae dictyota dichotoma and dictyota linearis, Phytochemistry 65 2025-2030.

36. K.C. Queiroz et al, Inhibition of reverse transcriptase activity of HIV by polysaccharides of brown algae. Biomedicine & pharmacotherapy 62 (2008) 303-307.

37. J. Vera et al, Seaweed polysaccharides and derived oligo saccharides stimulate defense responses and protection against pathogensin plants, Mar. drugs 9 (2011) 2514-2525.

38. M.A. Radwan et al, Synthesis and biologicalevaluation of new 3-substituted indole derivatives as potential anti-inflammatory and analgesic agents, Biol. med. chem.15 (2007) 3832–3841.

39. H.U. Yong-fang and W.U. Yong-jie, Effect of recombinant human basic fibroblast growth factor on acute inflammation in mice and rats,Acta. pharmacol. sin. 22 (2001) 357–379.

40. E. Tenório de souza et al, The antinociceptive and anti-inflammatory activities of caulerpin, a bisindole alkaloid isolated from seaweeds of the genus *Caulerpa,* Mar. drugs7 (2009) 689-704;

41. A. Cumashi et al, A comparative study of the anti-inflammatory, anticoagulant, antiangiogenic, and antiadhesive activities of nine different fucoidans from brown seaweeds,Glycobiology17 (2007) 541–552.

142. Guangling jiao et al, Chemical structures and bioactivities sulfatedpolysaccharides from marine algae,Mar. drugs9 (201 196-223.

143. S. Sakai et al, Inhibitory effect of carotenoids on the degranulatic of mast cells via suppression of antigen-induced aggregation high affinity IgE receptor. J. biol. chem. 284 (2009) 28172–2817

144. D. Vipul et al, Carrageenan: a natural seaweed polysaccharic and its applications, Carbohydrate polymers 105 (2014) 97–11

145. L. Chaudhary et al, Algae as a feedstock for Bioethan Production: New Entrance in Biofuel World, Internation Journal of chemTech research 6 1381-1389.

146. B.A. Yoza and E. M. Masutani, The analysis of macroalga biomass found around Hawaii for bioethanol production201 1860-1867

147. M. Okazaki et al, A study of calcium carbonate deposition i the genus Padina (phaeophyceae, Dictyotales), Br Phycol J. 2 (1986) 217-224.

148. G. Leilei, P. Wang, H. Mou, Study on saccharification techniqu of seaweed wastes for the transformation of ethanol, Renewab Energy36 (2011) 84–89.

149. S.J. Hornl M Aasen, K. Ostgaard, Ethanol production fror seaweed extract, Journal of Industrial Microbiology Biotechnology25 (2000) 249-254.

150. B. Kilinc et al, Seaweeds for food and industrial application http//dxdoi.Org/10.5772/53172

151. Seaweed Bioethanol Production in Japan. The Ocean Sunris Published in: OCEANS 2007 Date of Conference: Sept. 29 200 Oct. 4 2007 Page(s): 1 - 5 Conference Location: Vancouver, B DOI: 10.1109/OCEANS.2007.4449162 Publisher: IEEE

152. A. Demirbas and M.F. Demirbas, Importance of algae oil as source of biodiesel, 52 (2011) 163–170.

153. R. Maceiras et al,Raw material for biodiesel production, Applie energy 88(2011)3318-3323.

54. H.Van der Walet al, Production of acetone, butanol, and ethanol from biomass of the green seaweed Ulvalactuca128 (2013) 431-7. doi: 10.1016/j.biortech.2012.10.094.

55. T.Potts et al,The production of butanol from Jamaica bay macro algae,Environmental Progress & Sustainable Energy 31 (2012) 29–36.

56. F. Murphy et al,Biofuel Production in Ireland—An Approach to 2020 Targets with a Focus on Algal BiomassEnergies 6 (2013) 6391-6412.

57. Vivekanand, V.G.H. Eijsink,S.J. Horn,Biogas production from the brown seaweed *Saccharinalatissima*: thermal pretreatment and codigestion with wheat straw, 24 (2012) 1295-1301.

58. C. H. Vanegasand J. Bartlett,Green energy from marine algae: biogas production and composition from the anaerobic digestion of Irish seaweed species 2277-2283.

59. A. D. Hughes, Biogas from Macroalgae: Is it time to revisit the idea?, Biotechnology for Biofuels (2012) 5:86. doi:10.1186/1754-6834-5-86.

60. T. Matsui, Y. Koike, Methane fermentation of a mixture of seaweed and milk at a pilot-scale plant110 (2010)558-63.

61. N. Wei, J. Quarterman, Y.S. Jin, Marine macroalgae: An untapped resource for producing fuels and chemicals. Trends Biotechnol. 31 (2013) 70–77.

62. C. Jin etal,Progress in the production and application of n-butanol as a biofuel, 15(2011)4080–4106.

63 D. Chakraborty, Agar-agar manu facture from *Gracilaria confervoides*, Jour. Proc. Inst. Chem. 17 (1945) 188.

64. A.V. Rao, K.N. Patel, H.N.Shah, Manufacture of agar-agar from red seaweeds. Res. ind 10 (1965) 131-133.

65. C.S. Lobban and M.J. Wynne, The Biology of Seaweeds; ISBN: 9780520045859

166. K. Funaki and Y. Kojima. Studies on the properties of agar-aga from *Gracilaria confervoides*. Part-3, Bull. Jap. Soc. Sci. Fish. 1 (1951)401-404

167. G.T.A. Fenoradosoa et al, Extractionand characterization c an alginate from the brown seaweed *Sargassum turbinarioide*. 22 (2010) 131-137.

168. V. Rao and I.A.B. Thivy, Seaweed utilisation in India. Pro Symp. Algolog, (1960)ICAR, New Delhi. pp.345-365.

169. S.T. Moe, K.I. Draget, G. Skjåk-Braek, O. Smidsrød, Alginate Food polysaccharides and their applications,In A. M. Stephe (Ed.), New York: Marcel Dekker (1995) 245–286.

170. M. Rinaudo, Seaweed Polysaccharides. Comprehensiv Glycoscience In: Kamerling JP, editor Elsevier 2 (2007).

171. E. Fourest and B. Volesky, Contribution of sulfonate group and alginate to heavy metal biosorption by the dry biomass c *Sargassumfluitans*,Environ. Sci. Technol. 30 (1996)277–282.

172. G.T.A. Fenoradosoa et al, Extractionand characterization c an alginate from the brown seaweed *Sargassum turbinarioide*. 22 (2010) 131-137.

173. P. Vauchel et al, A New Process for Extracting Alginate from *Laminariadigitata,*Food and Bioprocess Technology (2008) 297 300.

174. E. Hill and E. Surrey, The polysaccharides of green, red an brown seaweeds: Their basic structure, Biosynthesis an function, Food and Bioprocess Technology 1 (2007) 297- 300.

175. A. Haug et al, Studies on the sequence of uronic acid residues i alginic acid, Acta Chem. Scand. 21 (1967) 691–704.

176. A. Haug, B. Larsen, B Smidsrod, Uronic acid sequence in alginat from different sources, Carbohyd. Res.32 (1974) 217–225.

177. A. Haug,B. Larsen, E. Baardseth, Comparison of the constitutio of alginates from different sources. Proc. int. Seaweed Symp (1969) 443–451.

78. K. Abe et al, In vivo studies on the synthesis of alginic acid in *Ishige okamurai*, Botanica mar. 16 (1973) 229–234.

79. T.Y. Lin and W.Z. Hassid, Isolation of guanosine diphosphate uronic acids from a brown alga,*Fucus gardneri*, Silva. J. Biol. Chem 241 (1966) 3283–3293.

80. J. Madgwick, A. Haug, B. Larsen, Polymannuronic acid 5-epimerase from the marine alga *Pelvetia canaliculata* (Dcne et Thur.), Acta Chem. Scand. 27 (1973) 3592–3594.

81. H. Ertesvåg and S. Valla, Biosynthesis and applications of alginates 59 (1998) 85–91.

82. H.A. Bernd,Alginates: Biology and Applications Springer Dordrecht Heidelberg London New York DOI: 10.1007/978-3-540-92679-5.

83. Title: **Production and utilization of products from commercial seaweeds** Division: Fishery and Aquaculture Economics and Policy Division ISBN: 9251026122 ISSN: 0429-9345.

84. H.J. Bixler and H. Porse, A decade of change in the seaweed hydrocolloids industry, 23 (2011) 321-335.

85. G. Bourgade, Improvement in treating marine plants to obtain gelatine.Journal of Applied Phycology(1871) 112,535.

86. E. Bjerre-Petersen, J. Christensen, P. Hemmingsen, Furcellaran. In Industrial gums. New York, Academic Press(1973)123-36.

87. M. Tutor-Ale, J.D. Mikkelson, A.S. Meyer,Important Determinants for Fucoidan Bioactivity: A Critical Review of Structure-Function Relations and Extraction Methods for Fucose-Containing Sulfated Polysaccharides from Brown Seaweeds, MarineDrugs 9 (2011) 2106-2130.

88. M. Tutor-Ale, J.D. Mikkelson, A.S. Meyer, Important Determinants for Fucoidan Bioactivity: A Critical Review of Structure-Function Relations and Extraction Methods for Fucose-Containing Sulfated Polysaccharides from Brown Seaweeds, Marine Drugs 10 (2014) 2130.

189. T. Thi-Thu-Thanhet al, Structure of Fucoidan from Brow Seaweed Turbinariaornata as Studied by Electrospray Ionizatic Mass Spectrometry (ESIMS) and Small Angle X-ray Scatterir (SAXS) Techniques, Mar. Drugs 11 (2013) 2431-2443.

190. B. Li et al, Fucoidan: Structure and bioactivity, Molecules 1 (2008) 1671–1695.

191. L. Chevolot et al, A disaccharide repeat unit is the major structu in fucoidans from two species of brown algae, Carbohydr. Re 330 (2001) 529–535.

192. A. Cumashi et al, A comparative study of the anti-inflammator anticoagulant, antiangiogenic, and antiadhesive activities (nine different fucoidans from brown seaweeds, Glycobiolog 17 (2007) 541–552.

193. M. Kusaykin et al,Structure, biological activity, and enzymat transformation of fucoidans from the brown seaweed Biotechnol. J. 3 (2008) 904–915.

194. J.B. Lee et al, Antiviral fucoidan from sporophyll (Undariapinnatifida (Mekabu). Chem. Pharm. Bull. 52 (200 1091–1094.

195. J. Trinchero et al, Antiretroviral activity of fucoidans extracte from the brown seaweed Adenocystisutricularis, Phytother. Re 23 (2009) 707–712.

196. H.R. Raghavendran, P. Srinivasan, S. Rekha, Immunomodulator activity of fucoidan against aspirin-induced gastric mucos damage in rats, Int. Immunopharmacol. 11 (2011) 157–163.

197. J. Wang et al, Potential antioxidant and anticoagulant capacit oflow molecular weight fucoidan fractions extracted fror *Laminaria japonica*, Int. J. Biol. Macromol. 46 (2010) 6–12.

198. A. Synytsya et al, Structure and antitumour activity c fucoidan isolated from sporophyll of Korean brown seawee *Undariapinnatifida*, Carbohydr.Polym. 81 (2010), 41–48.

99. Church et al, Antithrombin activity of function. The interaction of fucoidan with heparin cofactor II antithrombin III, nad thrombin, J. Biol-chem 264 (1989) 3618-3625.

00. G.E. Carlberg, E. Percival, M.A. Rhaman, Carbohydrates of the seaweeds, *Desmarestia ligulata* and *D. firma*., Phytochem. 17 (1978)1289-1292.

01. H. Kylin, Biochemistry of sea algae, H. Z. Physiol. Chem. 83 (1913) 171–197.

02. A. Beattie, E.L. Hirst, E. Percival, Studies on the metabolism of the Chrysophyceae, Biochem. J. 79 (1961) 531–537.

03. L. Rioux,S.L. Turgeon,M. Beaulieu, Structural characterization of laminaran and galactofucan extracted from the brown seaweed *Saccharinalongicruris*, Phytochemistry 71 (2010)1586-95.

04. E. Percival, Do the polysaccharides of brown and red seaweeds ignore taxonomy? In Modern approaches to taxonomy of red and brown algae (lrvine, D. E. G. and Price, J. H., editors) Systematics Association Special Volume No. 10.Academic Press, London and New York (1978) 47-62.

05. R.G.S. Bidwell, Photosynthesis and metabolism in marine algae. Products of synthesis in fronds of *Fucus vesiculosus* and their use in respiration. Can. or. Bot 45 (1967) 1557 1565.

06. R.S. Quatrano andP. Stevens, Cell wall assembly in *Fucus* zygotes. Characterisation of the polysaccharide components, Planta 58 (1976) 224-231.

07. A.O. Chizhov et al, Structural analysis of laminaran by MALDI and FAB mass spectrometry, Carbohydrate Research 310 (1998)203–210.

08. L.E. Riouxa, S.L. Turgeon, M. Beaulieu, Lausanne,Carbohydrate Polymers 69(2007)530–537.

09. G. Hocman, Prevention of cancer: vegetables and plants, Comp Biochem Physiol B. 93 (1989)201-212.

10. S. Abdussalam, Drugs from seaweeds, Med Hypotheses 32 (1990) 33-35.

211. N. Johnson, Seaweed and its synthetic analogues in obstetri and gynaecology. J R Soc Med. 83 (1990)387-389.

212. C. Aceves, B. Anguiano, G. Delgado, Is iodine a gatekeeper the integrity of the mammary gland?,J Mammary Gland Bi Neoplasia 10 (2005)189-196.

213. J. Teas et al, Dietary seaweed modifies estrogen and phytoestroge metabolism in healthy post-menopausal women. J Nutr. 13 (2009)939-944.

214. J.L. Frestedt, M.A. Kuskowski, J.L. Zenk,A natural seawee derived mineral supplement (Aquamin F) for knee osteoarthriti a randomised, placebo controlled pilot study, Nutr J. (2009) 8-

215. J.L. Frestedt et al, A natural mineral supplement provides reli from knee osteoarthritis symptoms: a randomized controlle pilot trial, Nutr J. (2008) 7-9.

216. Y. Yamori, A. Miura, K. Taira, Implications from and for foo cultures for cardiovascular diseases: Japanese food, particularl Okinawan diets, Asia Pac J Clin Nutr.10 (2001)144-145.

217. M. Krotkiewski et al, Effects of a sodium-potassium ioi exchanging seaweed preparation in mild hypertension, Am Hypertens. 4 (1991) 483-488.

218. J. Teas et al, Could dietary seaweed reverse the metaboli syndrome?, Asia Pac J Clin Nutr. 18 (2009)145-154.

219. S. Rajeshkumar et al, Seaweed-mediated synthesis gold nanoparticles using Turbinariaconoides and i characterization,Journal of Nanostructure in Chemistry 3 (201 44.

220. S.S Shankar, A. Rai, A. Ahmad, M. Sastry, Rapid synthesis of Au Ag, and bimetallic Au core Ag shell nanoparticles using Neer (Azadirachta indica) leaf broth, J. Colloid Interface Sci. 27 (2004) 496–502.

221. J.N. Liu et al, B cell stimulating activity of seaweed extracts, Ir J Immunopharmacol. 19 (1997)135-142.

22. B.E. Shan et al, Immunomodulating activity of seaweed extract on human lymphocytes in vitro, Int J Immunopharmacol. 21 (1999) 59-70.

23. J.Y. Song, B.S. Kim, Rapid biological synthesis of silver nanoparticles using plant leaf extracts, Bioprocess Biosyst. Eng 32 (2009) 79–84.

24. S. Nagarajan and K.A. Kuppusamy, Extracellular synthesis of zinc oxide nanoparticle using seaweeds of gulf of Mannar, India, Journal of Nanobiotechnology 11 (2013) 39.

25. P.K. Dhanalakshmi et al, Synthesis of silver nanoparticles using green and brown seaweeds, Phykos 42 (2012) 39-45.

26. M. Mahdavi et al, Green Biosynthesis and Characterization of Magnetic Iron Oxide (Fe_3O_4) Nanoparticles using Seaweed (*Sargassummuticum*) aqueous Extract, Molecules 18 (2013) 5954-5964.

27. W. Schramm. Editors Guiry M.D; Blunden G. Seaweeds for waste water treatment and recycling of nutrients. Book Seaweed resources in Europe: uses and potential(1991) 149-168.

28. J.H. Ryther et al, Physical models for integrated waste recycling-Marine polyculture system.Aquaculture 5 (1975)163-177.

29. J.Lust, The Herb Book, 1986, 16th. Impression, Bantam Publishing. ISBN 0-553-17273-5

30. M. Mitton, Herbal Remedies-Skin Problems. 1984. W.Foulsham ISBN 0-572-01254-3

31. M. Stuart, The Encyclopaedia of Herbs and Herbalism. 1986. Orbis ISBN 0-85613-700-6

32. B. Pramanick, K. Brahmachari, A. Ghosh. Effect of seaweed saps on growth and yield improvement of green gram, 8 (2013) 1180-1186.

33. W. Kahn et al, Seaweed extracts as biostimulants of plant growth and development, J Plant Growth Regul.27 (2009)270–279.

234. M. Povolny, The effect of the steeping of peat-cellulose flowerpo (Jiffypots) in extracts of seaweeds on the quality of toma seedlings, Proc Int Seaweed Symp. 8 (1981)730–733.

235. A. Jensen A, The nutritive value of seaweed meal fo domesticanimals, Proc Int Seaweed Symp. 7 (1972) 7–14.

236. M Casas-Valdezet al and E.A. Maina, Sargassum (Sargassaceae una alternativa tropical para la alimentation degrando caprin Rev Biol Trop 54 (2006) 83–92.

237. C. Rebours, E. Marinho-Soriano et al, Seaweeds: an opportuni for wealth and sustainable livelihood for coastal communities appl phycol 26 (2014) 1939–1951.

238. A.B. Zaixso, M. Ciancia, A.S. Cerezo, Seaweed resources i Argentina.World seaweed resources—an authoritative referenc system, Margraf publishers gmbh(2006), Germany.

239. H.J. Bixler, H. Porse, A decade of change in the seawee hydrocolloids Industry, J appl phycol. 23 (2011)321–335.

240. U.S.Murty, A.K. Banerjee, Seaweeds: the wealth of oceans. ir kim s.-k., editor. Handbook of marine macroalgae. John Wile & sons, ltd.; West sussex (2012) 36–44.

241. FAO fishery and aquaculture statistics 2010. Food and agricultur organization of the United Nations; Rome, Italy: 2012.

242. J. Dennis, McHugh, Worldwide distribution of commercia resources of seaweeds including Gelidium, Hydrobiologia 22 (1991) 19-29.

Index

D

G

M

N

O

P

Q

R

S

Y

Z